EX LIBRIS
**ST. JOHN FISHER
COLLEGE**

SEED BIOLOGY

PHYSIOLOGICAL ECOLOGY

A Series of Monographs, Texts, and Treatises

EDITED BY

T. T. KOZLOWSKI

University of Wisconsin
Madison, Wisconsin

T. T. KOZLOWSKI. Growth and Development of Trees, Volumes I and II – 1971

DANIEL HILLEL. Soil and Water: Physical Principles and Processes, 1971

J. LEVITT. Responses of Plants to Environmental Stresses, 1972

V. B. YOUNGNER AND C. M. MCKELL (Eds.). The Biology and Utilization of Grasses, 1972

T. T. KOZLOWSKI (Ed.). Seed Biology, Volumes I, II, and III – 1972

YOAV WAISEL. Biology of Halophytes, 1972

SEED BIOLOGY

Edited by

T. T. KOZLOWSKI

DEPARTMENT OF FORESTRY
UNIVERSITY OF WISCONSIN
MADISON, WISCONSIN

VOLUME III
Insects, and Seed Collection, Storage, Testing, and Certification

ACADEMIC PRESS   New York and London   1972

ACADEMIC PRESS, INC.
111 Fifth Avenue, New York, New York 10003

United Kingdom Edition published by
ACADEMIC PRESS, INC. (LONDON) LTD.
24/28 Oval Road, London NW1

LIBRARY OF CONGRESS CATALOG CARD NUMBER: 71-182641

PRINTED IN THE UNITED STATES OF AMERICA

CONTENTS

5   Essentials of Seed Testing

OREN L. JUSTICE

6   Seed Certification

J. RITCHIE COWAN

LIST OF CONTRIBUTORS

Numbers in parentheses indicate the pages on which the authors' contributions begin

G. E. BOHART, Entomology Research Division, Agricultural Research Service, USDA, Logan, Utah (1)

J. RITCHIE COWAN, Crop Science Department, Oregon State University, Corvallis, Oregon (371)

CHARLES R. GUNN, Plant Science Research, Division Agricultural Research Service, U.S. Department of Agriculture, Beltsville, Maryland (55)

JAMES F. HARRINGTON, Department of Vegetable Crops University of California, Davis, California (145)

R. W. HOWE, Pest Infestation Laboratory, Ministry of Agriculture Fisheries and Food, London Rd. Slough, Bucks, England (247)

OREN L. JUSTICE, Market Quality Research, Division Agricultural Research Service, U.S. Department of Agriculture, Federal Center Building, Fayettsville, Maryland (301)

T. W. KOERBER, Pacific Southwest Forest and Range Experiment Station, Forest Service, USDA, Berkeley, California (1)

PREFACE

Man's existence and health are directly or indirectly dependent on seeds. This fact has for many years pointed out the urgent need for a comprehensive coverage of information on seed biology. The importance of this work became even greater during the recent years of rapid population increases throughout the world. It was with these thoughts in mind that this three-volume treatise was planned to bring together a large body of important new information on seed biology.

The subject matter is wide ranging. The opening chapter outlines man's dependency on seeds as sources of food and fiber, spices, beverages, edible and industrial oils, vitamins, and drugs. Harmful effects of seeds are also mentioned. Separate chapters follow on seed development, dissemination, germination (including metabolism, environmental control, internal control, dormancy, and seed and seedling vigor), protection from diseases and insects, collection, storage, longevity, deterioration, testing, and certification. These books were planned to be readable and interdisciplinary so as to serve the widest possible audience. They will be useful to various groups of research biologists and teachers, including agronomists, plant anatomists, biochemists, ecologists, entomologists, foresters, horticulturists, plant pathologists, and plant physiologists. The work has many practical overtones and will also be of value to seed producers and users.

These volumes are authoritative, well-documented, and international in scope. They represent the distillate of experience and knowledge of a group of authors of demonstrated competence from universities and government laboratories in England, India, Israel, South Africa, and the United States. I would like to express my deep personal appreciation to each of the authors for his contribution and patience during the production phases. The assistance of Mr. W. J. Davies and Mr. P. E. Marshall in index preparation is also acknowledged.

T. T. KOZLOWSKI

CONTENTS OF OTHER VOLUMES

1

INSECTS AND SEED PRODUCTION

G. E. Bohart and T. W. Koerber

I. Introduction

Although the lives of plants and insects are interrelated in a multitude of ways, this chapter is limited to a discussion of the insects that directly affect seeds, seed-producing organs, or seed-bearing structures of plants. In general, it deals with broad generalizations concerning

1

relationships between insects and seed production and presents specific examples to illustrate them. We assume the reader is well aware that any factor that influences the health or vigor of a plant will, in turn, affect its reproductive capacity. Hence, many insects that feed on leaves or roots of plants affect seed production. Insects attacking seeds after maturity are discussed in Chapter 4 of this volume.

Insects have been around for a long time. The oldest insects, which lived 250 million years ago, have been identified as fossils in rocks of the Upper Carboniferous period (Carpenter, 1952). The first insects were rather generalized forms, some not too different from present-day roaches, and probably lived as scavengers in the extensive swampy forest lands of their day. Over successive eons a great variety of more specialized insects developed, until by the Jurassic period, 155 million years ago, many of the modern orders of insects existed. During the 100 million or so years during which insects were increasing in diversity, new types of plants evolved. During the Jurassic period, gymnosperms, of which present-day conifers, cycads, and ginkgo are surviving representatives, came to be dominant elements of the flora. Angiosperms began to appear in the Jurassic, but did not become abundant until the mid-Cretaceous period, 25 million years later (Darrah, 1960). Thus, the gymnosperms and insects evolved more or less simultaneously, but angiosperms did not appear until 130 million years after the first insects and evolved in a world where insects were a well-established, powerful force. Thus, it happened that the insects and plants, especially the angiosperms, by evolving together, have become mutually dependent.

Most insects exist by directly or indirectly exploiting plants. Many insects not only feed on plants, as do other animals, but also have developed intimate relationships far more complex than those between plants and most other animals. The small size of insects permits many of them to live within their host plants, some within a single seed. Many insects have developed colors and textures closely resembling those of parts of their host plants. We almost invariably find green caterpillars feeding on green leaves, some so closely matching the color and texture of the leaves as to defy detection. Other insects exhibit adaptations of form, habits, or both to resemble other parts of plants. The larvae of geometrid moths are characteristically elongated and colored to resemble twigs — a resemblance greatly reinforced by their habit of grasping a twig with their hindmost legs and extending their bodies rigidly at an angle from it. Adult geometrids are often patterned and colored to resemble closely tree bark or dead leaves. A few insects are even known to incorporate toxic compounds from their host plants into their bodies, thereby gaining a measure of protection from their predators.

To survive in the face of potentially unrestricted exploitation by insects, plants have developed various protective mechanisms. A plant may bear coverings of bark, wax, or hairs on its various surfaces and may have within it sticky fluids to impede insect attackers physically. It may further wage chemical warfare by producing compounds repellent or toxic to insects. These defense mechanisms do not provide immunity from insect attack but rather serve to restrict the kinds or numbers of insects which can successfully exploit a given species of plant.

Plants have also developed methods of exploiting insects, the most notable of which are the various systems by which angiosperms enlist the aid of highly mobile insects in their reproductive processes.* A substantial part of this chapter deals with the adaptations of plants for insect pollination. Nectar, produced in flowers to attract pollinators, is also produced in extrafloral nectaries, sometimes to the detriment of pollination. This apparent enigma has been explained, at least in part, by the discovery that such nectar is useful in attracting ants, which, in turn, protect the plants from potentially harmful insects.

II. Insect Pollination

A. Evolution of Insect Pollination

Since insect pollination (entomophily) is generally believed to be the primary condition in angiosperms, it can be said that the angiosperm flower evolved in response to the presence of insects as agents of pollen transfer. Since the gymnosperm orders Gnetales and Cycadales are apparently also primitively entomophilous, it appears that among seed plants, only the Coniferales and Ginkgoales were originally wind-pollinated (anemophilous). The angiosperm condition itself (protection of the ovules by the gynoecium wall) was probably an early development to prevent damage to the ovules by primitive pollen-feeding beetles. The hermaphroditic condition of reproductive strobili, an obvious adaptation for insect pollination, was exhibited by the Bennettitales, a Jurassic group sometimes postulated as ancestral to the angiosperms. Leppik (1960) suggested that these gymnosperms provided the emerging angiosperms with an existent anthophilous insect fauna. Although hermaphroditism in flowers undoubtedly presented problems in promoting cross-pollination, various structural modifications arose in flowers to keep pollen away from their own stigmas (at least initially), and the develop-

---

*Meeuse (1961), Percival (1965), Faegri and van der Pijl (1966), and Baker and Hurd (1968) give excellent treatments of the basic aspects of pollination, and Free (1970) does likewise for crop pollination.

ment of a style allowed for the development of several kinds of self-incompatibility systems (Baker, 1963).

Wind pollination, as it exists in the grasses, most Chenopodiaceae, many nut-bearing trees, and several genera of Compositae and other families, represents a secondary adaptation. The same is true of the many cases of bird pollination and the few of bat pollination. Secondary adaptations for pollen transfer by other agents, such as water currents, mollusks, and mammals also occur, but infrequently. Although insect pollination is primary, it can also be tertiary, as in the case of several genera of Cyperaceae which, having lost their perianth in a previous wind-pollinated stage, have more recently developed showy involucral bracts (Faegri and van der Pijl, 1966).

A number of plants are both anemophilous and entomophilous. *Salix* (willow) and *Castanea* (chestnut) are good examples. Since the percentage of wind-pollinated and self-pollinated plants increases with increasing latitude as well as with increasing exposure to wind, it appears that abiotic pollination arose in response to a scarcity of suitable animals and to the presence of suitable air currents.

Insects and other pollinating agents had an important influence on angiosperm evolution by facilitating sympatric reproductive isolation. Minor genetic changes (sometimes single gene mutations) can alter floral characteristics enough to change the pollinator complex and prevent full gene interchange (Baker, 1960; Grant, 1949). Within the same genus, different species may be pollinated by diverse animals such as bees, Lepidoptera, and birds (Grant, 1959).

B. Pollinating Insects

Pollinating insects occur primarily in the orders Coleoptera (beetles), Lepidoptera (butterflies and moths), Diptera (flies, midges, gnats, etc.), and Hymenoptera (sawflies, wasps, and bees). Coleoptera was a well-developed order in the early Cretaceous period when the major groups of angiosperms were developing, but only the more primitive families of Hymenoptera, Diptera, and Lepidoptera were present. Consequently, it is primarily the more advanced forms of the last three orders that are well adapted for flower visitation. Lepidoptera, being the most recent, has the highest percentage of flower-visiting forms. The order Thysanoptera (thrips) also contains many flower-visiting forms, but the role of these minute insects as pollinators appears to be minor (except, perhaps, in self-pollination) even for plants with correspondingly minute flowers.

Since flower-visiting beetles are usually polleniferous and crawling, plants with disc- or bowl-shaped flowers or inflorescences having abundant, freely exposed pollen, such as poppy, cactus, carrot, and many composites, are their preferred hosts. Flies, as well as beetles, eat pollen when it is easy to obtain and, consequently, tend to visit the same kinds of flowers. However, some genera of Bombyliidae, Acroceridae, Nemestrinidae, and Syrphidae have developed elongate mouthparts to obtain nectar from deep, narrow corolla tubes. Lepidoptera, which have long, slender tongues and appetites for nectar, prefer flowers with abundant nectar hidden in deep, narrow reservoirs where competition from other insects is minimal. Moths favor pale-colored flowers that shed their fragrance at night (e.g., evening primrose and gardenia). Many moths (e.g., sphingids and noctuids) hover while feeding and prefer flowers with a vertical opening and without landing structures. Butterflies prefer horizontal flowers or inflorescences on which they can alight, and, being among the few insects that distinguish red as a color, frequently visit red flowers or inflorescences (Faegri and van der Pijl, 1966). Hymenoptera, especially bees, are interested in both pollen and nectar and, as a group, visit a wide variety of flowers. All except a few highly specialized flowers are attractive to bees of some kind, but a number of forms normally pollinated by other animals may be poorly pollinated by bees even if visited by them (Fig. 1). The most "typical" bee flowers are bilaterally symmetrical types (e.g., those of Scrophulariaceae, Labiatae, and Papilionaceae) which require manipulation to expose their nectar and pollen. Bees are the principal insects that have the necessary tongue length, strength, and "dedication" to visit these flowers successfully.

C. Floral Adaptations for Insect Pollination

Most flowers attract insects by offering them pollen (Fig. 1) or nectar (Fig. 2) or both (Fig. 3) as food. Color, form, and perfume are merely signposts to enable the insects to find the primary attractants and return to the same food source. Some flowers satisfy other instincts than the ones to consume or store food, for instance, sex, oviposition, and territoriality. Several kinds of orchids, which mimic certain female bees or wasps in both form and odor, entice visitation by males of the model species. The flowers are pollinated while the insects are attempting to copulate with them (Kullenberg, 1961). There are flowers that offer places for oviposition and food for the developing larvae. On at least two types of flowers showing this kind of adaptation, the insects take what appears to be a purposeful (rather than accidental) role in pollination and make no attempt to feed or collect stores of pollen or nectar. The female fig wasp *(Blastophaga)* packs pollen into special pouches from the stamens of one fig inflorescence and places it on the pistils of another. After depositing her pollen, she oviposits in special gallflowers provided by the plant for the developing wasp larvae (Galil and Eisikowitch, 1969). The yucca moth *(Pronuba)* bears a similar relationship to the large flowers of Spanish bayonet. This moth scrapes pollen from the stamens with her specialized mouthparts and thrusts it upward into the dependent, invaginated stigma of another flower. Only then does she crawl to the ovary and insert therein a small number of eggs (Riley, 1892). A number of flowers with fetid odors attract flies interested primarily in oviposition, but in this endeavor, the flies are usually thwarted, even though the flower's purpose is accomplished. The territorial instinct of certain bees has been exploited by a genus of orchids. The flowers, which mimic the bees, wave back and forth on long, flexible stems in the manner of their hovering models. The male bees strike the flowers repeatedly in an attempt to drive them away, and thus accomplish pollination (Dodson and Frymire, 1961a). Even narcosis is employed by the orchid *Stanhopea* which attracts male bees of the genus *Eulaema* which have only to scratch the tepals with their chemoreceptive front tarsi to become "drunk" and eventually lose consciousness. Other orchids are similarly narcotic to other kinds of bees (Dodson and Frymire, 1961b).

D. Importance of Insect Pollination

The following discussion deals primarily with insect pollinators as they affect the production of seeds for the propagation of crop plants. It should be recognized, however, that by setting the stage for fertilization, insects are also significant in the production of seeds and fruiting structures used as food for humans and domestic animals, as spices, and in the production of medicines, fibers, oils, soaps, and many industrial chemicals. On the larger scene, insect pollinators are responsible for the origin and perpetuation of a large share of the earth's vegetation.

Although an undetermined percentage of uncultivated plants are capable of automatic self-pollination, they are dependent on at least occasional cross-pollination for the gene recombinations required to maintain the large gene pools they need to meet competition and changing conditions. Automatic self-pollinated crop plants, most of which are cross-pollinated only by insects, would be in a precarious position if they were incapable of cross-pollination. The wild ancestors of such crops are being increasingly suppressed and the frequently outcrossed varieties grown by primitive farmers are disappearing, thus causing the available gene pools for meeting future disease and insect problems to dwindle at an alarming rate. Gene banks supported by frequent crossings may be needed to meet this challenge (Chedd, 1970). Although most woody, rhizomatous, tuberous, and bulbiferous plants are vegetatively propagated, selections for new varieties are often made from the naturally occurring seedlings that result from cross-pollination. Hand pollination is frequently used for planned crosses, but insect pollinators are increasingly used to build up usable quantities of breeding stock.

The list in Table I, although limited as indicated by the heading illustrates the pervasive value of insect pollinators in our agricultural economy.

E. Degree of Dependency on Insects

Many of the above crops are almost entirely dependent on insects because of self-sterility, self-unfruitfulness, or inability to achieve automatic self-pollination. Others, such as cotton and safflower, may yield satisfactorily following automatic self-pollination, but under most circumstances, adequate insect pollination would increase their yields. The degree of dependence is often a question of the horticultural variety. In general, for obvious reasons, plant breeders have bred away from a dependence on insect pollination.

Increasing interest in the growing of hybrid seeds has resulted in recent years in a greater emphasis on insect pollination. Except in the case of wind-pollinated species, the dependence of hybrid seed crops on insects is total (unless the producer wishes to use the expensive alternative of hand pollination). Hybrid cottonseed, for example, cannot be grown without efficient pollination by bees. This is also true of cultivars such as tomato and garden pea which are normally independent of pollinating agents. It may become necessary to breed back into many of these plants the adaptations to cross-pollination and the features of attractiveness to pollinators which were largely lost in the development of autogamous varieties.

The foregoing discussion may indicate that the pollination biology of common wild plants and nearly all crop plants is well known. Actually, for only a tiny fraction of wild plants, is even the most rudimentary knowledge available, in spite of some excellent studies on unusual pollination systems by many botanists such as Darwin (1877) and van der Pijl (1954) and a few entomologists (Hurd and Linsley, 1963; Linsley, 1960). The more important insect-pollinated crops in this country have received some study, but in-depth studies are few, and the less common crops have scarcely been touched (Free, 1970).

F. Pollinators of Crop Plants

Throughout the world, there are hundreds of thousands of species of insects that habitually visit flowers and, thus, serve as pollinators. Some thousands of these benefit cultivated seed crops. For example, in the course of pollination studies at one location (Logan, Utah), 267 species of insects were found on onion flowers (Fig. 3) and 334 on carrot flowers (Bohart and Nye, 1960; Bohart et al., 1970). However, a considerably smaller number of insect species visit seed crops in sufficient numbers to

TABLE I

CROPS GROWN IN THE UNITED STATES (EXCLUDING HAWAII) KNOWN OR PRESUMED TO REQUIRE (OR BENEFIT FROM) INSECT POLLINATION (EXCLUDING ORNAMENTALS, MEDICINALS, CONDIMENTS, HYBRID SEEDS, AND RARELY GROWN TROPICAL FRUITS)ₐ

Anacardiaceae
(c) Mango, *Mangifera indica* L.
Cactaceae
(c) Prickly pear, *Opuntia ficus-indica* Mill.
Caricaceae
(c) Papaya, *Carica papaya* L.
Compositae
(ab) Safflower, *Carthamus tinctorius* L.
(ab) Sunflower, *Helianthus annuus* L.
Cruciferae
(a) Broccoli, *Brassica oleracea* L.
(a) Brussels sprouts, *Brassica oleracea* L.
(a) Cabbage, *Brassica oleracea* L.
(a) Chinese cabbage, *Brassica pekinensis* Rupr.
(a) Collards, *Brassica oleraceae* L.
(a) Cress (common) *Lepidium sativum* L.
(a) Cress (water) *Nasturtium officinale* R. Br.
(a) Cress (winter) *Barbarea verna* Aschers
(a) Kale (sea), *Crambe maritima* L.
(a) Kales, *Brassica oleracea* L., *B.* spp.
(a) Kohlrabi, *Brassica caulorapa* Pasq.
(a) Mustards, *Brassica* spp.
(a) Pakchoi, *Brassica chinensis* L.
(a) Radish, *Raphanus sativus* L.
(ab) Rape, *Brassica napus* L.
(a) Rutabaga, *Brassica napobrassica* Mill.
(a) Turnip, *Brassica rapa* L.
Cucurbitaceae
(ac) Cucumber, *Cucumis sativus* L.
(ac) Gourd, *Lagenaria siceria* Standl.
(ac) Muskmelon, *Cucumis melo* L.
(ac) Pumpkins, *Cucurbita* spp.
(ac) Squashes, *Cucurbita* spp.
(ac) Watermelon, *Citrullus vulgaris* Schrad.
Ebenaceae
(c) Persimmon (native), *Diospyros virginiana* L.
Euphorbiaceae
(b) Tung, *Aleurites fordii* Hemsl.

Fagaceae
(b) Chestnuts, *Castanea* spp.
Grossulariaceae
(c) Currants, *Ribes* spp.
(c) Gooseberries, *Ribes* spp.
Lauraceae
(c) Avocado, *Persea americana* Mill.
Liliaceae
(a) Asparagus, *Asparagus officinalis* L.
(a) Chive, *Allium schoenoprasum* L.
(a) Leek, *Allium porrum* L.
(a) Onion, *Allium cepa* L.
(a) Onion (Welsh), *Allium fistulosum* L.
(a) Shallot, *Allium ascolonicum* L.
Linaceae
(ab) Flax, Linseed, *Linum usitatissimum* L.
Malvaceae
(abc) Cottons, *Gossypium* spp.
Moraceae
(c) Fig (some vars.), *Ficus carica* L.
Papilionaceae
(a) Alfalfa (blue), *Medicago sativa* L.
(a) Alfalfa (yellow), *Medicago falcata* L.
(abc) Bean (broad bean), *Vicia faba* L.
(abc) Bean (lima), *Phaseolus limensis* Macf.
(abc) Bean (scarlet runner), *Phaseolus coccineus* L.
(a) Clover (alsike), *Trifolium hybridum* L.
(a) Clover (crimson), *Trifolium incarnatum* L.
(a) Clover (Dutch), *Trifolium repens* L.
(a) Clover (Egyptian), *Trifolium alexandrinum* L.
(a) Clover (Kura), *Trifolium ambiguum* Bieb.
(a) Clover (Ladino), *Trifolium repens* L. (var.)
(a) Clover (Lappa), *Trifolium lappaceum* L.
(a) Clover (large hop), *Trifolium procumbens* L.

(a) Clover (Persian) *Trifolium resupinatum* L.

(a) Clover (red), *Trifolium pratense* L.

(a) Clover (rose), *Trifolium hirtum* All.

(a) Clover (small hop), *Trifolium dubium* Sibth.

(a) Clover (strawberry), *Trifolium fragiferum* L.

(a) Clover (zigzag), *Trifolium medium* L.

(abc) Cowpea, *Vigna unguiculata* L.

(a) Crown vetch, *Coronilla varia* L.

(a) Indigo (hairy), *Indigofera hirsuta* L.

(a) Kudzu vine, *Pueraria thunbergiana* Benth.

(a) Lespedezas (some), *Lespedeza* spp.

(a) Lupin (blue), *Lupinus angustifolius* L.

(a) Lupin (white), *Lupinus albus* L.

(a) Lupin (yellow), *Lupinus luteus* L.

(a) Milk vetch, common, *Astragalus cicer* L.

(a) Milk vetch, sickle, *Astragalus falcatus*

(a) Rattleboxes, *Crotalaria* spp.

(a) Sainfoin, *Onobrychis viciaefolia* Scop.

(a) Sainfoin (Persian), *Onobrychis chorassinaca* Bunge ex Boiss

(a) Sweet clover (sour), *Melilotus indica* All.

(a) Sweet clover (white), *Melilotus alba* Desr.

(a) Sweet clover (yellow), *Melilotus officinalis* Lam.

(a) Trefoil (big), *Lotus uliginosus* Schkuhr.

(a) Trefoil (birdsfoot), *Lotus corniculatus* L.

(a) Trefoil (narrowleaf), *Lotus tenuis* Wald & Kit.

(a) Vetch (hairy), *Vicia villosa* Roth

Passifloraceae

(c) Passion fruits, *Passiflora* spp.

Pedaliaceae

(ab) Sesame, *Sesamum indicum* L.

Polygonaceae

(ab) Buckwheat, *Fagopyrum esculentum* Moench.

Punicaceae

(c) Pomegranate, *Punica granatum* L.

Rosaceae

(b) Almond, *Prunus amygdalus* Batsch.

(c) Apples, *Malus* spp.

(c) Apricot, *Prunus armeniaca* L.

(c) Blackberries, dewberries, *Rubus* spp.

(c) Cherry (sour) *Prunus cerasus* L.

(c) Cherry (sweet), *Prunus avium* L.

(c) Nectarine, *Prunus persica* Batsch.

(c) Peach, *Prunus persica* Batsch.

(c) Pear, *Pyrus communis* L.

(c) Plums, prunes, *Prunus* spp.

(c) Quince, *Cydonia oblonga* Mill.

(c) Raspberry (blackcap), *Rubus occidentalis* L.

(c) Raspberry (red), *Rubus idaeus* L.

(c) Strawberries, *Fragaria* spp. (hybrids)

Rutaceae

(c) Lemon, *Citrus limon* Burm.

(c) Tangelo, *Citrus reticulata paradisi* Macf.

(c) Tangerine, *Citrus reticulata* Blanco

Solanaceae

(ac) Eggplant, *Solanum melongena* L.

(ac) Pepper, *Capsicum annuum* L.

(ac) Pepper (tobasco), *Capsicum microcarpum* DC

Umbelliferae

(a) Anise (sweet), *Pimpinella anisum* L.

(c) Carrot, *Daucus carota* L.

(a) Celeriac, *Apium graveolens* L.

(a) Celery, *Apium graveolens* L.

(ac) Fennel, *Foeniculum vulgare* Mill.

(a) Parsley, *Petroselinum crispum* Nym.

(a) Parsnip, *Pastinaca sativa* L.

Vacciniaceae

(c) Blueberries, huckleberries, *Vaccinium* spp.

(c) Cranberry, *Vaccinium macrocarpum* L.

Vitiaceae

(c) Grapes (some), *Vitis* spp.

"Crop names preceded by (a) are grown for seed for propagation, those by (b) for seed for food or other uses, those by (c) for fruit for food or other uses.

be important. Bees predominate, although a few kinds of seed crops, representing only a small percentage of the total acreage, receive more benefit from other insects. There are over 25,000 species of bees alone, of which at least 1000 visit cultivated seed crops. Of these, several hundred are abundant enough at times and in places to be important pollinators (Bohart, 1967). Crops grown for their fruit or for seed used in other ways than for propagation are also pollinated primarily by bees. However, there are several important exceptions, such as cacao (by midges), fig (by *Blastophaga* wasps), and mango (mostly by flies) (Free, 1970). A relatively large percentage of uncultivated plants are pollinated by insects other than bees. This is especially true in tropical and cold-temperate regions where bees are relatively less common than in warm-temperate and arid regions.

G. Honeybee

Seed crops throughout the world undoubtedly benefit more from pollination by the honeybee (*Apis mellifera* L.) than from any other pollinator and, probably, from all others combined (Levin, 1967; Bohart and Todd, 1961). The genus *Apis* appears to have originated in southeastern Asia, and it is still represented there by the greatest number of species. *Apis mellifera* apparently originated in southeastern Europe or southwestern Asia and spread naturally throughout Europe, Africa, and much of western Asia. During the last few centuries, it has been introduced by man into eastern and southeastern Asia, Australia, New Zealand, and the New World. In most of southeastern Asia, other species of *Apis* (*A. florea* F., *A. dorsata* F., and *A. cerana* F.) are at the present time more abundant than *A. mellifera*, even in cultivated areas. Elsewhere, especially in cultivated areas, the honeybee predominates over all other bees.

The honeybee collects nectar and/or pollen from nearly every plant where either substance is found. In times of pollen scarcity, it even collects rust spores and grain millings. However, the honeybee tends to rank its host plants according to attractiveness, and the less attractive species may receive little or no visitation in the presence of more attractive species. In other words, the honeybee is a potential pollinator of nearly every kind of seed crop, but competition from other nectar and pollen sources may reduce or negate its effectiveness.

On some plants, the honeybee collects nectar and pollen separately (Figs. 4 and 5), and may rate such plants quite differently for the two substances. For example, the honeybee readily collects nectar from alfalfa, but only under certain conditions does it collect significant amounts of alfalfa pollen. However, it collects pollen from red clover more readily than it does nectar, probably because of the difficulty of obtaining a good supply of the latter (Bohart, 1960). Even on such relatively unspecialized flowers as those of onion and carrot, most of the foraging population is divided into nectar collectors and pollen collectors. On most plants, both forms of foraging result in pollination, and it is usually assumed to make little difference which substance the bees are collecting. Studies on carrots and onions have indicated, however, that nectar collectors stand higher on the flower heads and carry less pollen (see Fig. 6 for pollen collectors on onion). It is not known to what extent this affects their efficiency as pollinators (Bohart *et al.,* 1970). A similar problem exists with many varieties of apple on which the stamens and style are closely grouped in the center of the flower, thus allowing nectar-collecting bees to reach the nectaries at their base without contacting either anthers or stigma (Menke, 1951).

von Frisch (1947) discussed his results and those of others (mostly Russian) on the directing of honeybees to specific crops by "infusing" the hive with sugar syrup scented with flowers of the appropriate crop. Although he pointed out that the results were often only transitory and seemed to have little effect on ultimate seed or fruit yields, many Russian workers continued to report practical successes with this method. In 1958, Free reviewed the subject and reported his own essentially negative results. He concluded that, even if the method could be made reliable, it would probably only train nectar collectors which on some crops are less effective pollinators than pollen collectors.

On alfalfa, the nectar-collecting honeybees soon learn to avoid the pollination mechanism by approaching the flowers from the side (Fig. 7). They trip a few flowers (usually not over 1% of those they visit) by accident, but populations must be very high for this to result in commercially acceptable yields. Honeybees collect alfalfa pollen [with consequent "tripping" of the flowers (Fig. 8)] to a greater extent in the Southwest than they do elsewhere on this continent. Largely for this reason, they are rented for alfalfa pollination on a large scale in southern California and to a lesser extent in northern California and Nevada (Bohart and Todd, 1961). In the Northwest, some seed growers believe that honeybees inhibit pollination by reducing the attractiveness of the flowers to better pollinators, but there is little evidence to substantiate this viewpoint (Bohart *et al.,* 1967). Reasons for the trend toward decreasing alfalfa pollen collection from the Southwest toward the north and east are not known, though climatic factors affecting the condition of the plants (including the flowers) appear to be involved. Strangely enough, "competition" from other pollen sources appears to play little part, even though this is often cited as the prime factor (Bohart, 1954).

In the 1960s, the U.S. Department of Agriculture, Apiculture Research Laboratories at Logan, Utah, and Baton Rouge, Louisiana, developed a

FIG. 1. *Evylaeus aberrans* stripping pollen from stamens of evening primrose (pollination is by moths). (Photo by W. P. Nye, Federal Apiculture Research Laboratory, Logan, Utah.)

strain of honeybees that collects alfalfa pollen in northern Utah and western Idaho much more readily than "ordinary" strains (Nye and Mackensen, 1968) (Fig. 9). Although it is apparent that such a bee should be a much improved pollinator, the wide-ranging foraging flights of honeybees are such that a statistically valid demonstration of its value in terms of seed yields is extremely difficult. Private industry is now developing its own alfalfa pollen-collecting strain and may eventually bring about its large-scale use.

The honeybee has a definite aversion to cotton pollen and has seldom been seen to collect it, even though it is readily available In fact, nectar-collecting honeybees in cotton go to great trouble to scrape contaminating pollen grains off their bodies, and they soon become nearly bald in the process (McGregor, 1959). Furthermore, a large percentage of the nectar collectors are useless to the plant since they visit only the extrafloral nectaries. The ones that visit the floral nectaries sometimes effect pollination, but when the flowers are wide open, they often crawl in and out of the corolla and avoid the central sexual column (McGregor *et al.,* 1955) (Fig. 10). In future breeding for hybrid cottonseed production, it would probably be desirable to suppress the extrafloral nectaries and develop a partially closed flower.

Pollination of red clover by honeybees is poorly understood, in spite of extensive studies on this crop. Honeybees are notoriously poor pollinators of red clover in areas such as Scandinavia and England, which have cool, moist climates. The long corolla tubes of red clover, in which honeybees have difficulty reaching the nectar, have been cited as the principal deterrent to visitations by nectar-collecting honeybees (Bohart, 1957). The even greater reluctance of honeybees to visit tetraploid red clover, which has longer corolla tubes, supports this view. However, in the Intermountain region of the state of Washington, where the weather is warm and dry, honeybees gather less nectar than they do in the mid-

---

Fɪɢ. 2. *Peponapis pruinosa* taking nectar from pistillate flower of squash. (Photo by W. P. Nye, Federal Apiculture Research Laboratory, Logan, Utah.)

Fɪɢ. 3. Drone fly (*Eristalis tenax*) feeding on pollen and nectar from flower head of onion. (Photo by W. P. Nye, Federal Apiculture Research Laboratory, Logan, Utah.)

Fɪɢ. 4. Honeybee taking nectar from *Colutea*. (Photo by W. P. Nye, Federal Apiculture Research Laboratory, Logan, Utah.)

Fɪɢ. 5. Honeybee taking pollen from *Colutea*. (Photo by W. P. Nye, Federal Apiculture Research Laboratory, Logan, Utah.)

Fɪɢ. 6. Honeybee collecting pollen from head of onion. (Photo by W. P. Nye, Federal Apiculture Research Laboratory, Logan, Utah.)

Fɪɢ. 7. Honeybee collecting nectar from alfalfa. (Photo by W. P. Nye, Federal Apiculture Research Laboratory, Logan, Utah.)

Fɪɢ. 8. Honeybee collecting pollen from alfalfa. (Photo by W. P. Nye, Federal Apiculture Research Laboratory, Logan, Utah.)

western part of the country where it is more humid and rains are frequent. Nevertheless, honeybees are much more effective pollinators of red clover in the Intermountain region than in the Midwest because they collect pollen readily there, even in the presence of other attractive sources (Johansen, 1966). Apparently, intrinsic differences in the presentation of pollen by the plant (or in the pollen itself) under different growing conditions are at least as important in the extent of collection of red clover pollen by honeybees as differences in the nature and amount of competing pollen sources. Regardless of the reasons, honeybees appear to be unreliable red clover pollinators in Europe and New Zealand, intermediate in the Great Lakes region of the United States, and relatively reliable in Idaho, eastern Oregon, and eastern Washington.

Most seed crops are at least moderately attractive to the honeybee,[†] which, therefore, is the most useful species for their pollination. However, it is important to have enough colonies in or adjacent to the fields being pollinated and to have as few as possible competing floral sources in the area (Fig. 11). Since honeybees tend to be faithful to a particular plant species and to recruit their hive mates to it, apiculturists have recommended that colonies should not be brought to the target crop until blooming is well underway.

H. Bumblebee

Bumblebees (*Bombus* spp.) are best known as pollinators of red clover, but since they visit a wide variety of plants, they pollinate many other kinds of seed crops. They are, of course, especially valuable to crops for which the use of honeybees presents special problems. They also have a particular value in high latitudes because they tend to be abundant there (in comparison with other bees) and to work long hours during relatively low temperatures.

In Europe, bumblebees are generally thought to be the only effective pollinators of red clover. Other bees, such as *Eucera,* have been reported on red clover but, apparently, they are too scarce to be effective. In central France, I once observed an effective population of *Andrena ovatula* Kirby on a series of experimental red clover seed plots (unpublished). Perhaps this bee is overlooked at times because of its small size.

Bumblebees vary in tongue length and, in general, the species with long tongues are better pollinators of plants such as red clover, vetch, and sesame, which have long corolla tubes. The short-tongued species often cut holes in the bases of the corolla tubes and thus function more as "nectar thieves" than as pollinators (Fig. 12). They not only circumvent

---

[†]Exceptions include the horse bean (*Vicia faba* L.), which is only slightly attractive to honeybees, crown vetch (*Coronilla varia* L.), which honeybees rarely visit, and *Crotalaria* which have stiff flowers that are difficult for honeybees to manipulate.

pollination by this process, but the holes they cut are used by other species of bees, including honeybees (Free and Butler, 1959).

Knowledge of the nectar-thieving habit would have been useful 70 years ago when four species of bumblebees were introduced from England to New Zealand. Although red clover seed yields were dramatically improved, one of the species (*Bombus terrestris* L.) was a corolla cutter and, except when collecting pollen, was detrimental in the red clover seed fields. Nevertheless, for alfalfa pollination, it turned out to be the best of the four species (Palmer-Jones and Forster, 1965).

Bumblebees are excellent pollinators of *Crotalaria* and crown vetch, which have flowers that seem to be adapted to large bees. The long-tongued species are also reliable pollinators of vetch and horsebean. In the southeastern United States, bumblebees are among the few effective pollinators of cotton, although in the Southwest, they are seldom seen in cotton-growing areas. Several species of bumblebees collect pollen readily from tomato, and they might have a potential as pollinators of small hybrid tomato seed fields. Finally, bumblebees are well adapted for the pollination of several seed crops of ornamental plants such as snapdragon and pansy.

Although many persons have kept bumblebees for experimental purposes or as a hobby, their large-scale maintenance for the pollination of extensive acreages seems to be impractical. Most species never develop colonies with more than 100 individuals, and less than half of these are likely to be foraging at any one time (Fig. 13). A few species (for example, *Bombus terrestris*) develop as many as 1000 individuals in unusually prolific colonies (Hasselrot, 1960). However, bumblebee colonies require considerable care, and at least thirty better-than-average colonies per acre would be required to provide adequate pollination of red clover.

It would probably be more rewarding to improve the ecological conditions for bumblebees and protect them from predators than to keep colonies for pollination. Most bumblebee queens establish themselves in old rodent nests, especially those of field mice. However, since field mice often destroy young bumblebee nests, favorable conditions probably develop only as the mouse populations are declining. Undisturbed vegetation featuring clumps of perennial bunch grass are important as nesting areas for bumblebees. This was demonstrated by the general decline of bumblebee populations in the prairies of the United States and Canada following the destruction of perennial grassland conditions (Plath, 1934).

I. Alkali Bee

The alkali bee (*Nomia melanderi* Ckll.) (Fig. 14) is a nonsocial but gregarious, ground-nesting bee (Figs. 15 and 16) native to the Rocky Mountain and Pacific Coast states (United States) where it is an important

FIG. 9. Pollen trapped from honeybee colonies composed of high alfalfa pollen-collecting strain (top), low alfalfa pollen-collecting strain (middle), and "ordinary" strain (bottom). (Photo by W. P. Nye, Federal Apiculture Research Laboratory, Logan, Utah.)

alfalfa pollinator. Some of its nesting sites in moist alkaline soils harbor over a million nests per acre. In the 1950s, alfalfa seed growers in the Northwest became aware of its value and began protecting nesting sites and trying to enlarge them or create new ones. By 1958, growers began constructing "artificial" nesting sites to create controlled moisture conditions in soils where lateral and upward movement of water was difficult to bring about. These sites were built by lining excavations with plastic film, adding a layer of gravel to serve as a water reservoir, and backfilling with soil. Water supplied through pipes to the gravel layer seeped to the surface (Fig. 16). Salt added to the surface prevented excessive evaporation and inhibited growth of vegetation (Bohart, 1967). Newly prepared sites were stocked by transplanting cores of soil containing overwintering larvae (Fig. 17).

Although measures taken to protect and encourage alkali bees resulted in manyfold increases in both yields and acreage grown for seed, interest in these bees declined somewhat during the 1960s as a result of their decimation by untimely rains and insecticide applications, especially to neighboring crops such as mint. An equally important reason was the increasing reliance being placed on the newly arrived alfalfa leafcutter bee.

Although the alkali bee is useful primarily as an alfalfa pollinator, it collects pollen readily from several vegetable seed crops, such as carrot and onion. However, most of the vegetable seed fields in the Northwest are close to alfalfa seed fields and, consequently, do not attract many alkali bees. Probably for this reason, vegetable seed growers have not taken positive steps to increase alkali bees.

J. Alfalfa Leafcutter Bee

The alfalfa leafcutter bee (*Megachile rotundata* F.) (Fig. 18) was accidentally introduced onto the eastern seaboard in the late 1930s. By

FIG. 10. Honeybee collecting nectar from cotton flower (note some pollen adhering to bee). (Photo by W. P. Nye, Federal Apiculture Research Laboratory, Logan, Utah.)

FIG. 11. Apiary in alfalfa field. (Photo by W. P. Nye, Federal Apiculture Research Laboratory, Logan, Utah.)

FIG. 12. Scarlet sage flower with hole bitten into corolla tube by bumblebee. (Photo by W. P. Nye, Federal Apiculture Research Laboratory, Logan, Utah.)

FIG. 13. Nest of bumblebee (*Bombus griseocolles*). (Photo by W. P. Nye, Federal Apiculture Research Laboratory, Logan, Utah.)

FIG. 14. Adult female alkali bee at nesting site. (Photo by W. P. Nye, Federal Apiculture Research Laboratory, Logan, Utah.)

FIG. 15. Two cells of alkali bee in soil, one with pollen ball and egg, the other with a half-grown larva. (Photo by W. P. Nye, Federal Apiculture Research Laboratory, Logan, Utah.)

FIG. 16. "Artificial" nesting site for alkali bees with one area sheltered from rain. Note nest mounds. (Photo by H. Potter, Federal Apiculture Research Laboratory, Logan, Utah.)

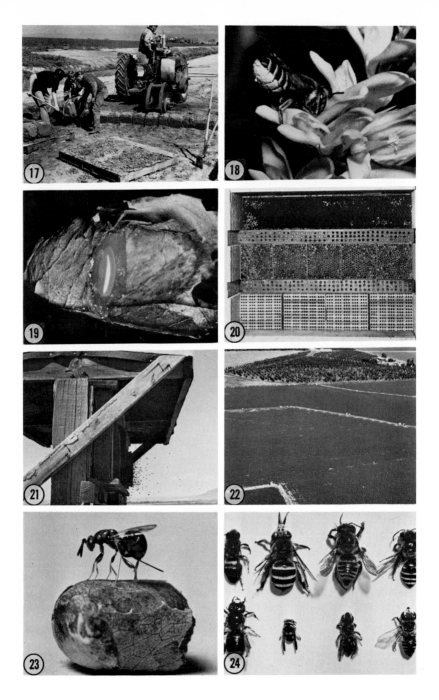

FIG. 17. Cutting and loading soil cores containing overwintering larvae of alkali bees. (Photo by H. Potter, Federal Apiculture Research Laboratory, Logan, Utah.)

1958, it had become abundant in the western states, and its potential as a new alfalfa pollinator was recognized. The 1960s marked the rapid development of a new industry devoted to increasing populations of the alfalfa leafcutter bee and using it in alfalfa seed fields.

The alfalfa leafcutter bee nests above ground in narrow, usually tubular cavities. It constructs a linear series of brood cells composed of leaf or petal pieces (Fig. 19). Seed growers soon started putting soda straws in containers and drilling holes in timbers to attract nesting bees (Fig. 20). During the last few years, many thousands of acres have been pollinated by millions of these bees, all of them raised in specially prepared nesting materials. Up to 5000 or more female bees per acre are recommended, and many growers have reached this figure. Since foraging by this species usually takes place within a few hundred feet of the nesting place, the nests and nesting materials are placed in shelters (Fig. 21) scattered through the fields at intervals of several hundred to 1000 ft (in the range of 3 to 10 acres per shelter) (Fig. 22). When the nearby flowers have been tripped, the shelters are often moved to less well-pollinated areas or to later-blooming stands of alfalfa (Bohart and Knowlton, 1967).

Although many of the factors involved in the survival and reproduction of *M. rotundata* can be placed under man's control, careful husbandry is required to control predators and parasites (Fig. 23) and to ensure the proper use of insecticides. Sometimes, even when the best-known techniques of management are used, there is a high mortality of immatures. The problem of this "unexplained mortality" must be solved before leafcutter bee management becomes truly reliable (Bohart, 1972).

Since the alfalfa leafcutter bee is manageable under relatively controlled conditions, it can be used successfully outside its natural areas of adaptation, as, for example, in Alberta. This fact, combined with the ease with which it can be transferred to and established in new areas, has

FIG. 18. Adult female alfalfa leafcutter bee "tripping" alfalfa florets. (Photo by W. P. Nye, Federal Apiculture Research Laboratory, Logan, Utah.)

FIG. 19. Egg of alfalfa leafcutter bee on food mass in leaf cell. (Photo by W. P. Nye, Federal Apiculture Research Laboratory, Logan, Utah.)

FIG. 20. Four kinds of nesting materials for alfalfa leafcutter bees (corrugated cardboard, drilled wood, soda straws, and grooved boards). (Photo by W. P. Nye, Federal Apiculture Research Laboratory, Logan, Utah.)

FIG. 21. Alfalfa leafcutters entering and leaving nests in soda straws. (Photo by W. P. Nye, Federal Apiculture Research Laboratory, Logan, Utah.)

FIG. 22. Shelters for alfalfa leafcutter bees along roadways through alfalfa seed fields. (Photo by H. Potter, Federal Apiculture Research Laboratory, Logan, Utah.)

FIG. 23. A wasp (*Monodontomerus obscurus*) laying eggs in leaf cell of the alfalfa leafcutter bee. (Photo by H. Potter, Federal Apiculture Research Laboratory, Logan, Utah.)

FIG. 24. Eight species of alfalfa pollinators collected in alfalfa fields in Iran. (Photo by W. P. Nye, Federal Apiculture Research Laboratory, Logan, Utah.)

resulted in recent worldwide interest in its utilization for alfalfa pollination. Among other seed crops for which the alfalfa leafcutter bee has been considered are red clover and crown vetch. It does not visit red clover readily and would probably be useful for pollination of this crop only in enclosures or in the absence of more favored pollen sources. Preliminary trials indicate that it would be an effective pollinator of crown vetch, although some degree of isolation from more attractive pollen sources might prove necessary.

Although it is generally believed that the honeybee is the logical species to use for any crop that it pollinates effectively, some growers claim that the alfalfa leafcutter, being more "domesticated" than the honeybee, should be considered for use on any crop that it pollinates effectively. In support of this thesis, they cite its tendency to stay on the field in which it is placed, its total lack of aggressiveness toward humans, and its complete hibernation (and, therefore, lack of care required) during most of the year. Insect-pollinated seed crops attractive to the alfalfa leafcutter include sweet clover, Dutch clover, and cole crops. However, it rarely visits some important seed crops such as red clover, onions, and carrots.

K. Other Bees

Although many bees pollinate seed crops, relatively few seem to have much potential for management as pollinators. The most useful would be those that pollinate crops for which the honeybee is not entirely satisfactory. For example, honeybees rarely visit tomatoes for pollen and are inefficient at it when forced to do so (as in a greenhouse). Nevertheless, it is known that in Peru, the homeland of tomatoes, many species of bees are specially adapted for collecting pollen from the cultivated tomato and its congeners (Rick, 1950). It is logical to assume that some of these bees could be successfully introduced to California or Arizona to assist in hybrid seed production. Production of hybrid tomato seeds in Peru is another logical possibility.

Although, as previously noted, we now have three manageable species of pollinators of alfalfa, other species would be helpful under some circumstances. For example, the alfalfa leafcutter bee declines in population about a month after the first generation emerges (by which time only a few of the second generation bees have appeared). A related bee, *Megachile concinna* Smith, was accidentally introduced into this country and has become well established in the Southwest. More recently, it has been found in Washington, where it performs much the same service as the alfalfa leafcutter bee. However, it emerges somewhat later and is still near its peak population during the downswing in the population of the alfalfa leafcutter bee (Eves, 1970). It appears that management techniques for the alfalfa leafcutter bee will also succeed with *M. concinna,* though minor modifications may be necessary. Many other pollinators of alfalfa are known in southeastern Europe and southwestern Asia (Fig. 24). Several of these show a special affinity for small-flowered legumes and at least a few, judging from their gregarious behavior and nesting habitat, could be managed successfully. These examples indicate that various segments of the seed industry could benefit from further research on the potential value of untried pollinators throughout the world and, particularly, on their suitability for management in seed-producing areas (Bohart, 1962).

L. Pollination in Enclosures

Plant breeders often enclose insect pollinators with plants to achieve various kinds of self- or cross-pollination without the introduction of unwanted germ plasm. Blowflies (Calliphoridae) are commonly used in small cages (usually less than 2 ft^3) and honeybees, bumblebees, or alfalfa leaf cutting bees in larger enclosures. Blowflies usually live better than bees in very small spaces and are readily available in large numbers. In Europe they are often obtained as pupae from companies supplying fish bait (Free, 1970). However, they cannot be used for plants that require special manipulation of the flowers, such as alfalfa or tomato.

Among the bees, honeybees have been used most commonly because of their universal availability. However, full-sized colonies collapse quickly and the large numbers of bees often damage the flowers. Small colonies or groups of bees require considerable care and are usually of no value after the pollination period is ended. Where contamination from outside pollen is no problem, colonies can be given entrances leading into the enclosure and into the open. In most cases, an hour or two a day of exposure to the bees results in adequate pollination (Shemetkov, 1960).

Bumblebees have small colonies that are well adapted to pollination in large field cages (Pedersen and Bohart, 1950; Horber, 1971). Bumblebee colonies are often difficult to obtain, however, and are not available until summer. For pollination in the spring, European workers sometimes use field-caught queens. Interestingly enough, Dutch workers, supposing that collecting healthy queens might reduce bumblebee populations, have used parasitic bumblebees (*Psithyrus*) and bumblebee queens parasitized by nematodes (Kraai, 1958).

Alfalfa leafcutter bees, now readily available as overwintering larvae from many suppliers in the western United States, are ideally suited for pollination in enclosures (Bohart and Pedersen, 1963). When confined, they even pollinate crops such as red clover and carrot which they rarely visit in the open. At the end of the pollination season, there is often a supply of overwintering larvae for use the next year.

Insects are also used for pollination of fruits and vegetables in large glasshouses. Since these glasshouses are usually much larger than cages or the glasshouse sections used for breeding work, they are better suited for at least small colonies of honeybees. Commercial glasshouse crops commonly pollinated by honeybees include strawberries, peaches, apricots, cucumbers, melons, and various flower seeds. Honeybees are also used to pollinate glasshouse tomatoes, even though they do not ordinarily visit tomatoes in the open.

III. Harmful Impact of Insects on Seeds and Seed Production

A. *Importance of Seed-Destroying Insects to Seed Production*

Estimates of economic losses to seed insects in a strict sense are elusive. Many seed crops are produced on small acreages either as part of an operation producing many kinds of seeds or as a sideline to some other agricultural operation. In some instances, the individual seed crops have not received the attention required to make estimates of losses resulting from insect attack.

Seeds of some of our more important crop plants are extensively grown. In these instances there are abundant data on seed yields and losses. However, the published figures usually are the result of several component factors. For example, in Nebraska in 1970, a loss of $36,090 was reported on 30,000 acres of alfalfa grown for seed production (Agricultural Research Service, 1970). The two main components used to arrive at this estimate were a cost of $24,750 for insect control operations and a reduction in seed yield valued at $11,340 on areas not treated for insect control. However, of the ten species of insects attacking the crop, only two, strictly speaking, were seed insects.

Voluminous data are available on increases in seed yield following insecticide applications, but it is usually impossible to separate the effect of the insecticide on insect pollinators and other insects from its effect on the seed insects. For example, Stone and Foley (1959) reported that the yield of dry lima beans was increased from 1650 lb/acre on untreated plots to 1910 lb/acre by properly timed application of DDT. In this case, it is probably safe to ascribe most of the increased yield of 260 lb/acre to a reduction in the population of two species of seed insects. However, each report of this nature must be analyzed with careful attention to the kinds of insects affecting the crop, the dependence of the crop upon insect pollinators, and the effect of the particular insecticide on the insects.

Thus, whereas this information is excellent for judging the effectiveness and economic benefits of insecticide applications, it is not quite adequate for judging the impact of seed insects.

In a few instances, intensive investigations on certain seed crops and the insects attacking them have provided a detailed picture of the impact of seed insects on their host plants. If one is willing to accept some risky generalizations, these studies will serve to illustrate the impact of seed insects on seed crops in general.

Lygus bugs (*Lygus* spp.) attacking the alfalfa seed crop are quite susceptible to contact insecticides. When DDT first became available, it was quickly discovered that properly timed applications greatly reduced the *Lygus* population, and seed yields were correspondingly increased. For example, Smith and Michelbacher (1946) reported that well-timed applications of DDT dust to alfalfa resulted in seed yield increases ranging from 163 to 328%.

In most instances, yields increased by more than 100 lb/acre. In an extreme case, when very dense *Lygus* populations were present for a long time, the yield increased from 15 to 411 lb/acre.

Carlson (1961) studied the effect of *Lygus* bug attack on yields of table beet seeds by enclosing known numbers of bugs in cages with developing beet seed heads. He reported reductions in seed yield ranging from 33 to 44%. The reduction in yield appeared to become economically significant at population densities exceeding 22 insects/plant. In addition to the reduction in the weight of seed produced, there was also a reduction in the number of viable germs per seed. At a population density of 22 *Lygus* bugs/plant, the viability of the seeds dropped below the acceptable minimum of 85%.

Hills (1943) obtained similar results when he caged individual *Lygus* bugs, or stink bugs (*Chlorochroa* spp.), on developing sugar beet seeds. The *Lygus* bugs reduced the number of viable seed balls produced from 425 in cages without insects to a range of 325 to 383 per cage with insects. The stink bugs destroyed most of the seeds in their cages, reducing the yield to 126 viable seed balls per cage.

Blickenstaff and Bauman (1961) tested insecticide-dipped pollen exclusion bags to control the corn earworm (*Heliothis zea* Boddie) in Georgia. They reported that 93% of the ears not covered by treated bags were infested by corn earworms and fall armyworms (*Spodoptera frugiperda* A. & S.) which damaged an average of 89.9 kernels/ear. Their most effective insecticide-dipped bags reduced the damage by 99.4%.

Buckley and Burkhardt (1962) confined known numbers of corn earworms on developing seed heads of grain sorghum. They found that a single larva destroyed 166 kernels or approximately 6% of the seed complement. Two larvae per head destroyed 10% and three larvae destroyed 13%.

Carlson (1967) investigated the effect of the sunflower moth (*Homeo-soma electellum Hbn.*). He found that the larvae destroyed an average of nine seeds each and yield reductions often reached 30–60%. Sunflowers protected by a series of insecticide treatments produced 3361 lb of seeds/acre, whereas unsprayed plants produced 2089 lb/acre.

Insects attacking tree seed crops exact an even greater toll. Koerber (1962) found that insects destroyed 79% of the Douglas fir seed on individual trees studied in northwestern California. The Douglas fir cone moth was responsible for two-thirds of the damage. The larvae of the Douglas fir cone moth destroyed approximately twenty seeds each, and three larvae per cone destroyed the entire seed complement.

Merkel (1967) studied insect infestations in the cones of fifteen slash pine trees for 3 consecutive years. Over the study period, the slash pine seed worm infested an average of 73% of the cones. Some individual trees had 90% or more of the cones infested each year.

The seeds of wild plants are at least as heavily attacked as those of crop plants. However, we are not usually much concerned with seed production of wild plants, so their seed insect problems have generally gone uninvestigated. Wild plants are not, however, unimportant to us. Vast areas of natural vegetation are managed to provide forest products, grazing for sheep and cattle, and wildlife habitat. Other large areas, such as parks and watershed lands, are also in natural vegetation. The quantity and quality of the products produced by these lands are partly dependent upon the composition of the vegetative communities covering them, and seed insects are one of the factors influencing the species composition of natural plant communities.

Natural plant communities such as forests and rangelands differ from croplands in that they are not replanted annually. In a forest, for example, the same plants occupy the land for tens or perhaps hundreds of years. As long as the land is fully stocked with desirable plants, the fate of the seed crop is of little consequence. Only a small amount of seed need survive to provide replacements for the individual plants that die. However, in the event of a disaster, such as a forest fire, the fate of the seed crop on the surviving trees and those on the edge of the devastated area may be of the utmost importance in determining the species composition of the new plant community for a long time after the fire.

There is also the possibility of slow shifts in species composition. Selective removal of desirable tree species or selective grazing of palatable range plants will tend to favor an increase in the proportion of less desirable species remaining. However, the ability of the less desirable species to increase and occupy the vacant spaces in the community is subject to the influence of seed insects, among other factors.

Much circumstantial evidence shows that seed insects influence the species composition of natural plant communities. However, clear, direct evidence is scanty. Gibson (1969) reported twenty-seven species of weevils of the genus *Curculio* which feed mainly on the seeds of oaks are found in the hardwood forests of North America. It seems reasonable that they might influence the proportion of oak in the forests. The cones of eastern white pine are destroyed by the beetle *Conophthorus coniperda* Hopkins. During one recent period in New England this insect was the principal cause of white pine trees failing to produce a good seed crop for 10 years in succession (Fowells, 1965). We may reasonably suppose that few white pine seedlings became established in that period, and proportionately more trees of other species are present in forest stands originating at that time. From a forestry standpoint, white pine is one of the most desirable trees growing in New England. It will become established under and grow up through a stand of thin-foliaged species such as birch (Toumey, 1919), but its growth rate is reduced in proportion to the competition for light.

In the Pacific Coast states, if desirable conifers do not become established within 3 to 5 years after fire or logging, they face serious competition from woody shrubs. A long-lived plant community dominated by *Ceanothus* and *Arctostaphylos* species, commonly called a brush field, is likely to develop and prevent or retard the establishment of trees for decades. The kind of plant community which develops is, of course, dependent on the quantity of various kinds of seed available, which is influenced by seed insects.

The seed insects of range plants have received some attention in recent years. Ferguson *et al.* (1963) found four species of insects destroying seeds of bitter brush (*Purshia tridentata*), a desirable range plant in Idaho. Nord (1965) reported two insects feeding on bitter brush seeds in California. Neither of these reports assesses the importance of the seed insects in regard to maintaining the bitter brush stand, but bitter brush grows in competition with less desirable plants such as sage (*Artemisia* spp.) and rabbit brush (*Chrysothamnus* spp.), and it seems reasonable to assume that after a fire, or in the event of overgrazing, the abundance of seeds and the insects destroying seeds will have a strong influence on the composition of the range plant community.

Undesirable range plants are also attacked by seed insects. Johnson (1970) found seven species of *Acanthoscelides* beetles destroying the seeds of loco weeds (*Astragalus* spp.) in Arizona, California, and Oregon. The loco weeds are notoriously poisonous to livestock, and the poison is present in high concentrations in the seeds. Evidently *Acanthoscelides* beetles have evolved an immunity to the toxic chemical in their food supply. It is interesting to note that another species of *Acanthoscelides* is a well-known pest of bean seeds.

Janzen (1972) has studied seed-destroying animals, including insects, in relation to the composition of plant communities. He regards seed insects as predators attacking a special sort of immobile prey, seeds. He reported that in tropical forests, seed predators, including insects, are very effective at finding and destroying nearly all the fruits or seeds of their host plants, resulting in plant communities characterized by marked species diversity and relatively great distances between individuals of the same species. In temperate zone forests, the seed predators are usually less effective, permitting interspecific competition to become more important in determining the species composition of the forests. When the community structure is set primarily by interspecific plant competition, the most competitive species tend to exclude all others with a similar life form, and species diversity is thus held to a low level.

B. *Destruction of Seeds by Insects*

Insects vary in the degree to which they specialize in attacking the reproductive structures of plants. We may consider the least specialized species to be facultative or opportunistic seed destroyers. These insects normally feed on something other than seeds, but the range of acceptable foods includes the reproductive structures of plants when they are available. Other insects are specialized feeders on buds, flowers, or fruits, and destruction of seeds is only incidental to their feeding on another portion of the reproductive system. The most specialized insects feed exclusively upon the seed or some other specific element of the reproductive system. These insects usually show remarkable adaptations to their specialized way of life and often attack only one species of plant. A great many of the common insect pests of our farms and gardens are facultative seed destroyers. The field cricket, *Acheta assimilis* F., feeds on a wide variety of plant and animal materials. When abundant, crickets will destroy the developing seeds of cereal grains, legumes such as peas, beans, and alfalfa, and will also eat the developing buds or fruits of squashes, tomatoes, and cotton (Dewey, 1970). Various species of aphids are often the most abundant insects found on crop plants. They grow and multiply with no difficulty on the vegetative parts of their host plants. However, when the grain aphid, *Macrosiphum granarium* (Kirby), is present on the developing seed heads of its small grain host plants, blasted seed heads may result. The pea aphid, *Macrosiphum pisi* (Kaltenbach), attacking legumes and the cabbage aphid, *Brevicoryne brassicae* L., attacking cruciferous plants, will also reduce the seed yields of their hosts when present in large numbers during the developmental stage of the seed. The Japanese beetle, *Popillia japonica* Newman, is a notorious opportunistic feeder. The larvae live in the upper layers of the soil, feeding on roots. The adults emerge from the soil in July and August and have been reported to feed on the foliage, flowers, and fruit of over 250 plants. The emerging silks of corn are a favorite food. When dense populations of Japanese beetles eat the corn silks, they, of course, interfere with pollination and greatly reduce the seed yield. Fortunately, the corn-growing areas of the Midwest have not yet been invaded. In forests of the western United States, a beetle, *Dichelonyx crotchii* Horn, also is a root feeder in the larval stage. The adults normally feed on the foliage of pines, but for a few weeks in the early summer, their favorite food is the tender young cones. Some of the young cones are entirely eaten and many more are damaged to the extent they later dry up and fall from the trees. Spruce budworms, *Choristoneura* spp., are well-known defoliators of firs, spruce, and Douglas fir. The larvae feed mainly on the foliage of the host tree, but they also consume the developing staminate and ovulate cones of the tree.

Although the foregoing species are not seed insects in the strict sense, their habits and other circumstances combine at times to permit them to seriously affect seed production, and they become *de facto* seed insects. There are a great many other insects the primary food of which is reproductive structures of plants. The remainder of this chapter is devoted to a close look at some representatives of these obligate seed insects.

Possibly the most ubiquitous members of this group are the thrips (Fig. 25). They are elongate, usually dark-colored insects ranging from 0.02 to 0.06 in. in length. Adults characteristically have two pairs of fringed wings which are held lengthwise over the back. Their mouthparts are basically of the chewing type but have been modified for piercing or rasping and sucking. A close examination of almost any flower will reveal a population of these tiny insects living among the floral parts. Here they feed by piercing or rasping away the tender surface cells and sucking up the exuding juices, or by emptying the contents of pollen grains. The females usually insert their kidney-shaped eggs into plant tissues. The eggs produce nymphs which resemble the adults except for being wingless and initially much smaller. They grow and reproduce rapidly and, under favorable circumstances, produce many generations per year. Several members of the genus *Frankliniella*, known collectively as flower thrips, are often abundant in the blossoms of deciduous fruit trees in the spring. Here they feed on the floral parts and developing fruits, causing blemished or malformed fruit to develop. Later in the season, they invade the blossoms of alfalfa, onions, and a wide variety of ornamentals, causing reduction of up to 50% in seed yield (Bailey, 1938). During the summer, a generation may be completed in 16 days. There are commonly five or six generations per summer. In warm climate areas, reproduction may continue at a reduced rate through the winter. The onion thrips, *Thrips tabaci* Lind., causes substantial yield reductions in onions grown for seeds. Elmore (1949) reports that thrips populations in the range of several hundred per umbel occurred in seed fields in various parts of California. Before the seed stalks appear, this thrips feeds on the vegetative portions of the plants. As soon as the bracts enclosing the flower umbel open, the insects invade the developing flower head and feed successively on the developing buds, flower pedicels, and anthers. Heavy feeding on the buds and pedicels prevents normal flower development and seed production. Another species of importance to seed producers is the composite thrips, *Microcephalothrips abdominalis* (Crawford), which is reported by Bailey (1938) to feed on the developing seeds of zinnias, marigolds, and other composites. Thrips also attack the reproductive structures of forest trees. The staminate cones of conifers are often heavily infested, and one species, *Gonothrips fuscus* (Morgan) (Fig. 25) is reported by Ebel (1961) to cause serious damage to the opening buds and young cones of slash pine. The surface damage permits beads of resin to exude from the cone. Up to 20% of the crop is lost as the damaged cones dry up and drop from the tree.

Several species of true bugs of the genus *Lygus* are widespread pests of many seed crops. *Lygus hesperus* Knight and *Lygus elisus* VanDuzee are western species; *Lygus lineolaris* (Polisat de Beauvais) is found in the East; and *Lygus campestris* L., a northern species, is sometimes important in the agricultural areas of Canada. These insects attack the seed pods of alfalfa, beans, carrots, beets, sugar beets, lettuce, cotton, safflower, and many ornamental plants.

The adult *Lygus* bugs are flattened oval insects about 0.25 in. long (Fig. 26). They are usually yellowish brown or greenish brown in color. In cold climates, the adults hibernate in sheltered locations in the winter, whereas in warmer areas, such as southern Arizona or California, they may be active all year.

The females insert their curved sausage-shaped eggs into stems or buds. The eggs hatch in 8 to 10 days, producing nymphs which at first are similar in appearance to green aphids but have a dark spot in the center of the abdominal dorsum (Fig. 27). The nymphs feed on plant juices and grow rapidly, passing through five developmental stages to become adults in 10 days to 2 weeks. The adults continue to feed and lay eggs, producing ever-increasing numbers of progeny as the summer advances. In most of our agricultural areas, there are four to six generations per year.

Both the nymphs and adults are active, and the adults fly readily. When early maturing crops, such as small grains, begin to ripen and become dry, they readily move to nearby fields of more succulent crops such as alfalfa. The mouthparts are designed to pierce the tissues of the host plant and suck fluids from them. They have two pairs of piercing stylets and two tubes enclosed in a jointed beak attached to the lower front of the head. One of the tubes is connected to the salivary gland and is used to inject saliva into the plant tissues. The other is connected to the esophagus and is used to suck up fluids. The feeding process of *Lygus* bugs has been intensively investigated by Strong and Kruitwagen (1968) and Strong (1970). They report that the insect inserts its stylets to a maximum depth of 2 mm, partly withdrawing and reinserting them to lacerate the tissues. At the same time, saliva containing polygalacturonase, a powerful digestive enzyme, is injected into the tissue. After a few seconds, the fluids released are sucked up, and the process is repeated until a mass of tissue up to 1 mm across is destroyed. *Lygus* bugs feed preferentially on the meristematic tissue and on the developing reproductive organs of their host. Depending on the structure attacked and the stage of its development, feeding by *Lygus* bugs may result in dropping of buds or fruits, formation of necrotic areas within fruits or seeds, or production of shriveled empty or embryoless seeds.

Sorenson (1939) studied two species of *Lygus* bugs in relation to alfalfa seed production. He noted several types of damage, depending on the stage of plant development. When *Lygus* bugs feed on alfalfa buds (Fig. 27), the buds are killed and in a few days they turn gray and drop off. Flowers are similarly affected and drop soon after *Lygus* feeding. The immature seeds within the pods are also attacked, causing them to become misshapen or shriveled (Fig. 28). Sorenson (1939) found the degree of injury to vary directly with the intensity of infestation. He estimated average *Lygus* infestations in the Uinta Basin of Utah to be about 311,000 bugs/acre.

Lygus bugs also attack beans. Elmore (1955) reports they prefer to feed on the flower pedicels and the small seed pods. Early feeding causes shedding of buds or blossoms. Small pods may be shed or they may be stunted or distorted. The number of beans in such pods is reduced and many of them are deformed. If the preferred young pods are scarce, the bugs feed on beans approaching maturity. They puncture the developing beans within the pod and destroy a spot of tissue immediately around the point of penetration. The spots of dead tissue show up as pits in the mature bean.

Lygus bugs cause a more subtle type of damage when they feed on developing seeds of umbelliferous plants such as carrot or celery. When feeding occurs during the flowering stage, some of the developing ovaries

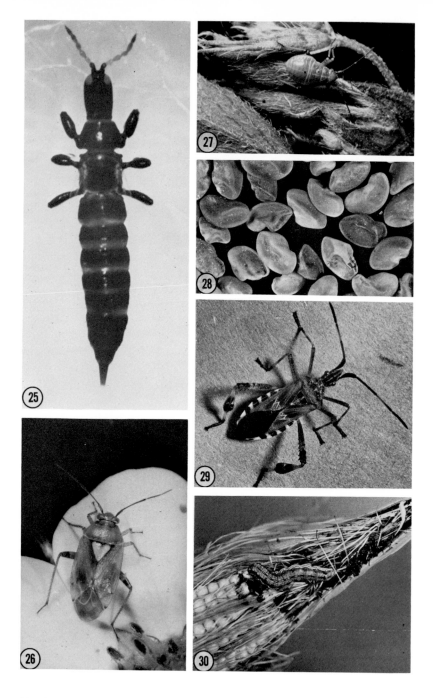

FIG. 25. A thrips, *Gonothrips fuscus,* from slash pinecones. (Photo by E. P. Merkel, S. E. Forest Experiment Station, Olustee, Florida.)

are destroyed, thus reducing seed yield. However, if *Lygus* bugs feed on developing seeds, they may kill the embryos but cause no apparent damage to the endosperm or fruit coat. This problem was studied by Flemion and Olson (1950), using the developing seed heads of coriander, dill, fennel, carrot, celery, parsley, and parsnip. They reported that in these plants the endosperm of the seed quickly matured and became firm, whereas the embryo remained immature for a relatively long time. When *Lygus* bugs feed on these seeds, they destroy the embryo, leaving a cavity where the embryo would normally be.

Lygus bugs also cause severe damage to the seed crops of both table and sugar beets. Hills (1950) reports that *Lygus* bugs feeding on sugar beet seeds cause a reduction in the percentage of viable seeds and in the number of sprouts per viable seed. By caging the insects on developing seed stalks at various stages of development, they found that there was no damage in the prebud, flower bud, or early stages of seed development. The greatest damage occurred when plants in the late bloom-to-soft seed stages were exposed to *Lygus* feeding. Carlson (1961), using similar methods, confirmed these findings and reported that *Lygus* populations equivalent to 12.7 bugs/table beet seed plant may be sufficient to cause a significant reduction in seed viability.

Another group, the pentatomids, or stink bugs, have some notably destructive representatives. The Say stink bug, *Chlorochroa sayi* Stal feeds on a great variety of crop plants. It is reported by Hills and Taylor (1950) to damage beet seeds in much the same way as *Lygus* bugs, but it is able to damage the mature seeds and may continue to damage them after the plants have been cut. Developing bolls of cotton are often attacked and heavy damage may cause shedding of the bolls. *Chlorochroa uhleri* Stal and *Chlorochroa ligata* Say were found by Nord (1965) to feed on the immature fruit of bitter brush, an important range plant. The seed produced was shriveled and did not germinate.

The developing seeds of forest trees, even those enclosed in large woody pine cones, are also subject to attack by true bugs. A leaf-footed

---

Fig. 26. Adult *Lygus* bug on strawberry blossom. (Photo by W. W. Allen, Department of Entomology, University of California, Berkeley, California.)

Fig. 27. Young *Lygus* bug nymph feeding on alfalfa buds. (Photo by W. P. Nye, Federal Apiculture Research Laboratory, Logan, Utah.)

Fig. 28. Alfalfa seed shriveled by *Lygus* bug feeding. (Photo by W. P. Nye, Federal Apiculture Research Laboratory, Logan, Utah.)

Fig. 29. Leaf-footed plant bug, *Leptoglossus occidentalis*. (Photo by T. W. Koerber, Pacific Southwest Forest & Range, Experiment Station, U. S. Department of Agriculture, Berkeley, California.)

Fig. 30. Larva of corn earworm, *Heliothis zea* in ear of sweet corn. (Photo, U. S. Department of Agriculture.)

bug, *Leptoglossus occidentalis* Heid. (Fig. 29), was reported by Koerber (1963) to feed on the developing seeds of various western coniferous trees. Debarr (1967) found a similar species, *Leptoglossus corculus* (Say), feeding on seeds of pines in the eastern United States. *Leptoglossus* bugs are large insects, almost an inch long, and have a proboscis capable of penetrating through 0.75 in. of cone scales to feed on the seeds within a pine cone. Studies by Krugman and Koerber (1969) showed that the stylets passed through the cone scales, causing only minimal damage. The nature of the damage to the seeds strongly suggests that the insect feeds by injecting digestive enzymes into the seed to break down or dissolve the tissues and then sucks up the resulting fluid. Seeds subjected to 1 to 5 hours of feeding by *L. occidentalis* showed disorganization of the cellular layers of the nucellus and endosperm and removal of the cell contents without complete destruction of the cell walls.

The larvae of various moths are prodigious destroyers of reproductive structures of plants. This group is typified by the familiar corn earworm, *Heliothis zea* (Boddie) (Fig. 30). The same moth is known as the cotton bollworm and tomato fruitworm, depending on the plant it is eating. Sorghum, peas, beans, and various other crops are also subject to attack. The adult moth is dull yellowish brown with darker brown lines and bands on the forewings. It has a wing span of about 1.5 in. The females are active at night when they fly from plant to plant, depositing their eggs singly on the silks of immature corn ears or the tender new growth of tomato, cotton, or other food plants. The eggs are white at first, darkening with age. They are slightly smaller than a pinhead and hemispherical in shape with a pattern of ribs radiating outward from the top center. The eggs hatch in 3 to 8 days, depending on temperature. The new larvae feed initially on the corn silks or tender new foliage. On corn they soon invade the silk end of the ear and begin to consume the developing kernels. On other host plants, they seek out the reproductive structures. On cotton, they bore into the base of the buds or immature bolls. On tomatoes, they bore into the stem end of the fruit, and on sorghum they invade the seed head and consume the developing kernels. Once they have bored into the reproductive structures, they are difficult to detect and very difficult or impossible to control with chemicals. The feeding activity of the larvae destroys part of the fruit and also furnishes an entry for yeasts, fungi, or other microorganisms which further damage the fruit.

The mature larva is variable in color, ranging from pale green to various shades of brown or almost black, and usually having lighter longitudinal stripes. The mature larva leaves the plant and burrows into the ground where it pupates and eventually becomes an adult moth. Individuals that reach the pupal stage in late summer or fall do not reach adulthood until the following spring. In the northern United States with its short summers, the corn earworm has two generations per year, whereas five or six generations per year may be produced in the South. For further information and color illustrations of the corn earworm, the reader is referred to the *Yearbook of Agriculture* (U.S. Department of Agriculture, 1952).

The corn earworm is not strictly a seed insect. It also causes severe reductions in the yield of cotton grown for fiber and is a serious quality control problem in canning tomatoes and sweet corn. In a seed production situation, it sometimes causes serious losses when the larvae feed inside the pollen exclusion bags used in production of inbred or hybrid seeds of grain, sorghum, and corn (Buckley and Burkhardt, 1962; Blickenstaff and Bauman, 1961). It appears that earworms inside the bag are partly protected from their natural enemies, and a high proportion survive to eat the developing seeds. Of course, the value of the seed has a bearing on the magnitude of the problem. A loss which might be tolerable in a feed grain production operation could be disastrous in a limited stock of a hand-pollinated hybrid line.

The pink bollworm, *Pectinophora gossypiella* (Saunders) is another lepidopterous pest of cotton bolls, but it is more specific to cotton than is the corn earworm. During its larval stage, the pink bollworm lives within an immature cotton boll, feeding primarily on the seeds. Heavily infested bolls may be shed, and more lightly damaged bolls contain many injured seeds and much damaged lint. During the summer generations, of which there are several in the southern United States, the mature larvae leave the bolls and pupate on the ground. During the winter, most of the larvae hibernate in infested bolls (Fig. 31) (Curl and White, 1952).

The seeds of trees are subject to destruction by the larvae of a group of moths of the genus *Dioryctrya*. Three species attack the cones of slash and longleaf pine in the southeastern United States (Ebel, 1963). The larvae, commonly called coneworms, are generally the most injurious insects attacking the cones of slash and longleaf pines. They mine indiscriminately through the cone, destroying seeds and cone scales alike. The *Dioryctrya* coneworms do not restrict their feeding to the cones but also attack staminate flower clusters and buds. Several generations per year are produced.

One species, *Dioryctrya abietella* (D. and S.), is a very widely distributed moth. It feeds in the cones of the pine and firs and including Douglas fir in most forested areas in the Northern Hemisphere. This species will sometimes kill a small area in a developing cone, allowing the remainder to grow and develop. The affected cones are sharply curved or distorted so that the cone scales on one side cannot open to release the seed. In this way, the insect causes the loss of much more seeds than it actually eats.

The cones of Douglas fir are attacked by a moth of more specialized habits, *Barbara colfaxiana* (Kearf.). The adult is a grayish brown moth with a wing spread of about 0.75 in. Its life cycle is closely linked to the development of its only host, the cones of Douglas fir, and its entire life cycle, except for the free flying moth stage, occurs on or within the Douglas fir cone.

The moths emerge from old cones in early spring just as the cone buds of Douglas fir are opening. They deposit their eggs on the bracts of the new cones. The eggs hatch in about 2 weeks and the new larvae bore into the cone between two scales or between a bract and a scale. They bore toward the center of the cone, feeding on the scales, bracts, and seeds. Pitch and frass accumulate in the irregular cavities excavated by the feeding larvae. Late in July, the larvae reach maturity and spin a papery cocoon within the cone. Pupation occurs upon completion of the cocoon, and the insects remain in the pupal stage until the following April (Fig. 32). Ten to 20% of the pupae do not transform to moths after the first winter but remain in the pupal stage for another year. These individuals serve to carry the population through years when the cone crop fails. Keen (1958) reports that one larva will destroy 45% of the seeds in a cone, and three larvae will destroy the entire seed complement.

Another relatively specialized insect is the notorious boll weevil, *Anthonomus grandis* Boheman (Fig. 33). The boll weevil is a grayish or brownish beetle about 0.25 in. long with a slender snout about half as long as its body. The adults emerge from their hibernating places in early spring. The boll weevil prefers the squares, as the buds of the cotton plant are commonly called, for both feeding and oviposition, although the bolls, which are the developing fruits, are also attacked. The eggs are deposited in a deep puncture made in a square, or boll, by the female weevil with her long snout. The egg hatches in a few days to produce a legless white grub with a brown head. The larva proceeds to eat the inside of the square or boll, reaching maturity in 1 to 2 weeks. Pupation occurs within the damaged square or boll. The pupal stage lasts for less than a week, after which the new adults chew their way out. After a few days of feeding, they are ready to start reproducing. The injured squares and smaller bolls usually drop from the plant, but the weevils complete their development, either on the plant or on the ground. In the warmer areas of the Southeast, there may be eight to ten generations per year. As winter approaches, the adult weevils seek out sheltered places under trash on the ground or loose bark of trees to hibernate through the winter. Those that survive the winter resume their attacks on the next crop of cotton. This insect is better known for its annual ravages of cotton fiber crops of the southeastern states than for damage to seed crops. However, it is an obligate destroyer of the reproductive structures of the cotton plant, and in a cottonseed production operation it is certainly a formidable seed insect. The weevils belong to a very large family which has many species of obligate seed insects. Adult weevils are characterized by the presence of a snout, and those that attack fruits and seeds often have unusually long ones. The legless larvae complete their entire development within a single fruit or seed. (For further information on the boll weevil and other weevils attacking crop plants, see Metcalf *et al.,* 1951.)

Another interesting group of beetles destroys the cones of pine trees. The ponderosa pine cone beetle, *Conophthorus ponderosae* Hopkins, is a good representative of these beetles. The adult cone beetle is a dark-brown cylindrical insect about 0.1 to 0.17 in. long. The female beetle bores into the base of a ponderosa pine cone and severs the water-conducting elements surrounding the core of the cone. The beetle then tunnels along the central axis of the cone. At intervals along the tunnel, it deposits white eggs in little pockets cut in the side of the burrow, usually adjacent to a developing seed. After depositing eggs, the beetle leaves the cone and seeks out another cone to repeat the process. When the water-conducting tissues of the cones are severed, they die and, thus, produce no seeds, regardless of whether the remainder of the insect's reproductive process is successfully completed. Miller (1915) reports that a similar species, *Conophthorus lambertianae* Hopkins, completes its boring and egg laying in 5 to 8 days. The period of time over which cones are killed suggests that each female beetle destroys six to eight cones. The eggs hatch in a few days, and the white legless larvae feed initially on the immature seed and later on the cone scales. The larvae mature in about a month and transform to pupae and, in turn, to adult beetles. The adults remain in the cone through the remainder of the summer and winter, emerging the following spring to attack the next cone crop. Miller (1915) reported that infested cones produced an average of 6.5 new beetles per cone. However, reproduction is often unsuccessful, and usually less than half of the cones that are killed produce any new cone beetles.

The gall midges (Cecidomyidae) are another group of insects of interest to the seed grower. The adults are very small mosquitolike flies. The larvae are small, elongate, often spindle-shaped maggots which are usually colored red, orange, yellow, or white. They feed on plant juices, frequently living in a small cavity within some part of the plant. Many species are extremely specialized. They may be restricted to a single plant host, live within a particular structure of the host, or even alter the biochemistry of the host to cause it to form a gall, enclosing and protecting the insect larvae. The larvae feed exclusively on liquids. In the case of species of interest to seed growers, these fluids are extracted from the reproductive structures of the plants.

The sorghum midge, *Contarinia sorghicola* (Coquillet), a representative of this group, attacks the seed of sorghums, broom corn, Johnson grass, and Sudan grass in the southeastern United States. The adult sorghum midge is about 0.06 in. long with an orange body. Walter (1941) reported that the female midge deposits eggs within the spikelets of developing sorghum heads, usually when the spikelets are in bloom. The eggs hatch in 2 to 3 days to produce a larva which remains within the spikelet and extracts plant juices from the developing seed, thus blighting it. The new larvae are nearly colorless, but as they grow, they become pink and finally dark orange. During the summer months, the larvae complete their development in 9 to 11 days. Depending on temperature conditions, the mature larvae either go into the pupal stage and transform to adults in another 3 to 4 days or spin a cocoon in which they remain dormant until the following spring. The adult midges live for only a day, during which they mate, and the females seek out host plants to deposit their eggs. Under favorable weather conditions, with a continuing supply of host plants in a susceptible state of development, there may be up to thirteen generations per year. The dormant larvae in cocoons are stimulated by spring rains to complete their development to the adult stage. The first spring brood often develops on Johnson grass, which reaches a susceptible stage of growth earlier than sorghum. Large populations may build up in Johnson grass and switch to sorghum when the sorghum spikelets come into bloom, causing serious losses.

Another specialized midge, the clover seed midge, *Dasyneura leguminicola* (Lintner), lives within the blossoms of red clover. Other clovers are sometimes lightly infested, but only red clover seed crops are seriously damaged. The adults are very small and gray-to-black, with a red abdomen. According to Creel and Rockwood (1932), the adults appear in April or May, and the females deposit their eggs in young clover heads. The eggs hatch in 3 to 5 days, and the young maggots enter the unopened florets, where they feed by extracting plant fluids, and destroy the developing ovules. The infested florets stop developing and never open, so that infested clover heads show green areas of infested florets among the normally developed red ones at the bloom stage. The larvae mature in about a month and leave the florets, usually during rain, and drop to the ground, where they spin cocoons, pupate, and transform to the adult stage. The new adults emerge in July, in time to attack the second crop of clover heads. In northern areas this generation of midge larvae leaves the plants in late summer and overwinters in cocoons in the surface layers of the soil, to emerge as adults the following spring. In the South, a third generation may be produced.

Various midges also attack the seed crops of trees, especially conifers. The Douglas fir cone midge, *Contarinia oregonensis* Foote, is a well-known destroyer of Douglas fir seed crops in the Northwest. This is another species which is very closely synchronized with the reproductive cycle of this host. The adult midges fly in the early spring, somehow managing to time their appearance to coincide with the opening of the cone buds of Douglas fir. A detailed account of the life cycle of this insect was presented by Hedlin (1961). The female midges seek out the open Douglas fir cone flowers and deposit their eggs between the cone scales which, for a short period, are open to receive the pollen. The eggs hatch in a short time, producing larvae which enter the tissues of the cone scale. The cone scale reacts to the presence of the larva by producing a gall, a roughly spherical, hard woody nodule enclosing the larva. The midge larvae live within the galls, extracting fluid from the cone scale. They are mature by late summer but remain within their galls until the cones are soaked by winter rains. Then the larvae drop to the ground, where they spin cocoons in the layer of litter under the trees. They remain in the cocoons until early spring, when they pupate and become adults in time to attack the new cone crop. The damage is caused by the formation of galls within the cone scale tissues. If only a few larvae are present, the galls may fuse some of the seed coats to the cone scale, preventing the removal of the ripe seed from the cone, even though the seeds are viable. When the cones are heavily infested, the galls become very numerous and the scales are swollen, misshapen, and the seeds are almost entirely displaced by the gall (Fig. 34).

Even more specialized seed insects such as seed chalcids feed only on seeds, to the exclusion of all other parts of the plant. Many insects of this group have reached a level of adaptation to their host plant which permits them to complete their entire life cycle within a single seed. The life cycles and habits of these insects are very often closely attuned to the reproductive cycle of their host plants. Specialization is often manifested by the restriction of a species to a single species of host plant.

Insects living within a single seed present serious problems to the seed-grower. Their damage may be almost impossible to detect while the crop is growing in the field. The infested seeds are sometimes very difficult to separate from sound seeds and may pass through the seed-cleaning process and move with seed shipments to previously uninfested territory. A number of destructive seed pests have attained worldwide distribution in this manner.

The alfalfa seed chalcid, *Bruchophagus roddi* Gussakowskii, an excel-

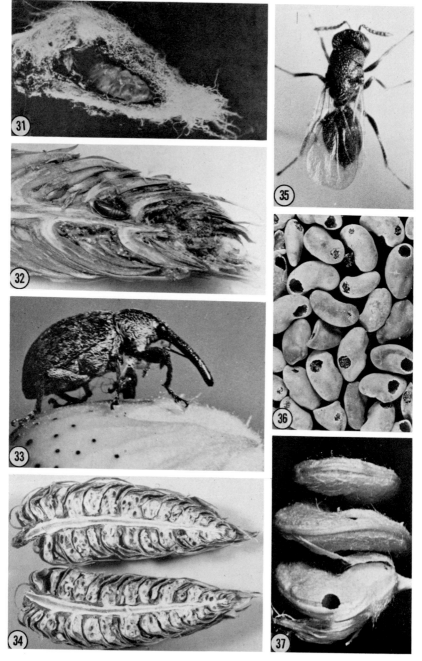

F1G. 31. Larva of pink bollworm, *Pectinophora gossypiella,* in cottonseed. (Photo, U. S. Department of Agriculture.)

lent example of a highly specialized seed insect, destroys alfalfa seed crops throughout North America. The adult is a tiny black wasp about 0.07 in. long (Fig. 35). The female is equipped with a hollow needlelike egg-laying apparatus which she uses to insert her eggs directly into the developing seed. The eggs are deposited in alfalfa seed within the seed pods before the seed coat hardens. The eggs hatch in a few days, and one larva develops to consume the contents of one seed in about 2 weeks. The mature larva pupates within the now hollowed-out seed and transforms to an adult. The adult wasp emerges to oviposit in more developing seeds. There are two generations per year in the northern part of the United States and three in the South, but the emergence of adults is spread over a long period of time so that adults are present in the fields throughout the summer. The alfalfa seed chalcid spends the winter as a mature larva in infested seeds in the fields (Fig. 36) and emerges in the spring in time to attack the next seed crop (Fig. 37).

The alfalfa seed chalcid is a difficult insect to deal with. The adults represent the only stage that lives outside the seed, and they are so small as to be practically invisible. Large populations may go unnoticed until the reduction of seed yields is noted. Most of its life is spent inside the seed where it is well protected from most potential enemies as well as from most types of insecticides. Alfalfa is so widely grown as a forage crop that there is almost always a readily available supply of host material to support chalcid populations near enough to infest seed fields. (For further information, see Bacon et al., 1964.)

Other seed chalcids attack the seeds of trees, especially those of true firs (Abies spp.) and Douglas fir. The Douglas fir seed chalcid (Megastigmus spermotrophus Wachtl) is typical of this group. The adult is a minute yellow wasp. The female uses her long ovipositor to insert eggs into developing seeds within the immature Douglas fir cone. When the eggs hatch, a single larva develops in each seed (Fig. 38). In early spring

FIG. 32. Longitudinal section of Douglas fir cone showing pupa of *Barbara colfaxiana*. (Photo by T. W. Koerber, Pacific Southwest Forest & Range Experiment Station, U. S. Department of Agriculture, Berkeley, California.)

FIG. 33. Adult cotton boll weevil, *Anthonomis grandis*, on cotton bud. (Photo, Clemson Agricultural College Extension Service, Clemson, South Carolina.)

FIG. 34. Longitudinal section of Douglas fir cone damaged by cone midge, *Contarinia oregonensis*. (Photo by T. W. Koerber, Pacific Southwest Forest & Range Experiment Station, U. S. Department of Agriculture, Berkeley, California.)

FIG. 35. Alfalfa seed chalcid, *Brucophagus roddi*. (Photo by W. P. Nye, Federal Apiculture Research Laboratory, Logan, Utah.)

FIG. 36. Alfalfa seed destroyed by alfalfa seed chalcid. (Photo by W. P. Nye, Federal Apiculture Research Laboratory, Logan, Utah.)

FIG. 37. Alfalfa seed pod with emergence hole of alfalfa seed chalcid. (Photo by W. P. Nye, Federal Apiculture Research Laboratory, Logan, Utah.)

the larva transforms to an adult which emerges from the seed in time to attack the next seed crop.

The pea weevil, *Bruchus pisorum* (L.), is another highly specialized seed insect. According to Brindley *et al.* (1946), the adult beetles appear in the field about the time that peas come into bloom. They feed on the pea plants, especially on the pollen and petals. When the pods begin to develop, the females deposit yellow or orange eggs on the surface of the immature pods (Fig. 39). Depending on the temperatures, the eggs hatch after 5 days to 2 weeks, and the larvae bore into the pod and enter the developing pea. Only one larva develops within each infested pea. After feeding for 4 to 6 weeks, the larvae transform to pupae which in another 2 weeks become adults. Some of the adult weevils may leave the peas and seek sheltered locations for hibernation, whereas the remainder remain in the pea (Fig. 40). In this condition, they may spend the winter in seed stocks in storage or in peas remaining in the fields after harvest. The adult pea weevils do not emerge from dry peas in storage and do not multiply in mature peas. If the embryo of the seed has not been destroyed by the weevil larva, infested peas may germinate if planted, but such seeds produce weak plants, due to the destruction of the food reserves in the cotyledons. The adult weevils emerge from infested seeds after they are planted, or from peas scattered in last year's pea fields, in time to attack the next pea crop.

Other species of weevils of the same family (Bruchidae) infest the seeds of other legumes, including those of beans, cowpeas, vetch, clover, lentils, and lupines. These infest developing seeds in the field, but some of them [for example, the bean weevil, *Acanthoscelides obtectus* (Say)] continue to breed in the stored seeds and are consequently discussed more fully in Chapter 4.

C. Reduction of Seed Quality by Insects

The outright destruction of seeds or seed-bearing organs of plants by insects is relatively easy to see and evaluate, but insects act in more subtle ways to reduce seed quality. Some of these effects have already been mentioned (Section A), but they are important enough to receive special attention.

Some insects, for example *Lygus* bugs feeding on beet seed, reduce the germinability of the seed. In effect, these nonviable seeds are dead. Such seeds may have the same physical characteristics as viable seeds and thus cannot be separated when the crop is processed. At its worst, this type of damage may render the whole seed lot unmarketable because of failure to meet minimum germination standards. More often the seed is marketable but the grower's reputation suffers when his customers do not get the stand of seedlings they expected. Another particularly serious aspect of this type of damage is that it often is not detected until the crop is harvested and processed, thus increasing production costs.

Insects also lower seed quality by causing low seedling vigor. In this instance, the seeds are viable but do not produce strong seedlings because insects have consumed part of the cotyledons or endosperm, or have interfered with normal seed development. Again, the end result is dissatisfied customers and damage to the grower's reputation. Unless the grower has an intensive seed-testing program which involves planting seed samples and evaluating the resulting plants, this defect may go undetected and, in any event, it is too late to correct the problem.

A number of insect pests cause their host plants to produce shriveled, shrunken, or undersized seed. It is, of course, possible to separate these seeds in the cleaning plant, but uniform, full-sized seed is always easier to process than a mixture of normal and undersized or defective seeds. For the grower, this problem necessitates additional time or expensive equipment to clean seeds and more time and attention by skilled personnel to produce seeds of high quality. Furthermore, as long as seeds are sold by weight, uniform, full-sized seeds will be more profitable to grow.

Insects, especially aphids and leafhoppers, are well-known carriers of plant diseases. In most cases, insect-transmitted plant diseases damage or kill the infected plants, thereby reducing seed yield, even though the insects are not, in the narrow sense, seed insects. There are a few insect-transmitted plant diseases, notably bean mosaic and lettuce mosaic, which also are seed-borne. That is, insects transmit the disease from plant to plant. The infected plants produce seeds which carry the disease organism and produce seedlings infected with the disease. The infected seeds cannot be separated from uninfected seeds and thus, constitute a very serious seed quality problem. There is also an additional hazard in that infected seeds may be shipped to previously disease-free areas. (For further discussion of seed diseases, the reader is referred to Chapter 5, Volume II, of this treatise.)

Insects may also cause problems of a strictly mechanical nature in harvesting and seed-processing operations. Some insects cause their host plants to produce deformed fruits, pods, etc. Often these do not open properly during processing and perfectly sound seeds trapped in the deformed structures are thus lost. For example, pine cones deformed by the fir coneworm may fail on drying to release 25–50% of their seeds. Insect-damaged pine cones may also exude resin, making them inconvenient and unpleasant to handle. Insect-damaged seeds, pellets of excrement, and the insects themselves are all contaminants which must somehow be removed in the seed-cleaning process.

FIG. 38. Larva of Douglas fir seed chalcid, *Megastigmus spermatotrophus,* in Douglas fir seed. (Photo by T. W. Koerber, Pacific Southwest Forest & Range Experiment Station, U. S. Department of Agriculture, Berkeley, California.)

The insects themselves are without question the most serious contaminant because of the very real problem of the spread of insect pests in seed shipments. The spread of insect pests in seed shipments is especially difficult to prevent in the case of insects that live within a single seed. The pea weevil and the alfalfa seed chalcid, for example, most certainly came to North America in infested seed shipments. The Douglas fir seed chalcid, native to western North America, has been exported to Europe and now destroys Douglas fir seed crops from Scotland to Poland.

A well-documented account of the spread of the pink bollworm, *Pectinophora gossypiella* (Saunders), was presented by Gains (1957). By means of the larvae hibernating in infested cottonseed, the pink bollworm has been spread around the world. This insect was first described from specimens collected in India in 1842. It is believed to have been spread by shipments of infested seeds to Egypt in 1906 and to Hawaii in 1909. Infested seeds were sent from Egypt to Mexico in 1911. The insect was further spread by shipment of infested seeds from the Laguna area of Mexico to Hearne, Texas, about 1917. Once established in the cotton-growing areas of the Rio Grande Valley, the pink bollworm was able to spread and now causes millions of dollars in losses to the cotton crop in the United States every year.

D. Beneficial Seed-Destroying Insects

It is probably safe to assume that seeds of all plants are subject to insect attack. Some of our most notorious seed insects, for example, the *Lygus* bugs, feed on the seeds of many plants and may be injurious or beneficial depending on one's viewpoint. If the plant happens to be a weed, the *Lygus* bugs destroying its seeds become beneficial insects. Unfortunately, the fate of weed seeds has not been of enough concern to justify systematic studies of their insect problems, so we are generally unaware of the influence of insects on the abundance of weeds. Probably the best-

---

FIG. 39. Eggs of pea weevil, *Bruchus pisorum,* on young pea pods. (Photo, U. S. Department of Agriculture.)

FIG. 40. Dried peas with emerging pea weevils. (Photo by W. P. Nye, Federal Apiculture Research Laboratory, Logan, Utah.)

FIG. 41. A big eyed bug, *Geocorus* sp., feeding on *Lygus* nymph. (Photo by Jack Eves, Entomology Department, Washington State University, Pullman, Washington.)

FIG. 42. A damsel bug, *Nabis* sp., feeding on nymphal *Lygus* bug. (Photo by Jack Eves, Entomology Department, Washington State University, Pullman, Washington.)

FIG. 43. An assassin bug, *Sinea diadema,* feeding on adult *Lygus* bug. (Photo by Jack Eves, Entomology Department, Washington State University, Pullman, Washington.)

FIG. 44. A hunting spider, *Lycosa* sp., feeding on nymphal *Lygus* bug. (Photo by Jack Eves, Entomology Department, Washington State University, Pullman, Washington.)

known insect associated with weed seeds is the milkweed bug, *Oncopeltus fasciatus* (Dallas), which feeds only on milkweed seeds (Andre, 1934). However, it is well known more because it is a convenient laboratory animal for research on insect physiology than because its role as a destroyer of milkweed seed is appreciated.

It has been shown only rather recently that many insects in the same family as the milkweed bug (Lygaeidae) are also seed feeders. Sweet (1960) reported that many species long believed to be predaceous actually feed on seeds. He suggested that insects in the Lygaeidae ought to be called "seed bugs" because the seed-feeding habit is so common among them.

The fact that insects destroy seeds of undesirable plants has stimulated interest in using insects as weed control agents. This is particularly true in instances where undesirable plants have been introduced to a new area and are flourishing in the absence of their normal complement of natural enemies.

The most successful introduction of a weed seed insect to date was that of a seed weevil to attack the seeds of puncture vine (*Tribulus terrestris* L.), a weed with spiny seed pods accidentally introduced from the Mediterranean region (Andres and Angalet, 1963). The weevil, *Microlarinus lareynii,* (Jacquelin du Val) deposits its eggs in the developing seed pods of the puncture vine, and the larvae develop within the pods, consuming the seeds. Another seed weevil, *Apion ulicis* Forster, which feeds on the seeds of gorse (*Ulex europaeus* L.) has been imported into California (Holloway and Huffaker, 1957).

E.   Control of Seed Insects

The term *insect control* is used here to include all measures taken to prevent or reduce the damage caused by insects. This broad definition includes everything that jeopardizes insect survival, reproduction, or spread.

It is difficult for most insects to survive even in the absence of control measures. Multitudes of potentially destructive insects are killed by unfavorable weather. Furthermore, all plant-feeding insects are preyed upon by other organisms and are, thus, subject to varying degrees of biological control. Insects are usually the most efficient control agents, especially where the pest is present in low-to-moderate numbers. Microorganisms and birds are most likely to become important when the insect population reaches outbreak proportions. Among the insects, those classed as predators usually feed on a wide variety of insects (Figs. 41 to 44), whereas those classed as parasites are usually more specific in their host selection. Most potentially destructive insects are actually prevented by one or more of their insect enemies from becoming abundant enough to cause noticeable damage.

Although the natural factors controlling insect populations generally are not subject to our manipulation, it is important to recognize these factors and understand how they influence potentially destructive insects. With a minimum of understanding, we can avoid interfering with natural factors that limit the abundance of potential pests.

Ideally, we should identify and learn to manipulate factors that regulate populations of insects commercially important. Normally, seed growers are not directly involved in such manipulations, but they may have the opportunity to support research programs which attempt to do so.

1. BIOLOGICAL CONTROL PROGRAMS

Biological control programs seek to exploit the ability of predatory or parasitic insects to limit the populations of pest insects. The most successful of these have usually involved importing parasitic or predatory insects to destroy an insect pest which previously had been inadvertently introduced. When an insect pest becomes established in a new area in the absence of its normal parasites and predators, it tends to multiply unchecked until it reaches the limit of its food supply. If the food supply happens to be one of our crop plants, the crop will be severely damaged. Effective biological control can best be achieved by studying the pest in its native habitat to find the factors that hold it in check there and by importing one or more effective natural enemies to control it in its new home. The parasite or predator considered for importation must first be thoroughly investigated to determine that it will effectively control the pest it is to combat, that it does not attack beneficial insects, and that it has a reasonable chance of adapting to the climatic conditions of the area to which it is imported.

The search for and importation of a natural enemy of an introduced pest is an expensive and complicated undertaking which normally does not involve seed growers. However, since a team of scientists involved in such a venture requires adequate funding and political support, it is in the interest of seed growers to support such projects. Although the initial expense of introducing a parasite or predator may be high, in the long run biological control is the most economical type of pest control known. Once a predator becomes established in its new habitat, it usually becomes a self-perpetuating, self-regulating force that continues to suppress the pest population year after year without further cost. Over a period of years, the growers of the affected crop automatically receive the benefit of free pest control and the initial cost of the project is likely to be recovered many times over.

2. CULTURAL CONTROL

Factors controlling the abundance or destructiveness of insects can also be manipulated by judicious cultural practices. Cultivation, weed control, planting, harvesting, and destruction of crop residues can all be manipulated to control insect pests in various stages of their life cycle.

Some insects, such as the corn earworm, pass through the pupal stage in upper soil horizons. Plowing or cultivating disrupts the soil surface so that some pupae are buried too deeply for adult moths to emerge, whereas others are exposed to birds or other predators.

Plowing is also an excellent way to dispose of crop residues which may harbor seed insects or provide sheltered places in which they may survive the winter. Both the pink bollworm and the cotton boll weevil pass through the winter in unharvested bolls or in trash on the ground. If cotton stalks are plowed under immediately after harvest, the insects remaining on the plants or in litter are buried before they can complete their development and disperse from the cotton field. Hence, the number of insects present at the start of the next growing season is greatly reduced.

Some of our most injurious seed insects feed on weeds in addition to (or as alternatives to) their crop plant hosts. *Lygus* bugs feed on a large number of noncrop plants. In areas having mild winters, these plants may sustain large insect populations during the winter when the crop plant hosts are not available. The alfalfa seed chalcid also infests bur clover, and the sorghum midge lives on Johnson grass. The weeds serving as alternative food supplies can be eliminated by plowing and cultivating, and when such practices are used in the fall, they insure that the insects will face a hard winter. Johnson grass is a particularly important alternative host for the sorghum midge because it supports the midge early in the spring before sorghum reaches the susceptible stage of growth. If large populations of midges are prevented from developing on Johnson grass, there will be few to attack sorghum when the spikelets come into bloom.

Other methods of destroying crop residues and weeds, such as burning or applying herbicide sprays, may also be effective in reducing insect populations. The particular method used and the appropriate time will depend upon the life cycle and habits of the insects involved, the operations necessary to produce the crop, and the climate of the area in question. As always, it is necessary to understand the habits of the pest insect in order to select the most effective method and time for a given cultural practice.

Insect damage can also be partly avoided by careful selection of the place to grow a seed crop. Most commonly, this involves crop rotation. By not growing the same crop year after year on the same field, it is possible to avoid a buildup of populations of insects which are adapted to that crop. Build up of specific disease organisms of crops and depletion of certain plant nutrients may also be avoided. Commercial seedgrowers in the Lompoc and Santa Maria areas of California have found it advisable to wait 3 years before repeating a given crop on the same field.

Some types of insect damage may be partly avoided by growing seed crops in places that are isolated from extensive commercial production of the same crop. It is well to remember, for example, that corn insects are likely to be abundant wherever corn is extensively grown.

Isolation of a seed crop is also helpful in reducing pollen contamination and outcrossing. Outcrossing is likely to be especially troublesome when specific crosses or inbred lines of plants pollinated by bees are being produced. Bees are likely to transport unwanted pollen for distances up to a mile. In the instance of bees cross-pollinating an inbred line, they become harmful insects in contrast to their usual beneficial role.

Small stocks of especially valuable plants may be grown in greenhouses. In this situation good sanitation within the greenhouse and screening over the vent openings will provide nearly perfect, if rather expensive, isolation.

Some insects such as *Lygus* bugs or the corn earworm are serious pests of several seed crops. Thus it is not advisable to follow one susceptible crop with another species which also is susceptible. This is another instance where a thorough knowledge of the habits of the insects attacking the crop is very useful. Occasionally it is also well to know something of the insects feeding on other crops in the area. For example, seedgrowers in the Lompoc area of California found that when asters were grown near artichoke fields, the flowers and seed heads of the asters became infested with larvae of the artichoke plume moth, *Platyptilia carduidactyla* Riley. The plume moth, which normally is found in artichoke buds and thistle heads, apparently does not develop self-sustaining populations on asters, but moths which have developed on artichokes deposit enough eggs on aster heads to cause 20–30% infestation.

Manipulation of planting and harvesting dates can help to reduce the amount of damage by certain insects. A number of important seed pests produce multiple generations in a growing season, with each generation more numerous than the previous one. In general, crops attacked by these insects suffer minimum damage when planted early and brought to maturity and harvested as quickly as possible. In this way, the period in which the crop is exposed to multiple generations of insects is minimized, and the crop-free period during which the insects will have difficulty surviving is maximized. Use of fertilizers and irrigation schedules to accelerate growth of crops also tends to decrease insect damage.

In crops such as alfalfa, which grow more or less continuously, harvesting operations may be scheduled to reduce populations of certain pests. For example, *Lygus* bug populations in alfalfa are greatly reduced if the crop is mowed before nymphs reach maturity. If both a hay and a seed crop are to be produced, the cutting of hay may be timed so that *Lygus* nymphs cannot withstand the sudden loss of the shelter and food provided by the growing crop, and since they have not yet matured and grown wings, cannot fly away to seek a more favorable environment. If the seed crop can be grown in large fields, or if neighboring growers can cooperate to the extent of following the same schedule, even adult bugs will perish as a result of being suddenly deprived of food and shelter. The seed crop can then start to grow in the presence of a low insect population and with a chance of maturing before the insect population recovers enough to affect seriously the seed yield. Again, successful manipulation of harvest schedules for insect control depends on an intimate knowledge of the habits of the insects affecting the specific crop at a specific time and place.

3. PLANT RESISTANCE

Plant breeders have managed to develop many crop varieties that are resistant to insect damage or various diseases. Farmers are generally advised to grow resistant varieties when they are available, and the same advice applies to seedgrowers. However, resistant varieties may be of little comfort to a seedgrower who must maintain a stock of the susceptible parent of a resistant cross.

4. LEGAL RESTRICTIONS

In many areas, certain cultural practices are required by law. Usually, these are legal ordinances requiring destruction of crop residues, or weed control measures, and are intended to insure that cultural controls are not rendered ineffective by a few careless individuals who fail to follow them.

On a broader scale, many countries and some states seek to protect their crops from potentially destructive insects by laws regulating the movement of products likely to harbor insect pests. These quarantine laws may require inspection of certain plant materials to determine whether they are free of insect pests, or they may entirely prohibit the importation of certain plant products from areas known to be infested by potentially destructive insects.

Seed growers, whose business inherently involves shipment of plant material across state and national boundaries, continually encounter these regulations and may often find them inconvenient. However, it is well to consider that there are hundreds of destructive insects in other parts of the world which we do not have to deal with as long as they can be kept out. The maintenance of a few thousand miles of ocean between a crop and its insect pests is one of the most effective and economical insect control methods available.

5. INSECTICIDES

Control of insects injurious to seed production is commonly (though not ideally) achieved with insecticides. The timing of the applications depends on the life histories of the insects involved and usually also on the flowering and fruiting cycles of the crops. For example, in the control of legume seed insects, there is usually (*1*) a bud stage application to ensure development of flowers, (*2*) an early bloom stage application to maintain the bloom, and (*3*) a late bloom stage application to protect the developing seeds. These applications are used primarily to control *lygus* bugs which are usually present and damaging from the bud stage until the seed contents become firm. Specific applications to control pentatomid bugs, which primarily affect the developing seed, are made only from mid-to-late bloom.

Prebloom applications of insecticides are often chosen for their ability to protect the crop over a long period. Previously, some of the chlorinated hydrocarbons were favored, but the persistence of their residues on crops and in the soil has discouraged their use in recent years. Although the danger of insecticide residues being transferred to livestock and subsequently to man via the food chain can be minimized by destroying (instead of feeding) the seed harvest residue, there is still danger that the insecticides that enter the soil will be incorporated in bales of hay the following year. Systemic insecticides, which are absorbed by and translocated in the crop plant, are now available for certain pests. These can supply the desired extended period of prebloom crop protection without remaining as dangerous residues into the harvest period.

Bloom stage applications (where insect pollinators are involved) are usually made with relatively selective materials. In general, bloom stage applications should be made only during hours when pollinators are not on the crop. Sometimes, materials which are highly toxic to pollinators but have a very short active life can be used safely when applied during evening hours. However, cool night temperatures usually increase the danger that these materials will retain enough activity to kill pollinators the next morning.

In many seed-growing areas, insecticide programs have been developed which control most of the injurious insects while causing a minimum of harm to beneficial insects, as well as vertebrate animals, including man. Information concerning these programs is usually available from seed companies, county agricultural agents, and in the form of state and federal leaflets and bulletins. In some districts, seed growers have organized to protect their pollinators by means of local ordinances regulating the use of insecticides on all crops attractive to bees within flight range of their fields.

REFERENCES

Agricultural Research Service. (1970). Cooperative Economic Insect Report No. 20, p. 806. Plant Prot. Div., U.S. Dept. Agr., Agr. Res. Serv., Washington, D.C.

Andre, F. (1934). Notes on the biology of *Oncopeltus fasciatus* (Dallas). *Iowa State Coll. J. Sci.* **9**, 73.

Andres, L. A., and Angalet, G. W. (1963). Notes on the ecology and host specificity of *Microlarinus lareynii* and *M. lypriformis* (Coleoptera: Curculionidae) and the biological control of puncture vine, *Tribulis terrestris. J. Econ. Entomol.* **56**, 333.

Bacon, O. G., Riley, W. D., Russel, J. R., and Batiste, W. C. (1964). Experiments on control of the alfalfa seed chalcid, *Bruchophagus roddi* in seed alfalfa. *J. Econ. Entomol.* **57**, 105.

Bailey, S. F. (1938). Thrips of economic importance in California. *Calif., Agr. Exp. Sta., Circ.* **346**, 1–77.

Baker, H. G. (1960). Reproductive methods as factors in speciation in flowering plants. *Cold Spring Harbor Symp. Quant. Biol.* **24**, 177.

Baker, H. G. (1963). Evolutionary mechanisms in pollination biology. *Science* **139**, 877.

Baker, H. G., and Hurd, P. D., Jr. (1968). Intrafloral ecology. *Annu. Rev. Entomol.* **13**, 385.

Blickenstaff, C. C., and Bauman, L. F. (1961). Treated bags for control of the corn earworm and fall armyworm. *J. Econ. Entomol.* **54**, 587.

Bohart, G. E. (1954). The effect of competing pollen sources on the number of honey bees collecting alfalfa pollen. *Rep. Alfalfa Improvement Conf., 14th, 1954* p. 24.

Bohart, G. E. (1957). Pollination of alfalfa and red clover. *Annu. Rev. Entomol.* **2**, 355.

Bohart, G. E. (1960). Insect pollination of forage legumes. *Bee World* **41**, 51 and 85.

Bohart, G. E. (1962). Introduction of foreign pollinators, prospects and problems. *Proc. Int. Symp. Pollination, 1st, 1960* Commun. No. 7, p. 181.

Bohart, G. E. (1967). Management of wild bees. *U. S., Dep. Agr., Agr. Handb.* **335**, 109.

Bohart, G. E. (1972). Management of habitats for wild bees. *Proc. Tall Timbers Conf., 3rd, 1971* (in press).

Bohart, G. E., and Knowlton, G. F. (1967). Managing the alfalfa leaf-cutting bee for higher alfalfa seed yields. *Utah State Univ. Ext. Leafl.* **104**, 1–7 (revised).

Bohart, G. E., and Nye, W. P. (1960). Insect pollinators of carrots in Utah. *Utah, Agr. Exp. Sta., Bull.* **419**, 1–16.

Bohart, G. E., and Pedersen, M. W. (1963). The alfalfa leaf-cutting bee, *Megachile rotundata* F. for pollination of alfalfa in cages. *Crop Sci.* **3**, 183.

Bohart, G. E., and Todd, F. E. (1961). Pollination of seed crops by insects. *Yearb. Agr. (U. S. Dep. Agr.)* p. 245.

Bohart, G. E., Moradeshaghi, M. J., and Rust, R. W. (1967). Competition between honey bees and wild bees on alfalfa. *Int. Beekeep. Congr., 21st, Prelim. Sci. Manage. Summ. Pap.* **32**, 66.

Bohart, G. E., Nye, W. P., and Hawthorn, L. R. (1970). Onion pollination as affected by different levels of pollinator activity. *Utah, Agr. Exp. Sta., Bull.* **482**, 1–57.

Brindley, T. A., Chamberlin, J. C., Hinman, F. G., and Gray, K. W. (1946). The pea weevil and methods for its control. *U. S., Dep. Agr., Farmers' Bull.* **1971**. 1–24.

Buckley, B. R., and Burkhardt, C. C. (1962). Corn earworm damage and loss in grain sorghum. *J. Econ. Entomol.* **55**, 435.

Carlson, E. C. (1961). Lygus bug damage to table beet seed plants. *J. Econ. Entomol.* **54,** 117.

Carlson, E. C. (1967). Control of sunflower moth larvae and their damage to sunflower seeds. *J. Econ. Entomol.* **60,** 1068.

Carpenter, F. M. (1952). Fossil insects. *Yearb. Agr. (U. S. Dep. Agr.)* p. 14.

Chedd, G. (1970). Hidden peril of the green revolution. *New Sci.* **22,** 171.

Creel, C. W., and Rockwood, C. P. (1932). The control of the clover flower midge. *U. S. Dep. Agr., Farmers' Bull.* **971,** 1–32.

Curl, L. F., and White, R. W. (1952). The pink bollworm. *Yearb. Agr. (U. S. Dep. Agr.)* pp. 505–511.

Darrah, W. C. (1960). "Principles of Paleobotany." Ronald Press, New York.

Darwin, C. (1877). "On the Various Contrivances by Which Orchids are Fertilized by Insects." Appleton, New York.

Debarr, G. L. (1967). Two new sucking insect pests of seed in southern pine seed orchards. *U. S., Forest Serv., Res. Note* **SE-78,** 1–3.

Dewey, J. E. (1970). Damage to Douglas-fir cones by *Choristoneura occidentalis. J. Econ. Entomol.* **63,** 1804.

Dodson, C. H., and Frymire, G. P. (1961a). Natural pollination of orchids. *Ann. Mo. Bot. Gard.* **49,** 133.

Dodson, C. H., and Frymire, G. P. (1961b). Preliminary studies in the genus *Stanhopea. Ann. Mo. Bot. Gard.* **48,** 137.

Ebel, B. H. (1961). Thrips injure slash pine female flowers. *J. Forest.* **59,** 374.

Ebel, B. H. (1963). Insects affecting seed production of slash and longleaf pines. *U. S., Forest Serv., Res. Pap.* **SE-6,** 1–24.

Elmore, J. C. (1949). Thrips injury to onions grown for seed. *J. Econ. Entomol.* **42,** 756.

Elmore, J. C. (1955). The nature of lygus bug injury to lima beans. *J. Econ. Entomol.* **48,** 148.

Eves, J. (1970). Washington State University, Pullman (personal communication).

Faegri, K., and van der Pijl, L. (1966). "The Principles of Pollination Ecology." Pergamon, Oxford.

Ferguson, R. B., Furniss, M. M., and Basile, J. V. (1963). Insects destructive to bitterbrush flowers and seeds in southwestern Idaho. *J. Econ. Entomol.* **56,** 459.

Flemion, F.,and Olson, J. (1950). Lygus bugs in relation to seed production and occurrence of embryoless seeds in various umbelliferous species. *Contrib. Boyce Thompson Inst.* **16,** 39.

Fowells, H. A. (1965). Silvics of forest trees of the United States. *U. S., Dep. Agr., Forest Serv.* **71,** 329.

Free, J. B. (1958). Attempts to condition bees to visit selected crops. *Bee World* **39,** 221.

Free, J. B. (1970). "Insect Pollinators of Crops." Academic Press, New York.

Free, J. B., and Butler, C. G. (1959). "Bumblebees." Collins, London.

Gains, J. C. (1957). Cotton insects and their control in the United States. *Annu. Rev. Entomol.* **2,** 319.

Galil, J., and Eisikowitch, D. (1969). Further studies on the pollination ecology of *Ficus Sycomorus* L. (Hymenoptera, Agaonidae). *Tydskr. Entomol.* **112,** 1.

Gibson, L. P. (1969). Monograph of the genus *Curculio* in the new world (Coleoptera: Curculionidae) Part I United States and Canada. Miscellaneous publication Ent. Soc. America **6,** 241.

Grant, V. (1949). Pollination systems as isolating mechanisms in angiosperms. *Evolution* **3,** 82.

Grant, V. (1959). "Natural History of the *Phlox* Family." Nijof, The Hague.

Hasselrot, T. B. (1960). Studies on Swedish bumblebees (genus *Bombus* Latr.): Their domestication and biology. *Opusc. Entomol., Suppl.* **17,** 60.

Hedlin, A. F. (1961). The life history and habits of a midge, *Contarinia oregonensis* Foote in Douglas-fir cones. *Can. Entomol.* **93,** 952.

Hills, O. A. (1943). Comparative ability of several species of lygus and the Say stinkbug to damage sugar beets grown for seed. *J. Agr. Res.* **67,** 389.

Hills, O. A. (1950). Lygus damage to beet seed in various stages of development. *Proc. Amer. Soc. Sugar Beet Tech.* **6,** 481.

Hills, O. A., and Taylor, E. A. (1950). Effect of the Say stink bug on maturing sugar beet seed. *Proc. Amer. Soc. Sugar Beet Tech.* **6,** 488.

Holloway, J. K., and Huffaker, C. B. (1957). Establishment of the seed weevil, *Apion ulicis* Forst., for suppression of gorse in California. *J. Econ. Entomol.* **50,** 498.

Horber, E. (1971). Bumble bees as pollinators in the breeding of alfalfa and red clover. *Rep. Centr. Alfalfa Improvement Conf., 12th, 1971* p. 17.

Hurd, P. D., Jr., and Linsley, E. G. (1963). Pollination of the unicorn plant (Martyniaceae) by an oligolectic, corolla-cutting bee (Hymenoptera: Apoidea). *J. Kans. Entomol. Soc.* **36,** 243.

Janzen, D. H. (1972). *Annu. Rev. Ecol.* **2** (in press).

Johansen, C. (1966). Pollination of clovers raised for seed in Washington. *Amer. Bee J.* **106,** 298.

Johnson, C. D. (1970). Biosystematics of the Arizona, California, and Oregon species of the seed beetle genus *Acanthoscelides* Shilsky. *Univ. Calif., Berkeley, Publ. Entomol.* **59,** 1–116.

Keen, F. P. (1958). Cone and seed insects of Western forest trees. *U. S., Dep. Agr., Tech. Bull.* **1169,** 1–167.

Koerber, T. W. (1962). Douglas fir cone and seed research. *U. S., Forest Serv., Pac. Southwest Forest Range Exp. Sta., Progr. Rep., 1959* pp. 1–37 (processed).

Koerber, T. W. (1963). *Leptoglossus occidentalis* (Hemiptera Coreidae), a newly discovered pest of coniferous seed. *Ann. Entomol. Soc. Amer.* **56,** 229.

Kraai, A. (1958). Bijen en hommels bij het veredelingswerk. *Meded. Dir. Tuinbouw (Neth.)* **21,** 291.

Krugman, S. L., and Koerber, T. W. (1969). Effect of cone feeding by *Leptoglossus occidentalis* on ponderosa pine seed development. *Forest Sci.* **15,** 104.

Kullenberg, B. (1961). Studies in *Ophrys* pollination. *Zool. Bidr. Uppsala* **34,** 1.

Leppik, E. E. (1960). Early evolution of flower types. *Lloydia* **23,** 72.

Levin, M. D. (1967). Pollination. *U. S., Dep. Agr., Agr. Handb.* **335,** 77.

Linsley, E. G. (1960). Observations on some matinal bees at flowers of *Cucurbita, Ipomoea,* and *Datura* in desert areas of New Mexico and southwestern Arizona. *J. N. Y. Entomol. Soc.* **68,** 13.

McGregor, S. E. (1959). Cotton-flower visitation and pollen distribution by honey bees. *Science* **129,** 97.

McGregor, S. E., Rhyne, C., Worley, S., and Todd, F. E. (1955). The role of the honey bee in cotton pollination. *Agron. J.* **47,** 23.

Meeuse, B. J. D. (1961). "The Story of Pollination." Ronald Press, New York.

Menke, H. F. (1951). Insect pollination of apples in Washington State. *Proc. Int. Beekeep. Congr., 14th,* 1951.

Merkel, E. P. (1967). Individual slash pines differ in susceptibility to seedworm infestation. *J. Forest.* **65,** 32.

Metcalf, C. L., Flint, W. P., and Metcalf, R. L. (1951). "Destructive and Useful Insects," 3rd ed. McGraw-Hill, New York.

Miller, J. M. (1915). Cone beetles: Injury to sugar pine and western yellow pine. *U. S., Dep. Agr., Bull.* **243**, 1–12.

Nord, E. C. (1965). Autecology of bitterbrush in California. *Ecol. Monogr.* **35**, 307.

Nye, W. P., and Mackensen, O. (1968). Selective breeding of honeybees for alfalfa pollen: Fifth generation and backcrosses. *J. Apicult. Res.* **7**, 21.

Palmer-Jones, T., and Forster, I. W. (1965). Observations on the pollination of lucerne (*Medicago sativa* Linn.). *N. Z. J. Agr. Res.* **8**, 340.

Pedersen, M. W., and Bohart, G. E. (1950). Using bumblebees in cages as pollinators for small seed plots. *Agron. J.* **42**, 523.

Percival, M. S. (1965). "Floral Biology." Pergamon, Oxford.

Plath, O. E. (1934). "Bumblebees and their Ways." Macmillan, New York.

Rick, C. M. (1950). Pollination relations of *Lycopersicum esculentum* in native and foreign regions. *Evolution* **4**, 110.

Riley, C. V. (1892). The *Yucca* moth and *Yucca* pollination. *Rep. Mo. Bot. Gard.* **3**, 99.

Shemetkov, M. F. (1960). Pollinating activities of bees in greenhouses. *Pschelovodstvo, Mosk.* **33**, 28.

Smith, R. R., and Michelbacher, A. E. (1946). Control of lygus bugs in alfalfa seed fields. *J. Econ. Entomol.* **39**, 638.

Sorenson, C. J. (1939). *Lygus hesperus* Knight and *Lygus elisus* VanDuzee in relation to alfalfa seed production. *Utah, Agr. Exp. Sta., Bull.* **284**, 1–61.

Stone, M. W., and Foley, F. B. (1959). Effect of time of application of DDT on lygus bug populations and yield of lima beans. *J. Econ. Entomol.* **52**, 244.

Strong, F. G. (1970). Physiology of injury caused by *Lygus hesperus*. *J. Econ. Ent* **63**, 808.

Strong, F. E., and Kruitwagen, E. C. (1968). Polygalacturonase in the salivary apparatus of *Lygus hesperus*. *J. Insect Physiol.* **14**, 1113.

Sweet, M. H. (1960). The seed bugs: A contribution to the feeding habits of the Lygaeidae (Hemiptera: Heteroptera). *Ann. Entomol. Soc. Amer.* **52**, 317.

Toumey, J. W. (1919). Relation of gray birch to white pine. *J. Forest.* **17**, 15.

U. S. Department of Agriculture. (1952). "Insects. The Yearbook of Agriculture, 1952." U. S. Dep. Agr., Washington, D. C.

van der Pijl, L. (1954). *Xylocopa* and flowers in the tropics. I-III. *Proc. Kon. Ned. Akad. Wetensch.* **57**, 413 and 541.

von Frisch, K. (1947). "Duftgelenkte Bienen in Dienate der Landwirtschaft und Imkerei Wien." Springer-Verlag, Berlin and New York.

Walter, E. V. (1941). The biology and control of the sorghum midge. *U. S., Dep. Agr., Tech. Bull.* **778**, 1–26.

2

SEED COLLECTING AND IDENTIFICATION

*Charles R. Gunn**

I. Seed Collecting by Man

If farmers took heed [worried] what would be lost, corn should never be cast upon earth (Panton and Donaldson, 1869–1874).

---

*Research Botanist, Plant Science Research Division, Agricultural Research Service, U.S. Department of Agriculture, Beltsville, Maryland.

A. *Premechanized Collecting*

Evidence of man's domestication of plants dates back only to about 9000 B.C. (Ucko and Dimbleby, 1969). It was during this time period, between 9000 and 7000 B.C. in the Old World, and 6000 B.C. in the New World, that some human populations gradually changed from a hunter–gatherer-oriented life of feast and famine to a more settled agrarian type of life with its balanced diet and rather secure food supply. This gradual change, labeled the Neolithic Revolution by Childe (1943), permitted the first great population expansion and the first step toward civilization (Bennett, 1965; Mangelsdorf, 1965).

Harris (1969) described the type of people that domesticated plants. They were generalized hunter–gatherers who formed local bands and had a small range in which they utilized elements of both the fauna and flora. Because of their small range and quasi-nomadic habit, they became more familiar with the various uses of animals and plants. Fussell (1965) speculated that within the group it was women who, at first, played the dominant role in harvesting plants and planting them close to their homes where harvesting would be more convenient. Selection started when one plant was chosen over another, and when its seeds were resown. Perhaps this is why ancient people worshipped a female goddess, under various names, as the giver of good bread.

The Second Book of Kings (2 Kings 19:29) sums up the process of plant domestication succinctly: "And this shall be a sign unto thee. Ye shall eat this year such things as grow of themselves; and in the second year that which springeth from the same; and in the third year sow ye and reap and plant vineyards and eat the fruits thereof."

There are two main groups of food-producing plants, seed-producing crop plants (seed culture) and vegetable-producing crop plants (vegeculture). Seed culture first appeared in the drier tropics and subtropics of the Old and New Worlds (Fig. 1). Vegeculture had its origins in the humid tropical lowlands of South America, Southeast Asia, and Africa.

At least three centers of seed domestication have been located in the Old World. Sauer (1952) circumscribed these three centers as north China, India (including the area to the eastern Mediterranean), and Ethiopia where the Abyssinian highlands adjoin the Sudan. In all three centers climate favored seed production but not vegeculture. "In each center a cultivated assemblage took form that included starchy seeds, seeds rich in protein, and those yielding vegetable oils." Important seed food plants for these three centers are presented in Table I.

The dominant seed food plants of North America and northern and extreme western South America were maize, beans, and squash or pumpkins. A unique feature of the culture of these seed foods is that they were

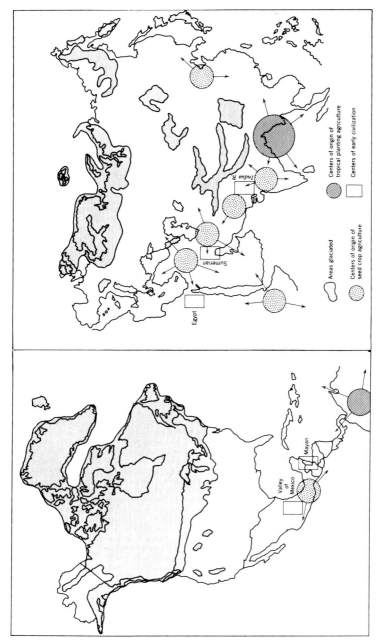

FIG. 1. Map showing relationship of glaciated areas to agricultural and civilization centers. (Reprinted from Dasmann, 1968, Fig. 28. Copyrighted 1968 by John Wiley and Sons, Inc. Used by permission.)

TABLE I
OLD WORLD SEED FOOD PLANTS LISTED BY CENTERS OF ORIGIN[a]

North China	India (eastern)	India (western)	Ethiopia
Millets[b]	Millets[c]	Wheat[d]	Millets[e]
Soybeans	Chick-pea	Peas	Sorghum
Velvet bean	Pigeon pea	Horsebean	Barley
Adzuki bean	Bean		Cowpea
			Sesame

[a] After Sauer (1952).
[b] Including true and foxtail millets.
[c] Including *Paspalum* and *Eleusine*.
[d] Introduced later.
[e] Including *Pennisetum, Eleusine,* and *Tef.*

grown together in a symbiotic complex. Maize grew rapidly and gave support to the early producing bean plants. These, in turn, enriched the soil by supporting nitrogen-fixing bacteria in their roots. Squash or pumpkin vines produced a late-maturing fruit crop.

Prehistoric man's adventures with agriculture took place in areas that met the following requirements: the vegetation was easily cleared, alluvial soils were near a freshwater source, and progenitors of crop species were present. The plant centers mentioned above had these characteristics and are the hearths of many of our present-day seed crops (de Candolle, 1886; Vavilov, 1949–1950).

In evaluating the "centers of origin" concept, Smith (1969) pointed out:

> Certain areas of the world's surface, through a combination of latitude, altitude, soils and climate, were the natural areas of geographical distribution of a series of species of plants attractive to man as a source of food. Whatever might have been the reasons for his movements, man sooner or later found those geographical areas in which the gathering of plant foods was the least work. One point which the anthropologists have largely overlooked is that these areas of plentiful plant resources were not the province of man alone, but that other animals dependent upon the gathering of food plants for existence also found these areas. They were, thus, available to be hunted and, eventually, some of them to become domesticated. Because the wild progenitors of the crop plants are rarely locally restricted endemics, but are frequently quite widely distributed wherever a suitable habitat is available, we have every reason to believe that many of them were brought into cultivation in a number of different places.

For centuries man has collected and selected seed crops. These crops, so painstakingly sown, cared for, and harvested, are the food of the present, and the thread of life for the future. Through trial and error, man learned to channel the energy and nutrients of seed plants to his well-being. In all its diverse aspects, agriculture is, indeed, the basis for civilization and the life we now know.

B. Mechanized Collecting

Hand tools, associated with seed production since Neolithic times, have been improved slowly over the centuries. But man's strength was not enough, so he turned to animal power. Man and his draft animals tilled the soil and harvested seed foods in the Old World since 3000 B.C. Farming practices and equipment in use during the Graeco-Roman period essentially were the same ones used by farmers through the seventeenth century. The results of farming during this time period were to feed a relatively few people, to improve and perpetuate crop species, and in some regions to exhaust and erode the soil.

A new pressure was placed on agriculture in the nineteenth century when the population increased as a result of the Industrial Revolution. Caught up in this revolution, the population of Europe increased by 200 million. Many people were employed by industry, not by agriculture. More food seeds were needed which meant that farmers needed more mechanization to increase food production. During the 1830's, Cyrus McCormick invented the reaper, and portable steam power became available as a supplement for horsepower, first as steam-driven threshers and later in a limited way as power for plowing.

Gasoline-powered tractors and allied general and specialized power equipment have revolutionized seed production and processing. Mechanized American farmers produced over 12 billion lbs of crop, vegetable, flower, and tree seeds in 1969. The tremendous increase in seed production, brought about by mechanization, better seeds and fertilizers, increased use of weed and insect controls of various types, and soil conservation, has kept pace with our growing population but not with the world's population growth.

The world's current population of 3.5 billion may double by the year 2000. If this happens, we shall need at least twice as much food as we now produce. Although a major breakthrough in human uses of marine-life proteins is possible, the chances are that increased food production will come primarily from conventional seed crops. Some projections of our current surpluses indicate that they will be exhausted in 1985 (Hardin, 1969). The President's Panel on the World Food Supply (President's Science Advisory Committee, 1967) arrived at the following basic conclusions:

1. The scale, severity, and duration of the world food problem are so great that a massive, long-range, innovative effort unprecedented in human history will be required to master it.

2. The solution of the problem that will exist after about 1985 demands that programs of population control be initiated now. For the immediate future, the food supply is critical.

3. Food supply is directly related to agricultural development and, in turn, agricultural development and over-all economic development are critically interdependent in the hungry countries.

4. A strategy for attacking the world food problem will, of necessity, encompass the entire foreign economic assistance effort of the United States in concert with other developed countries, voluntary institutions, and international organizations.

A burgeoning population can cause irreparable harm, as it has in the past, to agriculture, and subsequently to our civilization. As the need for seed food increases, more intensive use is made of our agricultural land, and marginal agricultural land is pressed into service. This, coupled with general apathy for agricultural and environmental problems, tends to weaken agriculture to the extent that our food supply may be imperiled. These types of problems have brought ruin to other civilizations and have done irreparable damage to the soil through water and wind erosion. Although these problems have been faced successfully and unsuccessfully before, we face more complex problems which, if not solved, will bring devastation. Among these new problems are environmental pollution, vast acreages of genetically similar cultivars, and narrow genetic bases for our seed food crops.

Because of the recent publicity given to environmental pollution, the public is generally aware of dangers of misuse and overuse of agricultural chemicals (Galston, 1970), the plant-damaging capacity of air pollutants (Agricultural Research, 1970), and the pressures on our land (Humphrey, 1970). However, little publicity has been given to another major danger to seed crops. Stearn (1965) has pointed out that "extinction of the old, varied, economically inferior cultivars and their replacement by a much smaller number of more uniform cultivars in the course of agricultural and and horticultural progress may lead to a situation where, to use Frankel's phrase, they hold no genetic reserve in store." Genetic bankruptcy would prevent us from having resources to breed new cultivars that would be resistant to new diseases or pests, or be better adapted to our needs.

In 1944 the National Research Council recommended that the U.S. Department of Agriculture (USDA) establish a facility for preservation of valuable germ plasm. This led to establishment of the National Seed Storage Laboratory at Fort Collins, Colorado. Its primary function is to store genetically useful seed stocks for plant breeders. To maintain viability, the seeds are stored at low temperature and low humidity. Periodic germination tests are made on each seed stock. When germination of a lot decreases to about 50%, arrangements are made to have it regrown, and a new supply of fresh seeds is stored.

The National Seed Storage Laboratory is under the direction of the

New Crops Research Branch (NCRB),* which, in addition to maintaining the germ-plasm bank, is responsible for exploring for new crop plants and germ plasm of established crops, and for developing new crops (Creech, 1970; Hyland, 1970).

Although the USDA has seen the need for a plant germ-plasm bank, most countries do not have this type of centralized organization. Fisher (1969) observed that Argentina, Australia, Ghana, India, Israel, Republic of South Africa, and USSR are the only countries which maintain organizations that systematically conduct plant introduction and exchanges with other countries. Brazil and Japan have the mechanics for handling introductions, but they are not organized on a national basis.

C. *Nonmechanized Collecting*

In this age of power and emphasis on mass production, it is interesting to note that much of our agricultural research is based on hand-harvested seeds. Many people still harvest seeds as our ancestors did, viz., agronomists, horticulturists, plant breeders, plant explorers, producers of specialized seed crops, gardeners, and hobbyists. They do this because they are working on too small a scale for mechanization or because they need to evaluate seed production of individual plants.

Seed collecting can be a rewarding hobby, whether it be gathering seeds to be planted the following year, or collecting them as one would collect stamps or coins. Seeds should be collected when they are ripe and kept in paper (not plastic) envelopes until thoroughly dried. They should be sprayed with an insecticide and periodically checked for insect damage. If the seed collection is to be kept as a hobby, then it is necessary to take field notes at the time the seed collections are made. These notes should include data that are ordinarily recorded when collecting herbarium specimens. If it is possible, a herbarium specimen or photograph should be made. Smith (1971) gives details about collecting plants and recording field data. Once the seeds are cleaned, they are ready to be identified and filed. Identification and filing are discussed in Section III of this chapter. For additional information on seed collecting as a hobby, the reader is referred to Hutchins (1965), Parker (1952), Quinn (1936), Ricker (1961), Rockcastle (1961), Russell (1961), Selsam (1957), and Vinal (1919).

II. Seed Collecting by Agents Other Than Man

Man is a recent collector of seeds when compared to other living seed collectors such as other mammals, birds, and ants, or when compared to

---

*U.S. Department of Agriculture, Plant Industry Station, Beltsville, Maryland.

nonliving seed collectors such as ocean currents, air currents, and time. Collecting seeds should not be confused with seed dispersal discussed in Chapter 4, Volume I of this treatise. Although some seeds may be accidently dispersed while they are being collected, the bulk of the seeds is intended for storage to be eaten by animals doing the collecting. For example, pigs are dispersal agents when they root under oak trees for acorns and accidently bury some acorns. Tree squirrels that pick up acorns and bury them for winter use are collectors and storers, and usually unintentional dispersal agents, because not all of the acorns will be eaten.

A. Mammals

Most seed-storing mammals are rodents. Chickarees, chipmunks, flying and ground squirrels, and various species of mice in the family Cricetidae store seeds in their nests. Tree squirrels bury nuts and acorns "shallowly in the soil in scattered spots throughout their home range, seldom placing more than one nut in a digging, and never more than two or three. When the tree squirrel needs food, it wanders over its storage plot, sniffing here and there, until its acute sense of smell enables it to locate a hidden store" (Jackson, 1961).

B. Birds

Birds are well-known seed dispersal agents, but only a few are true seed storers. Haftorn (1956) recorded the seed-storing habit of several species of tits and crows in Norway. These birds have a strong storage instinct. Some store seeds, then move them to other storage places. The seeds are usually stored under lichens, in bark crevices, and between pine needles. Other birds that store seeds are jays, nutcrackers, nuthatches, titmice, and woodpeckers.

Perhaps the most famous seed-storing bird is the California woodpecker. This bird is also known as the California acorn-storing woodpecker because of its unusual habit (almost a mania) of storing acorns in trees, telephone poles, gables, cornices, and, in fact, any wooden structure that is near an acorn source.

The California woodpecker drills a hole of suitable size and depth and then inserts an acorn, base out, into the hole. The acorn is then driven in until its base is flush or countersunk with the bark (Fig. 2). Acorns that are not driven flush or countersunk are vulnerable to pilferage by rodents and other birds. In addition to acorns, the California woodpecker stores almonds, hazelnuts, dates, maize, English and California walnuts, pecans, eucalyptus fruits, and stones which may or may not resemble acorns.

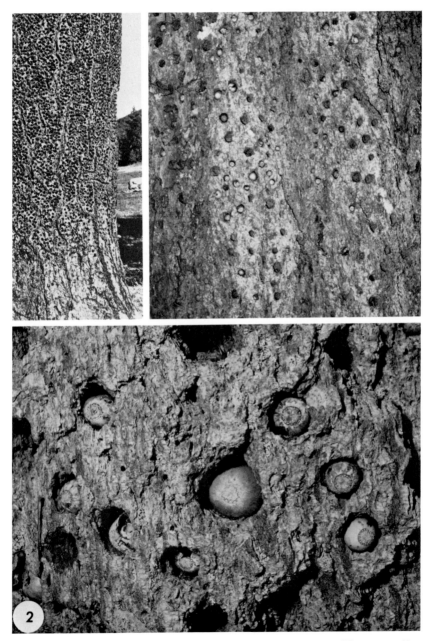

FIG. 2. Acorn granary tree of the California woodpecker shown at different magnifications. (Courtesy of Mr. and Mrs. N. G. Johannesma, Novato, California.)

About half the diet of the California woodpecker consists of acorns and the other half insects. The birds return to their wooden granaries to feed on the acorns when the supply from the trees becomes low or unavailable.

The holes seldom penetrate the cambial layer of trees. Therefore, it appears that these holes do not harm trees. An estimated 50,000 acorns have been imbedded in a large yellow pine and about 20,000 acorns in a large sycamore (Ritter, 1929, 1938).

Cavities containing half a dozen acorns from three species of oaks were found in the heartwood of a redwood in 1942. The 1080 tree rings between the cavities and the outside of the tree placed the age of the acorns at 802 A.D. The holes were clearly made by woodpeckers, probably the California woodpecker (R. C. Miller, 1950).

C. Ants

The ants and the grasshopper (Aesopus, 1850):

Once in winter the ants were sunning their seed store which had been soaked by the rains. A grasshopper saw them at this, and being famished and ready to perish, he ran up and begged for a bit. To the ant's question, "What were you doing in summer, idling, that you have to beg now?" he answered, "I lived for pleasure then, piping and pleasing travellers." "O, ho!" said they, with a grin, "dance in winter, if you pipe in summer. Store seed for the future when you can, and never mind playing and pleasing travellers."

The granivorous habit of some ant species was known for centuries but never was completely accepted by entomologists until Sykes (1835) studied *Phsidole providens* in India. His work and that of Moggridge (1873) and Wheeler (1926) have clearly proven that harvesting ants collect seeds which, upon being carried to the nest, are cleaned, and then stored in underground granaries (Figs. 3 and 4). Some species of ants are said to nip the radicle end of the seed to prevent germination. All studied species moved moist seeds outside on sunny days to dry them in an attempt to prevent germination. Sprouted seeds are not eaten but removed and discarded at the periphery of the kitchen midden. The stored seeds are crushed and fed to the larvae during periods when food is scarce. Moggridge counted seeds from eighteen plant families while studying the harvesting ants at Menton, France. He found about a quarter of a liter of seeds in an average-size hill.

D. Ocean and Air Currents

The ocean currents are huge reservoirs of buoyant seeds produced by terrestrial plants. Some of these seeds may be transported by ocean currents for thousands of miles in a year or two and may be washed ashore

FIG. 3. Ant carrying a flax seed. (Reprinted from Hutchins, 1961, p. 77. Copyrighted 1960 by Dodd, Mead and Company. Used by permission.)

FIG. 4. Ants tending *Paspalum* seeds in their underground granary. (Reprinted from Hutchins, 1960, p. 77. Copyrighted 1960 by Dodd, Mead and Company. Used by permission.)

in a viable condition (Fig. 5). This, in part, is the way some plants have attained a pantropic status and colonize newly formed islands (Gunn, 1968; Guppy, 1917; Muir, 1937).

Although there are no known studies on seeds accumulated by air currents, it is logical to assume that air currents, like ocean currents, do accumulate seeds which eventually become trapped in depressions and crevices.

E. Time

Time, the great collector of all things, has accumulated a treasurehouse of seeds. Recent seeds collected from arable soils have been studied by several scientists, including Jensen (1969). Seed deposits from the Neolithic period (Fig. 6) studied by archeologists have made definite contributions to our knowledge of the history of agriculture (Bertsch, 1941; Godwin, 1956; Helbaek, 1953, 1954; Katz et al., 1965; Smith, 1969). Plant phylogeny is rooted in fossil evidence, some of it from seeds (Fig. 7) uncovered by paleobotanists and geologists (Chaney et al., 1938; Elias, 1932, 1935, 1942, 1946; Katz et al., 1965; MacGinitie, 1941; Reid and Chandler, 1926, 1933).

III. Seed Identification

A. Importance

The seed is one of the distinctive features of the spermatophytes which sets them apart from the so-called lower plants. However, despite the economic importance of seeds, there are few data available on comparative seed morphology. Both internal and external structures of seeds have been neglected by morphologists and taxonomists, and this has made identification of isolated seeds more difficult than it should be.

Seed identification is a necessary part of seed testing, crop improvement, wildlife management, archaeology, paleobotany, and taxonomy. The importance of seed identification in seed testing and crop improvement is discussed in Chapter 5 of this Volume. Identification of seeds found in crops of birds (Fig. 8) and stomachs of other animals provides useful information for selecting food and cover crops to plant in game management areas. Seeds and pollen are the two main sources of information for archeologists and paleobotanists concerning past vegetation, climate, and agriculture. Fossil seeds have contributed to the history of spermatophytes.

Both internal and external seed characteristics are remarkably stable; therefore, they provide reliable criteria for positive identification of unknown seeds, such as the example shown in Fig. 9. They also increase our

FIG. 5. Seeds and fruits like these have been carried by the Gulf Stream from the Caribbean region to northern Europe, a distance of 4000 miles in 1 to 2 years.

FIG. 6. Top: 800-year-old charred seeds of bitter vetch recovered from an excavation in Turkey. Bottom: bitter vetch seeds collected in 1969 from plants growing in Turkey.

FIG. 7. *Carya* endocarps, ca. 45 million years old, from the Clarno formation of central-northern Oregon. (Courtesy of the Bones Collection, Smithsonian Institution.)

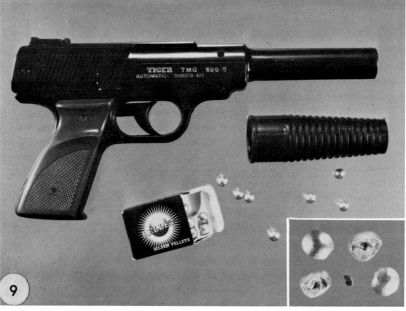

Fig. 8. Seeds removed from the crops of Rosy Finch chicks nesting on Amchitka Island, Alaska. (Magnification × 7.)

Fig. 9. Toy pellet gun imported from Japan shoots clay pellets which were formed around seeds of Japanese Barnyard grass shown in the insert as it was removed from a pellet. (Natural size.)

understanding of plant phylogeny. Although seed characteristics need not be given more recognition than other valid taxonomic characters, they should be used on a broader scale, especially by taxonomists. Examples of how seed characteristics may be used, with or without other plant parts, are cited in Section III,E.

Incorrect identification of seeds can be a health hazard. For example, the jequirity bean *(Abrus precatorius)* is deadly poisonous, but because of its colorful seed coat it is used extensively in jewelry and trinkets made in the tropics and subtropics (Figs. 10 and 11). *Rhynchosia pyramidalis* seeds which are nonpoisonous are similar to poisonous jequirity beans (Fig. 12). The seeds of the two species may be distinguished because the hilum of the jequirity bean is always in the black portion of the vermilion and black seed coat, whereas the hilum of the *R. pyramidalis* seed is always in the vermilion portion of the vermilion and black seed coat.

Following publicity about the danger of jequirity beans (Gunn, 1969), the Food and Drug Administration stopped sale of items containing these seeds. The castor bean apparently is not as deadly poisonous as the jequirity bean, but it does adversely affect humans by making them ill if ingested or causing, in some, an anaphylactic reaction if handled (Lockey and Dunkleberger, 1968).

B. Organizing and Maintaining a Seed Collection

1. ORGANIZATION

Small and specialized seed collections do not present the same problems inherent in organizing a large, general collection. However, both require that seed samples be stored so that they can be rapidly located. Additional requirements include finding a convenient method of handling various sizes of seeds, retaining collateral data, expanding the collection, and revising labels.

a. RAPID RETRIEVAL AND FILING. If the only objective is to locate known seed samples, an alphabetical system based on scientific or common names is useful. Often the objective of retrieval is to find a seed the characteristics of which match the seed to be identified. In order to do this rapidly, the seeds must be filed systematically by orders and families. It is essential in identifying an unknown seed to recognize possible family or order characteristics. With these characteristics in mind, the search for similarly identified seeds is greatly facilitated.

The most popular, though not necessarily the best, plant family system based on phylogeny is that of Engler and Prantl, popularized in their twenty-volume work, *Die naturlichen Pflanzenfamilien* (1887–1899), and numbered by Dalla Torre and Harms in their *Genera Siphonogamarum*

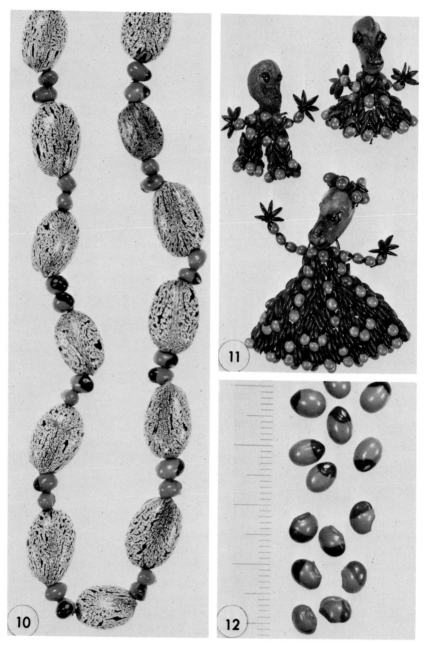

FIG. 10. A colorful souvenir necklace composed of deadly poisonous *Abrus precatorius* L. seeds and dangerously poisonous *Ricinus communis* L. seeds.

ad Systema Englerianum Conscripta (1900–1907). Lawrence (1951) briefly discussed other phylogenetic systems, but none of these are in wide use and none offer the advantage of the Engler and Prantl system. The naturalness of the Engler and Prantl system was questioned by Cronquist (1968) who proposed an alternative system. The thirty-seven families at the end of this chapter are presented according to the Cronquist system for comparative purposes.

Once a family system has been adopted for a seed collection, the families should be numbered. These family numbers may be used on trays and samples as a filing aid. The family numbers presented in Table II are used in the NCRB seed collection and at the U.S. National Herbarium, Smithsonian Institution. The families are presented alphabetically in Table III.

TABLE II

NUMERICAL LIST OF THE FAMILIES OF FLOWERING PLANTS BASED ON DALLA TORRE AND HARMS *Genera siphonogamarum ad systema Englerianum conscripta* (1900–1907)[a]

1.	Cycadaceae	23.	Araceae	45.	Musaceae
2.	Bennettitaceae	24.	Lemnaceae	46.	Zingiberaceae
3.	Cordaitaceae	25.	Flagellariaceae	47.	Cannaceae
4.	Ginkgoaceae	26.	Restionaceae	48.	Marantaceae
5.	Taxaceae	27.	Centrolepidaceae	49.	Burmanniaceae
6.	Pinaceae	28.	Mayacaceae	50.	Orchidaceae
7.	Gnetaceae	29.	Xyridaceae	51.	Casuarinaceae
8.	Typhaceae	30.	Eriocaulaceae	52.	Saururaceae
9.	Pandanaceae	31.	Rapateaceae	53.	Piperaceae
10.	Sparganiaceae	32.	Bromeliaceae	54.	Chloranthaceae
11.	Potamogetonaceae	33.	Commelinaceae	55.	Lacistemaceae
12.	Najadaceae	34.	Pontederiaceae	56.	Salicaceae
13.	Aponogetonaceae	35.	Philydraceae	57.	Myricaceae
14.	Juncaginaceae	36.	Juncaceae	58.	Balanopsidaceae
15.	Alismataceae	37.	Stemonaceae	59.	Leitneriaceae
16.	Butomaceae	38.	Liliaceae	60.	Juglandaceae
17.	Hydrocharitaceae	39.	Haemodoraceae	61.	Betulaceae
18.	Triuridaceae	40.	Amaryllidaceae	62.	Fagaceae
19.	Gramineae	41.	Velloziaceae	63.	Ulmaceae
20.	Cyperaceae	42.	Taccaceae	64.	Moraceae
21.	Palmae	43.	Dioscoreaceae	65.	Urticaceae
22.	Cyclanthaceae	44.	Iridaceae	66.	Proteaceae

FIG. 11. Figures made in Jamaica from seeds of *Abrus precatorius* L., *Leucaena leucocephala* (Lam.) de Wit, and *Anacardium occidentalis* L. (Magnification × 0.5.)

FIG. 12. Top: *Abrus precatorius* L. seeds (hilum on black portion of seed). Bottom: *Rhynchosia pyramidalis* Urb. seeds (hilum on vermilion portion of seed). Space between each line, 1 mm.

TABLE II *(Continued)*

67. Loranthaceae	112. Droseraceae	158. Celastraceae
68. Myzodendraceae	113. Podostemonaceae	159. Hippocrateaceae
69. Santalaceae	114. Hydrostachyaceae	160. Stackhousiaceae
70. Grubbiaceae	115. Crassulaceae	161. Staphyleaceae
71. Opiliaceae	116. Cephalotaceae	162. Icacinaceae
72. Olacaceae	117. Saxifragaceae	163. Aceraceae
73. Balanophoraceae	118. Pittosporaceae	164. Hippocastanaceae
74. Aristolochiaceae	119. Brunelliaceae	165. Sapindaceae
75. Rafflesiaceae	120. Cunoniaceae	166. Sabiaceae
76. Hydnoraceae	121. Myrothamnaceae	167. Melianthaceae
77. Polygonaceae	122. Bruniaceae	168. Balsaminaceae
78. Chenopodiaceae	123. Hamamelidaceae	169. Rhamnaceae
78a. Didiereaceae	124. Platanaceae	170. Vitaceae
79. Amaranthaceae	125. Crossosomataceae	171. Elaeocarpaceae
80. Nyctaginaceae	126. Rosaceae	172. Chlaenaceae
81. Bataceae	127. Connaraceae	173. Gonystylaceae
82. Theligonaceae	128. Leguminosae	174. Tiliaceae
83. Phytolaccaceae	128a. Pandaceae	175. Malvaceae
84. Aizoaceae	129. Geraniaceae	176. Triplochitonaceae
85. Portulacaceae	130. Oxalidaceae	177. Bombacaceae
86. Basellaceae	131. Tropaeolaceae	178. Sterculiaceae
87. Caryophyllaceae	132. Linaceae	179. Scytopetalaceae
88. Nymphaeaceae	133. Humiriaceae	180. Dilleniaceae
89. Ceratophyllaceae	134. Erythroxylaceae	181. Eucryphiaceae
90. Trochodendraceae	135. Zygophyllaceae	182. Ochnaceae
91. Ranunculaceae	136. Cneoraceae	183. Caryocaraceae
92. Lardizabalaceae	137. Rutaceae	184. Marcgraviaceae
93. Berberidaceae	138. Simaroubaceae	185. Quiinaceae
94. Menispermaceae	139. Burseraceae	186. Theaceae
95. Magnoliaceae	140. Meliaceae	187. Guttiferae
95a. Degeneriaceae	141. Malpighiaceae	188. Dipterocarpaceae
96. Calycanthaceae	142. Trigoniaceae	189. Elatinaceae
97. Lactoridaceae	143. Vochysiaceae	190. Frankeniaceae
98. Annonaceae	144. Tremandraceae	191. Tamaricaceae
99. Myristicaceae	145. Polygalaceae	192. Fouquieriaceae
100. Gomortegaceae	146. Dichapetalaceae	193. Cistaceae
101. Monimiaceae	147. Euphorbiaceae	194. Bixaceae
102. Lauraceae	148. Callitrichaceae	195. Cochlospermaceae
103. Hernandiaceae	149. Buxaceae	196. Koeberliniaceae
104. Papaveraceae	150. Coriariaceae	197. Canellaceae
105. Cruciferae	151. Empetraceae	198. Violaceae
106. Tovariaceae	152. Limnanthaceae	199. Flacourtiaceae
107. Capparaceae	153. Anacardiaceae	200. Stachyuraceae
108. Resedaceae	154. Cyrillaceae	201. Turneraceae
109. Moringaceae	155. Pentaphylacaceae	202. Malesherbiaceae
110. Sarraceniaceae	156. Corynocarpaceae	203. Passifloraceae
111. Nepenthaceae	157. Aquifoliaceae	204. Achariaceae

TABLE II *(Continued)*

205. Caricaceae	229. Cornaceae	255. Nolanaceae
206. Loasaceae	230. Clethraceae	256. Solanaceae
207. Datiscaceae	231. Pyrolaceae	257. Scrophulariaceae
208. Begoniaceae	232. Lennoaceae	258. Bignoniaceae
209. Ancistrocladaceae	233. Ericaceae	259. Pedaliaceae
210. Cactaceae	234. Epacridaceae	260. Martyniaceae
211. Geissolomataceae	235. Diapensiaceae	261. Orobanchaceae
212. Penaeaceae	235a. Theophrastaceae	262. Gesneriaceae
213. Oliniaceae	236. Myrsinaceae	263. Columelliaceae
214. Thymelaeaceae	237. Primulaceae	264. Lentibulariaceae
215. Elaeagnaceae	238. Plumbaginaceae	265. Globulariaceae
216. Lythraceae	239. Sapotaceae	266. Acanthaceae
217. Sonneratiaceae	240. Ebenaceae	267. Myoporaceae
217a. Crypteroniaceae	241. Styracaceae	268. Phrymaceae
218. Punicaceae	242. Symplocaceae	269. Plantaginaceae
219. Lecythidaceae	243. Oleaceae	270. Rubiaceae
220. Rhizophoraceae	244. Salvadoraceae	271. Caprifoliaceae
221. Combretaceae	245. Loganiaceae	272. Adoxaceae
222. Myrtaceae	246. Gentianaceae	273. Valerianaceae
223. Melastomataceae	247. Apocynaceae	274. Dipsacaceae
224. Onagraceae	248. Asclepiadaceae	275. Cucurbitaceae
224a. Trapaceae	249. Convolvulaceae	276. Campanulaceae
225. Haloragaceae	250. Polemoniaceae	277. Goodeniaceae
225a. Hippuridaceae	251. Hydrophyllaceae	278. Stylidiaceae
226. Cynomoriaceae	252. Boraginaceae	279. Calyceraceae
227. Araliaceae	253. Verbenaceae	280. Compositae
228. Umbelliferae	254. Labiatae	281. Incertae sedis

[a] Prepared by D. H. Nicolson, U. S. National Herbarium, Smithsonian Institution, 1969.

TABLE III

ALPHABETICAL LIST OF THE FAMILIES OF FLOWERING PLANTS BASED ON
DALLA TORRE AND HARMS *Genera siphonogamarun ad systema Englerianum
conscripta* (1900–1907)[a]

266. Acanthaceae	13. Aponogetonaceae	93. Berberidaceae
163. Aceraceae	157. Aquifoliaceae	61. Betulaceae
204. Achariaceae	23. Araceae	258. Bignoniaceae
272. Adoxaceae	227. Araliaceae	194. Bixaceae
84. Aizoaceae	74. Aristolochiaceae	177. Bombacaceae
15. Alismataceae	248. Asclepiadaceae	252. Boraginaceae
79. Amaranthaceae	73. Balanophoraceae	32. Bromeliaceae
40. Amaryllidaceae	58. Balanopsidaceae	119. Brunelliaceae
153. Anacardiaceae	168. Balsaminaceae	122. Bruniaceae
209. Ancistrocladaceae	86. Basellaceae	49. Burmanniaceae
98. Annonaceae	81. Bataceae	139. Burseraceae
247. Apocynaceae	208. Begoniaceae	16. Butomaceae

TABLE III *(Continued)*

149. Buxaceae	95a. Degeneriaceae	224a. (Hydrocaryaceae)
210. Cactaceae	235. Diapensiaceae	17. Hydrocharitaceae
148. Callitrichaceae	146. Dichapetalaceae	251. Hydrophyllaceae
96. Calycanthaceae	78a. Didiereaceae	114. Hydrostachyaceae
279. Calyceraceae	180. Dilleniaceae	162. Icacinaceae
276. Campanulaceae	43. Dioscoreaceae	281. Incertae sedis
197. Canellaceae	274. Dipsacaceae	44. Iridaceae
47. Cannaceae	188. Dipterocarpaceae	60. Juglandaceae
107. Capparaceae	112. Droseraceae	36. Juncaceae
271. Caprifoliaceae	240. Ebenaceae	60. Juglandaceae
205. Caricaceae	215. Elaeagnaceae	196. Koeberliniaceae
183. Caryocaraceae	171. Elaeocarpaceae	254. Labiatae
87. Caryophyllaceae	189. Elatinaceae	55. Lacistemaceae
51. Casuarinaceae	151. Empetraceae	97. Lactoridaceae
158. Celastraceae	234. Epacridaceae	92. Lardizabalaceae
27. Centrolepidaceae	233. Ericaceae	102. Lauraceae
116. Cephalotaceae	30. Eriocaulaceae	219. Lecythidaceae
89. Ceratophyllaceae	134. Erythroxylaceae	128. Leguminosae
78. Chenopodiaceae	181. Eucryphiaceae	59. Leitneriaceae
172. Chlaenaceae	147. Euphorbiaceae	24. Lemnaceae
54. Chloranthaceae	62. Fagaceae	232. Lennoaceae
193. Cistaceae	199. Flacourtiaceae	264. Lentibulariaceae
230. Clethraceae	25. Flagellariaceae	38. Liliaceae
136. Cneoraceae	192. Fouquieriaceae	152. Limnanthaceae
195. Cochlospermaceae	190. Frankeniaceae	132. Linaceae
263. Columelliaceae	211. Geissolomataceae	206. Loasaceae
221. Combretaceae	246. Gentianaceae	245. Loganiaceae
33. Commelinaceae	129. Geraniaceae	67. Loranthaceae
280. Compositae	262. Gesneriaceae	216. Lythraceae
127. Connaraceae	4. Ginkgoaceae	95. Magnoliaceae
249. Convolvulaceae	265. Globulariaceae	202. Malesherbiaceae
150. Coriariaceae	7. Gnetaceae	141. Malpighiaceae
229. Cornaceae	100. Gomortegaceae	175. Malvaceae
156. Corynocarpaceae	173. Gonystylaceae	48. Marantaceae
115. Crassulaceae	277. Goodeniaceae	175. Malvaceae
125. Crossosomataceae	19. Gramineae	260. Martyniaceae
105. Cruciferae	70. Grubbiaceae	28. Mayacaceae
217a. Crypteroniaceae	187. Guttiferae	223. Melastomataceae
275. Cucurbitaceae	39. Haemodoraceae	140. Meliaceae
120. Cunoniaceae	225. Haloragaceae	167. Melianthaceae
1. Cycadaceae	123. Hamamelidaceae	94. Menispermaceae
22. Cyclanthaceae	103. Hernandiaceae	101. Monimiaceae
82. (Cynocrambaceae)	164. Hippocastanaceae	64. Moraceae
226. Cynomoriaceae	159. Hippocrateaceae	109. Moringaceae
20. Cyperaceae	225a. Hippuridaceae	45. Musaceae
154. Cyrillaceae	133. Humiriaceae	267. Myoporaceae
207. Datiscaceae	76. Hydnoraceae	57. Myricaceae

TABLE III (Continued)

99. Myristicaceae	250. Polemoniaceae	160. Stackhousiaceae
121. Myrothamnaceae	145. Polygalaceae	161. Staphyleaceae
236. Myrsinaceae	77. Polygonaceae	37. Stemonaceae
222. Myrtaceae	34. Pontederiaceae	178. Sterculiaceae
68. Myzodendraceae	85. Portulacaceae	278. Stylidiaceae
12. Najadaceae	11. Potamogetonaceae	241. Styracaceae
111. Nepenthaceae	237. Primulaceae	242. Symplocaceae
255. Nolanaceae	66. Proteaceae	42. Taccaceae
80. Nyctaginaceae	218. Punicaceae	191. Tamaricaceae
88. Nymphaeaceae	231. Pyrolaceae	5. Taxaceae
182. Ochanaceae	185. Quiinaceae	186. Theaceae
72. Olacaceae	75. Rafflesiaceae	82. Theligonaceae
182. Ochnaceae	91. Ranunculaceae	235a. Theophrastaceae
213. Oliniaceae	31. Rapateaceae	214. Thymelaeaceae
224. Onagraceae	108. Resedaceae	174. Tiliaceae
71. Opiliaceae	26. Restionaceae	106. Tovariaceae
50. Orchidaceae	169. Rhamnaceae	224a. Trapaceae
261. Orobanchaceae	220. Rhizophoraceae	144. Tremandraceae
130. Oxalidaceae	126. Rosaceae	142. Trigoniaceae
21. Palmae	270. Rubiaceae	176. Triplochitonaceae
128a. Pandaceae	137. Rutaceae	18. Triuridaceae
9. Pandanaceae	166. Sabiaceae	90. Trochodendraceae
104. Papaveraceae	56. Salicaceae	131. Tropaeolaceae
203. Passifloraceae	244. Salvadoraceae	201. Turneraceae
259. Pedaliaceae	69. Santalaceae	8. Typhaceae
212. Penaeaceae	165. Sapindaceae	63. Ulmaceae
155. Pentaphylacaceae	239. Sapotaceae	228. Umbelliferae
35. Philydraceae	110. Sarraceniaceae	65. Urticaceae
268. Phrymaceae	52. Saururaceae	273. Valerianaceae
83. Phytolaccaceae	117. Saxifragaceae	41. Velloziaceae
6. Pinaceae	257. Scrophulariaceae	253. Verbenaceae
53. Piperaceae	179. Scytopetalaceae	198. Violaceae
118. Pittosporaceae	138. Simaroubaceae	170. Vitaceae
269. Plantaginaceae	256. Solanaceae	143. Vochysiaceae
124. Platanaceae	217. Sonneratiaceae	29. Xyridaceae
238. Plumbaginaceae	10. Sparganiaceae	46. Zingiberaceae
113. Podostemonaceae	200. Stachyuraceae	135. Zygophyllaceae

[a] Prepared by D. H. Nicolson, U. S. National Herbarium, Smithsonian Institution, 1969.

Within most families the seed samples may be conveniently filed alphabetically by genus and then by species. For large families, such as Compositae, Gramineae, Leguminosae, and Rosaceae, seed samples should be filed systematically by tribes and then either alphabetically or systematically by genus. Within a genus, the species should be filed alphabetically. If the genus name is known, but the family name is unknown, consult Airy-Shaw (1966).

Another filing aid is to number the trays or boxes which hold seed samples. If enough expansion room has been built into a seed collection, a simple numbering sequence can be used, such as 1 to 100 or 1 to 1000. However, if expansion is a problem, a family–tray number combination can be used; for example, 13.12 represents the twelfth tray in the thirteenth family. The next tray may be 14.1. If additions are made to the thirteenth family and there is no room, then a new tray, 13.13, can be inserted without renumbering. The samples within the existing trays can be adjusted to fill partially the new tray.

b. CONTAINERS FOR SEED SAMPLES. Seed samples should be placed in containers that are clear enough to permit the seeds to be studied without pouring them out, or mounted so they are fully visible. The most expensive, yet most convenient, way to store average-size seeds is in a 60×14 mm screw-capped glass vial or a 50×12 mm cork-topped glass vial (Fig. 13). A screw-capped vial is several times more expensive than a corked one. However, a plastic or metal cap with a cushioned, slick paper inner sealer forms a permanent barrier against insect penetration, and there is little danger of the cap shaking loose or popping off. Corks, on the other hand, may not be a permanent barrier to insects and occasionally shake loose, allowing the seeds to spill out. One disadvantage of a screw-capped vial is the constriction (collar) where the glass is threaded. This collar, only 12 mm in diameter, restricts the size of seeds that can be inserted into the 14 mm diameter barrel. Seeds with appendages may be forced through the collar, expand in the barrel, and then be difficult to remove, except by breaking the vial. Other sizes of screw-capped vials are available—the 80×21 mm vial should be considered because the collar is nearly absent. However, only about half as many of these vials can be fitted into the same space as the 60×14 mm vials.

Care should be taken in placing labels on vials. In Fig. 14 one label was placed on a vial along the long axis. This permits easy reading of long scientific names or notations. Unfortunately, the label completely blocks the view of the seeds so that the vial must be rotated (label turned down) or pulled out for seed viewing. This can be quite time consuming when scanning trays. A label placed around the vial will permit reading of most of the data. More importantly it will allow the seeds to be visible at all times.

To avoid gluing and to use both sides of the label, some collectors may prefer an inside label to an outside label. The drawback of gluing has been solved by the excellent pressure-sensitive labels. Data on the back side of an inside label may not be readily visible because of the seeds, and the front of an inside label may become buried behind a layer of seeds. Extra data may be recorded on index cards and filed in a collateral card file.

Another way to store seeds is in clear polyethylene bags which, regardless of size, are inexpensive, and are not harmed by insecticide sprays or *p*-dichlorobenzene (PDB). Their disadvantage is that they conceal seeds when bags are filed. Therefore, bags must be handled during the scanning process. Bags have proven to be quite satisfactory, especially for seeds which are too large for vials. Figure 15 shows two convenient bag sizes, a 16 × 8 cm nongusseted bag for regular-size seeds and a 20.5 × 10 × 5 cm gusseted bag for larger seeds. The backing card which holds the bag stiff and bears seed sample data is file-card stock. Half of a 3½ × 5 in. file card is ideal for the small bag. A 4 × 6 in. card with the two lower corners rounded (to prevent the pointed corners from tearing the bag when it is inserted) is ideal for the large bag.

Care should be taken to be certain that seams are completely sealed and that the polyethylene film is at least 1.5 mils but no more than 2 mils thick (to insure transparency) and at least 95% pinhole free. Static electricity is a drawback when filling polyethylene bags with small seeds. When bags are being filled, they should not be rubbed.

Two specially built seed sample containers are shown in Fig. 16. The box type is used by the Canadian Seed Research Laboratory, and the plastic bottle type is used by the Danish State Seed Testing Laboratory. Other less conventional ways of storing a small or specialized seed collection include embedding the seeds in plastic (Fig. 17), gluing tiny seeds to glass slides with a cellulose acetate plastic (Terrell, 1968), placing seeds in display-type cases (Figs. 18 and 19), or attaching small plastic vials containing the seeds to 3 × 5 in. index cards.

c. CABINETS. Most seed collections are kept in shallow-drawer metal cabinets (Fig. 20) or herbarium cases (Fig. 21). The drawers can be fitted with wooden or cardboard dividers which hold seed vials in place. A convenient way to make wooden seed vial holders is to bind two pieces of wood of proper dimensions together and drill properly spaced holes centered on seams between the boards. When the bindings are removed, each board will bear semicircular grooves in which vials can rest. If herbarium cabinets are used to house a seed collection, boxes with dividers can be used (Fig. 22). These boxes can be stacked six to the pigeonhole. Deeper boxes with dividers can be used to hold larger seeds in polyethylene bags. An advantage to using herbarium cases is that seeds of various sizes can be stored in the same case (Fig. 21).

d. LABELING. The content of a label is governed by label size and need for information. Required information includes family name or number for ease in filing, scientific name, and some notation about sample source. The latter notation is important if the seed collection is to be

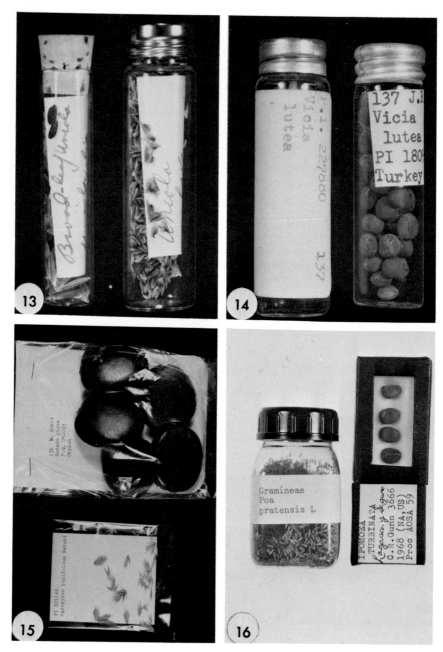

FIG. 13. Two popular-size glass seed vials: a 50 × 14 mm corked vial and a 60 × 14 mm screw-capped vial, both with inside labels.

meaningful. A sample vouchered by a herbarium specimen is intrinsically a better sample than one of which the origin and original identification are not known.

e. CARD FILE. In most seed collections collateral data, viz., notes, citations, annotations, and photographs are lost because there is no place to record and file them. A collateral card file, which can house these valuable adjuncts, should be maintained (Fig. 23).

f. PREPARING SAMPLES FOR ACCESSIONING. No seed sample should be accessioned until it is thoroughly dried and treated with an insect control. If possible, all seed samples should be fumigated with methyl bromide under 3 lb pressure at 80°F for 2 ½ hours. This will kill external insects as well as those borne within the seeds. It will not affect subsequent germination unless moisture content of the seeds exceeds 10% at time of treatment or unless the treatment is extensive. Other compounds have been used, including carbon bisulfide by the Australian plant introduction laboratory, Phostoxin in Europe and South Africa, hydrogen cyanide, and ethylene oxide. The 3-day treatment needed for Phostoxin may be inconvenient. The USDA, which in the past used hydrogen cyanide, now uses methyl bromide. In addition to being an insecticide, ethylene oxide is a fungicide and reduces the germination capacity of seed samples. Care should be taken in handling any of these fumigants, especially ethylene oxide.

The final step prior to placing seeds in labeled containers is their cleaning. This must not be done haphazardly. The person doing the cleaning should be familiar with peculiarities of the seeds being cleaned. Too often a sample is meticulously cleaned with removal of all plant parts, which are not seeds, and even removal of underdeveloped seeds. This type of cleaning will create a pretty, but not always a useful, sample.

The goal of the cleaning process should be to save a representative seed sample along with those plant parts which might be useful in making the seed identification. In preparing grass samples for filing, entire disseminule, floret, spikelet, and if possible complete (or a portion thereof) fruiting head should be saved. For families in which fruits and seeds are separate entities, both should be saved. If the fruit is too large for storing,

FIG. 14. Vertical outside label gives maximum exposure to the data but hides seeds, whereas horizontal outside label shows some data but does not hide seeds.

FIG. 15. Two popular-size polyethylene bags, 1.5 mils thick: a 16 × 8 cm nongusseted bag and a 20.5 × 10 × 5 cm gusseted bag.

FIG. 16. A plastic bottle seed storage container used by Danish State Seed Testing Laboratory and a box type used by Canadian Seed Research Laboratory.

a section or piece of it should be saved. Fruits should be sectioned prior to filing to serve as identification aids. Special care should be taken to include the two, or occasionally three, types of seeds that some plants produce.

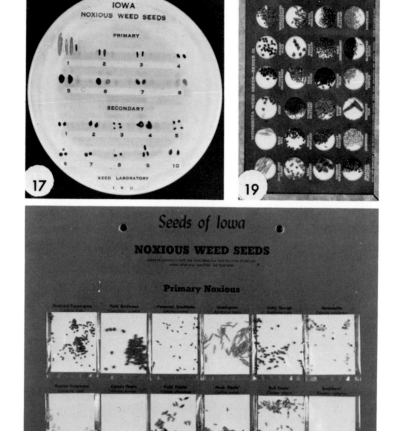

FIG. 17. Iowa noxious weed seeds embedded in clear plastic molded in a petri dish.

FIG. 18. Minnesota weeds in a glass-covered, 5 × 7 display case. One of a series of four.

FIG. 19. One-half of a spiral-bound display folder designed to exhibit Iowa noxious weed seeds. One of a series of four.

2. MAINTENANCE

In general, maintenance of a seed collection is not complicated or difficult. However, a few problems must be considered, including insect control, clouding of vials, relabeling, and accessioning new samples.

a. INSECT CONTROL. The best procedure is to fumigate each seed sample before it is accessioned. Samples that are not fumigated should be checked during the first 6 months or so for insect damage. Insect frass (Fig. 24) can usually be seen at the bottom of a vial along with larval casts and perhaps live larvae or adults. The sample should be cleaned of insects and their residue before being treated with PDB. A chunk of PDB placed in the vial will prevent any recurrence of insects. Another method of controlling insects, which has not been fully tested, is to freeze the sample for 24 hours in a deep freezer. Ordinarily insects should not be a problem in a seed collection where each sample is isolated from the next and housed in insectproof cases.

The room where the seed collection is housed should be kept clean, the floor swept regularly, and the floor and bases of the cases sprayed periodically with an all-purpose insecticide.

b. CLOUDING OF VIALS. Perhaps the most aggravating maintenance problem is the gradual clouding of some vials (Fig. 25) which is caused by exudates from the seeds. This is more a problem in some families, such as Berberidaceae, Cruciferae, Euphorbiaceae, Labiatae, Lauraceae, Palmae, Papaveraceae, Punicaceae, Rhamnaceae, Rosaceae, Rutaceae, and Verbenaceae, than in other families. The only solution is to replace cloudy vials. This problem will probably occur in time with polyethylene bags.

c. RELABELING. Some labels must be revised. Notes can be placed in the card file until time permits relabeling, if the relabeling is extensive. Usually relabeling merely involves updating of scientific names, and this can be done by making pencilled corrections on the labels. These labels can be used until new labels are prepared.

d. ACCESSIONING NEW SAMPLES. This is usually a routine matter. One problem is how many samples should be kept for one species. Obviously there are no hard and fast rules, but there are guidelines which will help to solve this problem.

Some species produce seeds the characteristics of which are superficially and basically quite stable, whereas other species have seeds the characteristics of which are superficially different but basically stable. Unfortunately superficial characteristics usually catch the eye. *Vicia*

FIG. 20. Seed collection (drawer type) at Federal Seed Laboratory, Grain Division, Consumer and Marketing Service, Agriculture Research Center, Beltsville, Maryland.

FIG. 21. One of the twenty-four herbarium cases housing New Crops Research Branch seed collection, Plant Science Research Division, Agricultural Research Service, Plant Industry Station, Beltsville, Maryland.

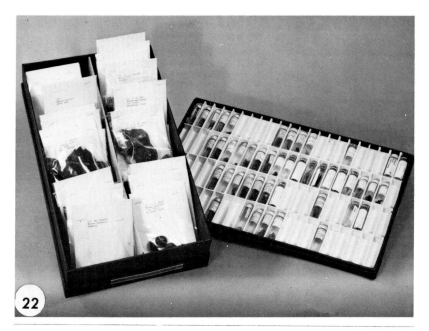

128
Vicia americana Muhl. ex Willd.
American Vetch

＊C. R. Gunn 2566, Story Cp., Iowa, Aug. 27, 1962 (ISC).
Hilum silvery, seed coat not densely mottled.

Mary Ryman -- taken from a V. villosa seed lot grown at
Sheridan, Ore., 1963. Hilum color of seed coat, seed coat
densely mottled.

C. R. Gunn (2566) Story Co., Iowa, 7.24.63. Black color of
legumes caused by *Alternaria* tenuis auct. sensu Wiltshire
according to Martin Kulik.

＊C. R. Gunn 2631, Idyllwild, Calif., June, 1963, legumes bright
straw, seeds dark (ISC)

23 ＊ Support of A H 392.
See Card Two

FIG. 22. Boxes designed to fit into a herbarium case pigeonhole — one for large seeds and
one for average-size seeds.
FIG. 23. Sample card showing the type of collateral data which is desirable to keep but
which cannot be recorded on a small label.

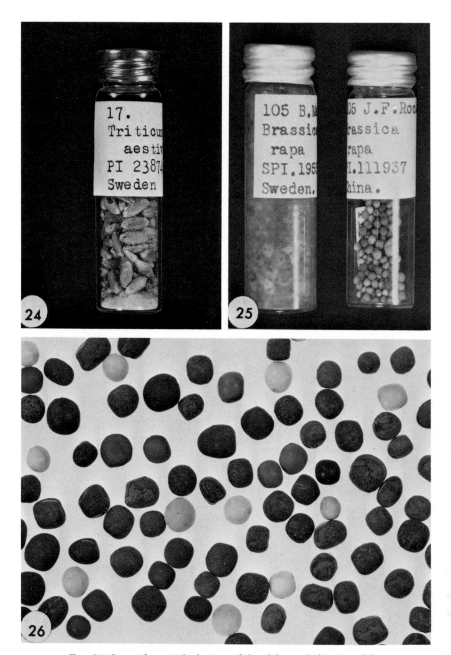

Fig. 24. Insect frass at the bottom of the vial reveals insect activity.

sativa seeds (Fig. 26) serve as an example of seeds with superficial variations within a species. It has automatically self-pollinating flowers. Outcrossing occurs less than 1% of the time. The superficial seed characteristics — size, color, and shape — are perpetuated by the self-pollination mechanism. These seed lines breed true, so that fifty or so samples would be required to show the complete range of variation. However, for an outcrossing vetch, such as *Vicia villosa,* of which the self-pollination rate is low and of which the seeds are uniform, only a few seed samples would be needed, and these selected from the range of the species. An effort should be made to have more than one sample of each species and to select for characteristics not shown in the samples already in the collection.

This is also the time to upgrade the collection. If a sample in the collection is not vouchered by a herbarium specimen, and the new one is, then consider discarding the one in the collection, if it does not add any information.

C. Seed Identification Centers

In most countries the main seed identification center is part of an official or commercial seed testing laboratory the main interest of which is identification of crop and weed seeds. These laboratories may or may not have the staff and the means to identify seeds for outsiders. Many countries which have government-operated seed testing laboratories belong to the International Seed Testing Association (ISTA). In North America there are two seed laboratory organizations. Federal, state, and province laboratories belong to the Association of Official Seed Analysts (AOSA); commercial laboratory personnel are members of the Society of Commerical Seed Technologists (SCST).

Unfortunately, there is no list or organization for the few seed identification specialists who work for museums, botanical gardens, universities, or federal agencies. Their work may be in diverse disciplines, such as archaelogy, paleobotany, and taxonomy.

Two major organizations not involved in seed testing which have trained, full-time seed identification specialists are the N. I. Vavilov All-Union Institute of Plant Industry, Leningrad, and New Crops Research Branch, U. S. Department of Agriculture. The Vavilov Institute

FIG. 25. *Brassica rapa* L. seeds placed in a vial in 1907 have clouded the glass, whereas those placed in a vial in 1935 have not caused clouding.

FIG. 26. *Vicia sativa* L. seeds (natural size) exhibit a variety of superficial characteristics (color, size, and shape) as well as stable characteristics (hilum characteristics and distance from hilum to lens) which permit accurate seed identification.

has a collection of 175,000 samples of seeds, including a germ-plasm bank, and an active seed identification program (Brezhnev, 1970). The NCRB collection of about 90,000 samples is separate from the 85,000 samples in the germ-plasm bank seed collection located at the National Seed Storage Laboratory (see Section I,B).

D. Methods

Seed identification methods are usually quite simple, because most isolated seed identifications are made by memory or comparison, either with known seed samples or with illustrations. A well-trained experienced seed analyst may be able to identify by sight about 500 species of seeds; perhaps as many as 1000 species. Others, such as agronomists, archeologists, morphologists, plant quarantine inspectors, and taxonomists, who make routine seed identifications, rely on their memories. When memory fails, one consults a seed collection or available literature.

Most workers, who make visual identifications, find it difficult to enumerate characteristics which they use in making the identifications. This is partially due to a deficiency of descriptive words and to a lack of standardization of seed terminology. Thus, drawings and photographs, which enchance descriptions, are important and are emphasized in seed identification publications.

Although there are several classification schemes for fruits, there apparently are only three published classifications of seeds. The classification by MacMillan (1902) has not been accepted by seed workers, because information about the ontogeny of individual seeds is required. The work of Martin (1946) is discussed under Internal Topography below. The most recent phylogenetic seed classification (Smirnova, 1965) is in Russian. Smirnova divided phanerogamous seeds into five groups based on the ratio of embryo size to food-storage tissue. This ratio is expressed as embryo length to total vertical length of the ripe seed.

The "polyclave" system perfected by Duke (1964a,b, 1965b, 1969a,b) is an information-retrieval system designed primarily to facilitate identification of unknowns, among them isolated seeds. The system, described by Duke (1969c), consists of a family-name underlay with symbols for 480 spermatophyte families and a series of transparent overlays which may be superimposed singly or in combinations over the underlay. Family symbols for families exhibiting the character called for on an overlay are not printed (families not exhibiting the character have their symbols printed) so that symbols for families with the character may be read from the underlay when the overlay is in place. By using a series of overlays, the number of family choices may be radically reduced. An unknown isolated seed for which the family characteristics are not recognized may be rapidly identified (at least to family) by using this system. Although the polyclave has not wide acceptance, it is quite useful and is recommended.

Aids that are useful in seed identification are discussed below under External Topography, Internal Topography, Chemical Analysis, and Seedlings.

1. EXTERNAL TOPOGRAPHY

With experience it is often possible to determine the family of an isolated seed by external topography alone. Important features of seeds are shape, size, seed coat surface, placement of hilum, and presence or absence of associated parts such as arils, caruncles, and elaiosomes. Studies of seed coat surfaces may be facilitated by using cellulose acetate peels (Stoddard, 1965) or acrylic polymer emulsion peels (Horanic and Gardner, 1967). Although developed to study leaf surfaces, these peels should work as well on seeds as the celluloid membrane used by Harada (1934) on seeds from the Amaranthaceae, Caryophyllaceae, and Portulacaceae.

The discussion of seed characteristics of thirty-seven selected families at the end of this chapter provides an example of how external characteristics can be used in making identifications of plant families. Other publications emphasizing external seed identifications at the family level include Anghel *et al.* (1959, 1965), Beijerinck (1947), Berggren (1969), Bertsch (1941), Bouwer (1927), Brouwer and Stahlin (1955), Dobrokotov (1961), Fong (1969), Heinisch (1955), Hubner (1955), Isely (1947), Kiffmann (1955–1960), Korsmo (1935), Lhotska (1957), Martin and Barkley (1961), McClure (1957), Murley (1951), Musil (1963), and Scurti (1948).

2. INTERNAL TOPOGRAPHY

The gross internal organization of seeds has been long neglected. Martin (1946) wrote: "Not since the time of Gaertner (1788–1805), a century and a half ago, has there been any extensive study in the internal organization of seeds. This neglect has served as a handicap not only in identification of seeds and their storage and germination, but also in formulation of correct concepts of plant phylogeny and classification, since in plants, as in animals, embryos are of fundamental value in denoting relationships." This statement is part of the introduction of Martin's study of the internal morphology of seeds of 1287 genera from 155 families. In addition to describing and depicting the internal morphology of these seeds, he organized a classification system and a phylogenetic classification of spermatophytes based on the internal morphology of seeds. The principal internal characteristics are type, size, placement of the embryo, quantity and quality of the food reserve, and size of seed. Martin's publication is indispensable to anyone who identifies seeds.

One problem in making freehand seed sections (with a razor blade or hacksaw) to display the embryo is that the embryo may be difficult to locate, especially if it is embedded in a nonstarchy endosperm. Martin found it advantageous to moisten the sections with water, whereas others have used softening solutions, such as a mixture of 74% distilled water, 25% methyl alcohol, and 1% dioctyl sodium sulfosuccinate (Aerosol OT). After soaking for 3 minutes or more, the embryo can be teased out with a dissecting needle. For a seed with a starchy endosperm, a dilute solution of iodine will darken the endosperm but not the embryo, permitting easier study of the embryo. A recently developed chemical, 2,3,5-triphenyl tetrazolium chloride, known as tetrazolium, has become quite useful in making rapid determinations of seed viability. The colorless tetrazolium solution is oxidized in the presence of dehydrogenase enzymes to its colored elementary form (see Chapter 5 of this volume). The brightly colored embryo (Fig. 27) is easily seen and its dissection facilitated. For instructions on the use of tetrazolium, see Grabe (1970).

Embryos can also be studied by X-raying seeds. Figure 28 is a radiograph of a *Pinus palustris* seed. The embryo with its several cotyledons can be clearly seen, and the seed is intact. X-Ray techniques and applications are described by Gustafsson and Simak (1963), Milner *et al.* (1952, 1953), Nicholson *et al.* (1953), and Swaminathan and Kamra (1961).

Netolitsky (1926) reviewed the literature on seed coat anatomy, as well as internal morphology of the seed. Because of the specialized equipment and training required, anatomical studies of the seed coats are seldom used in seed identification. Although gross external and internal morphologies are more frequently used, the importance of seed coat structure should not be overlooked. A survey of seed structures is given by Harz (1885), and Indian research on seed characteristics is summarized by Singh (1964).

G. Davis (1966) summarized the ovule position in 315 families of angiosperms (Fig. 29). Two families, Balanophoraceae and Loranthaceae, do not form ovules. Ovule morphology is constant in 248 families, of which 204 have an anatropous form, 20 an orthotropous form, 13 a hemianatropous form, 5 a campylotropous form, 4 an amphitropous form, and 1 each an anacampylotropous or circinotropous form. The remaining 67 families have two or more ovule types. The most variable taxon is the Lotoideae (a subfamily of Leguminosae) which exhibits four ovule types.

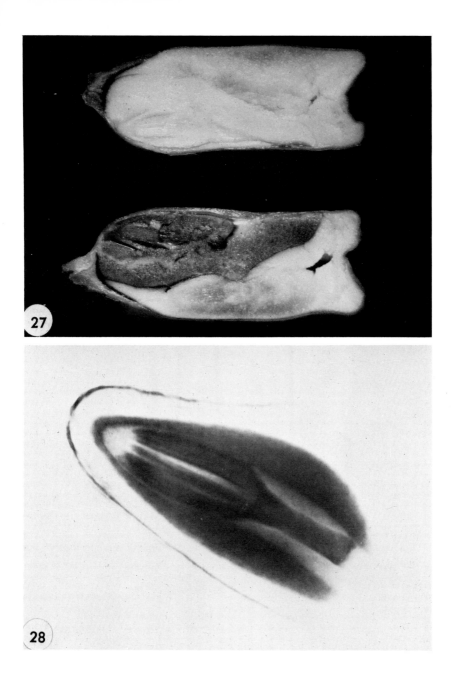

FIG. 27. Dead embryo of maize (top) remained white, whereas live embryo (bottom) turned red in a 1% aqueous solution of 2,3,5-triphenyl tetrazolium chloride.

FIG. 28. Radiograph of a *Pinus palustris* Mill. seed showing a straight, axile embryo with several cotyledons. (Magnification ca. × 8.)

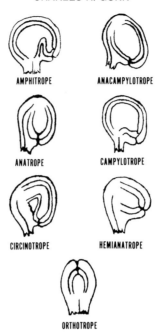

AMPHITROPE ANACAMPYLOTROPE

ANATROPE CAMPYLOTROPE

CIRCINOTROPE HEMIANATROPE

ORTHOTROPE

FIG. 29. Diagrams defining ovule morphological terms used in family descriptions.

3. CHEMICAL ANALYSIS

Chemical assays of seeds may be useful in formulating phylogenetic concepts at various taxonomic levels or in identifying seeds. Too often erroneous data and conclusions have found their way into the biochemical systematic literature through careless identifications of the seed stocks or through faulty or premature interpretations of relationships of the chemical substances.

Two examples will serve to illustrate the usefulness of seed assays and point up some dangers. Free amino acids and related compounds have been extracted and analyzed from seeds in *Lathyrus* and *Vicia*. These two genera are closely related members of the Leguminosae. Although the morphological differences used, both in the seeds and plants, to separate *Vicia* and *Lathyrus* are indefinite, Bell (1966) was able to separate seeds of fifty-two species of *Lathyrus* from seeds of forty-two species of *Vicia* by chemical analysis. Other papers germane to this subject are Bell and O'Donovan (1966) and Hanelt and Tschiersch (1967). The most striking chemical differences are the presence of C_6 guanidino compounds or canavanine in *Vicia* seeds, and C_7 guanidino compounds (including lathyrine) or α, γ-diaminobutyric acid in *Lathyrus* seeds. Within each genus the species can be grouped into chemical alliances. At the species level the chemical analysis of a partially mature seed of *Vicia menziesii*, an extinct Hawaiian endemic, was used as an independent line of evidence to show that this species was a member of the genus *Vicia* and not *Lathyrus*. Chemical data were also used to establish the position of this species in the genus *Vicia* and its relationship to other species (Gunn, 1970a).

Care must always be exercised in interpreting biochemical data. For example, E. A. Bell, in a personal communication (1971) about the *Vicia* data, observed that he thought canavanine must arise from homoserine by transguanidation. This suggests a basic difference in his *Vicia* alliances based on the ability or lack of ability to synthesize homoserine. However, if a mutation of the transguanidase gene occurred, there might be a buildup of homoserine and arginine or hydroxyarginine. Thus, a seemingly wide biochemical difference may in fact arise from a single gene mutation.

Correlations, such as those given above, are not limited to the Leguminosae. The results of seed oil analyses in the genus *Lesquerella,* in the Cruciferae, indicate that the auriculate-leaved alliance has a relatively high amount of C_{18} hydroxy acids but no C_{20} hydroxy acids. The latter chemical is found in *Lesquerella* species which do not belong to the alliance (Barclay *et al.,* 1962).

In discussing the distribution of fatty acids in plant lipids extracted from seeds, Shorland (1963) concluded "Although the data on the types and distribution of fatty acids do not provide an unequivocal guide to the classification of plants, many correlations of taxonomic significance have become apparent in spite of the small number of species examined up to now."

Although there is yet to be a complete fusion of chemistry and systematics, it is evident that both will benefit from a mutual exchange of data. A biochemical test to identify an isolated seed is not possible at present. Nevertheless, biochemistry will hopefully become increasingly useful in seed identification.

4. SEEDLINGS

Seedling identification is beyond the scope of this chapter. However, there are times when identification of a viable isolated seed can only be made by germinating the seed and consulting references such as Chancellor (1959), Duke (1965a, 1969c), Krummer (1951), Lubbock (1892), and Vasil'chenko (1965). If the seedling cannot be identified, then it should be grown to flowering stage when identification is possible. This procedure assumes the seed is alive and the flowering stage will be reached reasonably soon.

E.   Seed Characteristics of Selected Families

Spermatophyte families currently recognied by taxonomists are the product of multiple correlations of floral, and to a lesser extent, of vegetative characters applied on a trial and error basis. Although family parameters have become fairly stable, their relationship to each other has not been stabilized. Seeds do not have a priori importance in supporting or refuting a family's parameter. However, it is significant that seed characters do reinforce prevailing concepts of family parameters.

Walters (1961) recognized families of spermatophytes as "definable" or "indefinable." When these terms are applied to seed characteristics, the Amaranthaceae, Labiatae, and Orchidaceae are considered definable families, and the Liliaceae, Ranunculaceae, and Rosaceae are indefinable families.

Several excellent references consulted in preparing the family descriptions are not referred to with each entry but are cited here with deep appreciation. They made the descriptions much easier to prepare, and should be available to serious students. The terminology classifying the positions of ovules (Fig. 29) follows G. Davis (1966). The classification of embryos (Fig. 30) was adopted from Martin (1946). The works of

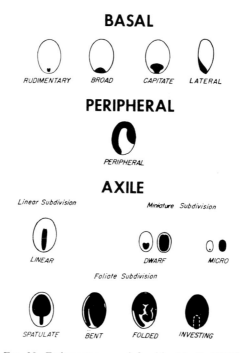

FIG. 30. Embryo types as defined by Martin (1946).

Isely (1947) and McClure (1957) were most helpful in preparing descriptions of some families. The general works of Gunderson (1950), Hutchinson (1960), Le Maout and Decaisne (1876), and Lubbock (1892), and seed references compiled by Barton (1967) were also used.

The families are presented according to the phylogenetic system of Cronquist (1968) which is based on modern data and current thinking, as well as traditional evidence. Table IV will facilitate the location of an individual family.

TABLE IV

ALPHABETICAL LISTING OF FAMILIES[a]

Amaranthaceae, 10	Leguminosae (Caesalpinoideae), 17(a)
Asclepiadaceae, 21	Leguminosae (Lotoideae), 17(c)
Betulaceae, 6	Leguminosae (Mimosoideae), 17(b)
Boraginaceae, 24	Liliaceae, 35
Cactaceae, 7	Malvaceae, 12
Caryophyllaceae, 8	Onagraceae, 18
Chenopodiaceae, 9	Orchidaceae, 37
Compositae, 30	Palmae, 34
Convolvulaceae, 23	Papaveraceae, 3
Cruciferae, 15	Pinaceae, 1
Cucurbitaceae, 14	Plantaginaceae, 27
Cyperaceae, 32	Polygonaceae, 11
Euphorbiaceae, 19	Ranunculaceae, 2
Fagaceae, 5	Rosaceae, 16
Gramineae (Festucoideae), 33(a)	Rubiaceae, 29
Gramineae (Panicoideae), 33(b)	Scrophulariaceae, 28
Iridaceae, 36	Solanaceae, 22
Juglandaceae, 4	Umbelliferae, 20
Juncaceae, 31	Verbenaceae, 25
Labiatae, 26	Violaceae, 13

[a] The number following the family name may be used to locate the family in the text.

The key to rapid identification of unknown isolated seeds is family recognition. Once the family is known, genus and, perhaps, species identification are possible.

Seed characteristics of 37 families important in the Northern Hemisphere are summarized. Characteristics of seeds of Leguminosae and Gramineae are presented by subfamilies because of the size and complexity of the families as a whole. The family text includes discussions of disseminule classification, external and internal descriptions of disseminules, and selected examples of how disseminule characteristics have been used at the family or genus levels. A pictorial summary of disseminule characteristics accompanies the text. Inserts of individual disseminules, selected to emphasize familial characteristics, are used when necessary. These disseminules are not typical or average specimens; rather, they illustrate one or more important characteristics. Magnifications are noted in the captions. Large disseminules were excluded when preparing the layouts, especially for the Leguminosae, Palmae, and Pinaceae, because of the room they occupied and focusing problems they presented.

1. PINACEAE (Fig. 31)

Disseminule: seed. No fruit in the Pinaceae.

External: seed oblong, ovate, or ovoid usually acute at base, or triangular; 4–21 mm excluding wing. Seed coat with adnate or articulate wing, or wing seldom rudimentary or absent; resinous dots or resin vesicles may be present; usually dull, black, gray, or brown. Hilum quite inconspicuous at base of seed, seldom noted.

Internal: seed coats 2; outer usually thickened, inner membranous. Embryo axile in anatropous ovule; usually straight or nearly so; linear to almost spatulate; cotyledons 2–18; usually shorter than inferior radicle. Perisperm abundant; horny or fleshy farinaceous, or oily.

Notes. isolated seeds may be identified to genus and species (Dallimore and Jackson, 1966; Forest Service, 1948).

Abies seed—ovoid or oblong, winged, outer seed coat soft with several resin vesicles; cotyledons 4–10; perisperm fleshy.

Cathaya seed—with broad wing; cotyledons 3–4.

Cedrus seed—irregularly triangular, with broad wing, outer seed coat soft, oily; cotyledons 9–10; embryo bent at base of cotyledons; perisperm oily.

Keteleeria seed—like *Abies* except only 2 cotyledons.

Larix seed—small, triangular, winged; outer seed coat crustaceous; cotyledons 6.

Picea seed—small, oblong, acute at base, with large wing; outer seed coat hard; resin pits lacking; cotyledons 4–15.

Pinus seed—usually ovate or oblong, acute at base, usually with a wing (in some species soon deciduous); outer seed coat hard, cotyledons 3–18. Uyeki (1927) has prepared a key, descriptions, and illustrations of *Pinus* seeds.

Pseudolarix seed—white small, triangular ovate, with large wing, cotyledons 5–7.

Pseudotsuga seed—like *Picea* with large wing; cotyledons 6–12.

Tsuga seed—ovate to oblong, compressed, nearly surrounded by a large wing; outer seed coat with resin vesicles; cotyledons 3–6.

FIG. 31. Pinaceae. (Magnification × 3.)

FIG. 32. Ranunculaceae. (Magnification × 3.) Inserts: *Delphinum ajacis* L. and *Ranunculus bulbosus* L. (Magnification × 4.)

2. RANUNCULACEAE (Fig. 32)

Disseminule: seed or achene.

External: seed from a follicle (a capsule in *Nigella*) or rarely from a berry in *Actaea* and *Hydrastis;* ovoid or flattened and oblong; ranging from angular to rounded. Seed coat smooth or wrinkled; usually with a prominent raphe which may be obscured by well-developed wrinkles. Hilum inconspicuous, seldom noted, at one end of elongated seed; in *Ranunculus* hilum and fruit scar are adjacent. Achene usually flattened and ovate, occasionally angular; usually tipped by a persistent style (occasionally plumose) with fruit scar at opposite end.

Internal: seed coats 1 or 2; outer seed coat of free seed crustaceous, whereas within achene it is thin and pliable. Embryo rudimentary to linear in anatropous ovule (hemianatropus in *Ranunculus*); less than one-tenth of seed length except in some species of *Ranunculus* and *Nigella;* generally difficult to locate because of its size and similarity to endosperm. Cotyledons 2, rarely 1; appressed or divergent. Endosperm abundant, generally watery–fleshy, or oily.

Notes. The Ranunculaceae may be subdivided on the basis of seed and fruit characteristics. The tribe Helleboreae (Helleboraceae of Hutchinson, 1960, 1969) has more than one ovule per carpel and has a fruit that is a follicle, rarely united into a capsule, or rarely a berry. The tribe Anemoneae has only one ovule per carpel, and fruit is an achene, or rarely baccate. The Anemoneae may be subdivided into Ranunculeae, Clematideae, and Anemoneae partially on the basis of achene characteristics. The Ranunculeae has an erect seed with the hilum located just beneath the scar area. In the other subtribes the seeds are pendulous and the hilum is located under the beak of the style, opposite the scar area. Wiegand (1895) surveyed the fruit and seed characteristics and placed the genera in eight groups.

3. PAPAVERACEAE (Fig. 33)

Disseminule: seed; indehiscent fruit segments in *Platystemon.*

External: seed globose to subreniform, slightly larger at one end. Seed coat smooth to reticulate, usually with a crest arising from raphe, or raphe smooth or arillate; dull or shiny and dark colored, occasionally brown or whitish. Hilum inconspicuous, seldom noted, at small end of seed or at base of raphe on concave surface.

Internal: seed coat thick or thin; brittle. Embryo rudimentary to linear; axile in anatropous ovule; less than half seed length. Cotyledons 2; appressed or divergent. Endosperm abundant, soft, watery–fleshy, opaque–white, with an oily food reserve.

Notes. Seed characteristics have been little used in the classification or

FIG. 33. Papaveraceae. (Magnification × 3.) Insert: *Papaver somniferum* L. (Magnification × 11.)

FIG. 34. Juglandaceae. (Magnification × 1.)

identification of the taxa in the Papaveraceae. Seeds are useful in defining the family and its genera, and may be useful at the species level.

4. JUGLANDACEAE (Fig. 34)

Disseminule: drupe or nut without wings, or winged nutlet.

External: seed dominated by 2 large, corrugated, two-lobed cotyledons usually separate except where joined to the radicle. Seed coat brown, smooth, veined. Hilum quite inconspicuous, seldom noted.

Internal: seed coats 2; outer membranous, inner quite thin. Embryo straight, investing in orthotropous ovule dominated by cotyledons with cerebriform or cordate bases; radicle quite small, superior. Cotyledons 2; often two-lobed; unusually large; fleshy and oily. Endosperm absent.

Notes: Hutchinson (1960, 1969) recognized eight genera with the following fruit characteristics.

Alfaroa fruit — ellipsoidal, softly hispid, not winged (Manning, (1949).

Carya fruit — a large drupe without wings; exocarp splitting at maturity; endocarp smooth, often angled.

Cyclocarya fruit — not recorded.

Engelhardtia fruit — a winged nutlet; wing formed by three-lobed foliaceous bract.

Juglans fruit — a large drupe without wings; exocarp indehiscent; endocarp sculptured or rugose.

Oreomunnea fruit — a winged nutlet; wings formed by three-lobed, large and rigid bract.

Platycarpa fruit — a winged nutlet.

Pterocarya fruit — a winged nutlet; wing formed by bracteoles.

5. FAGACEAE (Fig. 35)

Disseminule: nut with or without an involucre or capsule.

External: seed filling nut cavity, thus shape of nut. Seed coat thin; brown. Hilum quite inconspicuous, seldom noted. Nut ovoid, globose (occasionally with a flattened side) or cylindrical; fruit leathery, brown, shining, with large fruit scar; or, if triangular, fruit thin, brown, fruit scar small.

Internal: nut coat leathery to hard, usually thin. Seed coats 2; both thin. Embryo investing in anatropous ovule; straight, or folded in *Fagus;* cotyledons dominant; radicle small, superior. Cotyledons 2; straight or folded or sinuous; fleshy. Endosperm absent.

Notes: Hutchinson (1960, 1969) recognized six genera with the following nut characteristics.

Fagus nuts — triangular, within a two- to four-lobed, regularly splitting, two-seeded involucre, surrounded by prickly, subulate, or bractlike appendages.

FIG. 35. Fagaceae. (Magnification × 1.)

FIG. 36. Betulaceae. (Magnification × 4.) Insert: *Betula lutea* Michx. f. (Magnification × 5.)

Nothofagus nuts — triangular, within a two- to four-lobed, regularly splitting, one- to three-seeded involucre, transversed by an entire or toothed scale.

Castanea nuts — rounded, partially within a spiny involucre that does not split, or if so, then splitting irregularly.

Castanopsis nuts — like *Castanea.*

Lithocarpus nuts — rounded, 1 in a nonspiny cupule.

Quercus nuts — rounded, 1 in a cupule.

The nuts of *Lithocarpus* and *Quercus* are described by Camus (1948).

6.   BETULACEAE (Fig. 36)

Disseminule: nut or nutlet; wings present or absent, or enclosed in a foliaceous lobed, laciniate, or bladderlike involucre.

External: seed filling nut or nutlet cavity, thus shape of nut. Seed coat thin, brown, shiny or dull, usually reticulate. Hilum quite inconspicuous, seldom noted. Nut ovoid-compressed to compressed, basal scar inconspicuous except for *Corylus* where basal scar is prominent.

Internal: nut or nutlet coat thin to thick. Seed coat 1; membranous, thin. Embryo investing in anatropous ovule; straight; dominated by the cotyledons; radicle small, superior. Cotyledons 2; fleshy, oily, flat. Endosperm minute, restricted to radicle end of seed; transparent.

Notes. The family is divided into two tribes, Betuleae and Coryleae. Some authors recognize these tribes as families. The seeds and fruits would appear to possess useful characters at the genus and species levels. However, little general information is available about them.

7.   CACTACEAE (Fig. 37)

Disseminule: seed.

External: seed flattened, semicircular to nearly circular, bent, or straight; smooth, angular, or indented; occasionally with margin almost winged; mostly between 1 to 5 mm in length. Seed coat smooth, punctate, reticulate, papillate, or foveolate; dull or shiny; black, brown, white, gray, or reddish. Hilum small to large; generally circular; basal but may appear sublateral or lateral.

Internal: seed coats 2; outer crustaceous to nearly bony; inner thin. Embryo peripheral in circinotropous ovule; circular, bent, or straight, usually dominated by well-developed hypocotyl, less so by cotyledons; radicle stout, inferior. Cotyledons 2; seldom well-developed, usually small and knoblike. Endosperm, when present, ranging from abundant to barely present, usually floury–starchy, or viscid in *Rhipsalis.*

Notes. The seeds are contained in a juicy or occasionally a dry berry, or a capsule in one species. Seed characteristics have seldom been used

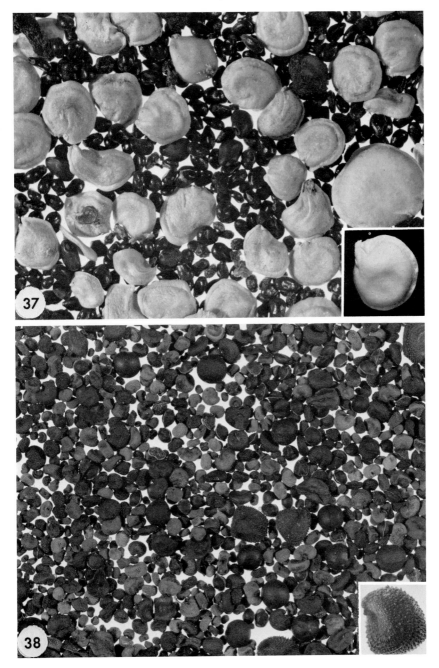

FIG. 37. Cactaceae. (Magnification × 3.) Insert: *Opuntia basilaris* Engelm. & Bigel. (Magnification × 2.5.)

FIG. 38. Caryophyllaceae (Magnification × 4.) Insert: *Agrostemma githago* L. (Magnification × 6.)

below the family level, and information about seeds is widely scattered. Seed discussions in Benson (1969) and Britton and Rose (1963) were particularly helpful in preparing this discussion.

8. CARYOPHYLLACEAE (Fig. 38)

Disseminule: seed.

External: seed reniform, globose, obovoid, or scutiform. Seed coat tuberculate, muricate, or smooth and shining, rarely winged; usually dark colors; outermost cells of some species with sinuate margins with humped centers. Hilum at narrow end of elongate seed *(Dianthus* and *Tunica)* or lateral in reniform seed type and flush or in a notch and surrounded by a collar or ridge.

Internal: seed coat 1; crustaceous. Embryo well-developed; peripheral (axile in *Dianthus* and *Tunica)* in campylotropous to hemianatropous ovule; usually semicircular to nearly circular, to spiral in *Spergula* and *Drypis.* Cotyledons 2; appressed and usually half or more of the embryo's length. Perisperm abundant; mostly centrally placed; floury or rarely sub-fleshy, semitranslucent, or white and soft. Endosperm reduced to a thin sheath around radicle.

Notes. Seed characteristics have been little used in the classification or identification of the taxa in the Caryophyllaceae. Duke (1961) used seed characteristics to advantage in his study of *Drymaria.* Kowal (1966) proposed sections within *Spergula* and *Spergularia* based solely on seed characteristics and found it relatively easy to identify species by seeds. However, Kowal and Wojterska (1966) found that seed characteristics of *Dianthus* were of limited value in making species identifications.

9. CHENOPODIACEAE (Fig. 39)

Disseminule: utricle or rarely a berry with or without a modified or un-changed calyx; rarely a seed or a conglomerate ("seed ball").

External: seed lenticular and notched or reniform, or rarely thick obconic with lateral spiral furrows in *Salsola.* Seed coat minutely pitted to smooth; usually black or brown, shiny or dull; usually obscured or partially so by a cellular scarious pericarp. Hilum quite inconspicuous, at or near the marginal notch formed by the end of the cotyledons and adjacent radicle.

Internal: seed coats, usually 2; outer crustaceous, inner membranous, or if 1, membranous. Embryo well-developed; peripheral in campylotropous ovule; usually curved into a ring or a horseshoe shape, or spirally coiled. Cotyledons 2; appressed and usually half or more of the embryo's length. Perisperm abundant; centrally placed; farinaceous, rarely subfleshy; absent in *Anabasis, Salicornia, Salsola, Sarcobatus;*

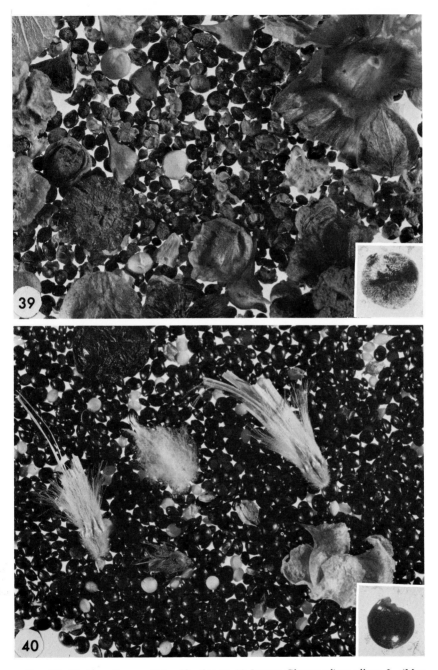

FIG. 39. Chenopodiaceae. (Magnification × 4.) Insert: *Chenopodium album* L. (Magnification × 8.)

FIG. 40. Amaranthaceae. (Magnification × 3.5.) Insert: *Amaranthus graecizans* L. (Magnification × 10.)

or divided in two portions in some Spirolobeae. Endosperm reduced to a thin sheath around radicle.

Notes. The Chenopodiaceae may be divided into two tribes, Cyclolobeae and Spirolobeae, on the basis of seed characteristics. In Cyclolobeae the embryo is circular or horseshoe-shaped and surrounds an abundant perisperm. In Spirolobeae the embryo is spiralled and the perisperm is scant or absent.

Some species of *Atriplex, Axyris,* and *Chenopodium* have dimorphic or trimorphic fruits. These differences include variations of shape, size, and color. The fruit of *Atriplex* is enclosed in foliaceous, valvate bracts, whereas in *Spinacia* the bracts are hardened and completely fused at maturity forming a nutlike, indehiscent disseminule. Flowers of the genus *Beta* are borne in glomerules. Their calyces become corky thickened and adnate at maturity forming a several-seeded seed-ball which is the disseminule.

The seed and associated plant parts offer useful characters at genus and species levels. Several seed treatments are available, especially for seeds of *Chenopodium* (Baranov, 1969; Cole, 1961; Guinet, 1959; Herron, 1953; Kowal, 1953).

10.   AMARANTHACEAE (Fig. 40)

Disseminule: seed usually free of pericarp and calyx; enclosed by a hairy calyx in *Froelichia.*

External: seed lenticular, rounded, or elongated slightly; usually with a distinct marginal rim; notched. Seed coat black, shiny, smooth. Hilum quite inconspicuous, at or near marginal notch.

Internal: seed coats usually 2; outer crustaceous, inner membranous, or if 1, thin membranous. Embryo well-developed; peripheral in anacampylotropous, campylotropous, or circinotropous ovule; usually curved into a ring or partially overlapping. Cotyledons 2; appressed and usually half or more of embryo's length. Perisperm abundant; centrally placed; hard and translucent to granular and whitish. Endosperm reduced to a thin sheath around radicle.

Note. Seeds of the Amaranthaceae are remarkably similar and of little value below the family level. Based on seed characteristics, the Amaranthaceae and Chenopodiaceae appear to be closely related.

11.   POLYGONACEAE (Fig. 41)

Disseminule: achene with or without associated perianth parts.

External: achene trigonous or lenticular. Achene coat smooth to faintly pitted or roughened, occasionally shiny; brownish to black. Seed coat dull brown. Hilum quite inconspicuous, seldom noted.

Internal: achene coat crustaceous. Seed coat 1; thin; free from achene

FIG. 41. Polygonaceae. (Magnification × 3.5.) Inserts: *Rumex fenestratus* Greene and *Polygonum coccineum* Muhl. (Magnification × 3.)

FIG. 42. Malvaceae. (Magnification × 2.) Insert: *Abutilon theophrasta* Medikus. (Magnification × 3.)

coat. Embryo peripheral in orthotropous ovule; curved, bent, convoluted, or straight; linear or spatulate; radicle superior. Cotyledons 2; linear to broad and thin. Endosperm usually abundant; outer layer forming a distinct external sheath, an aleurone layer, rich in protein rather than starch as rest of endosperm; hard and semitransparent, occasionally crystalline–granular, or whitish and soft to firm; rarely divided into two portions by the embryo.

Notes. The achenes and their associated perianth parts are useful at the family and genus level and in some cases at the species level. Marek (1954, 1958) and Martin (1954) used seed characteristics at the genus and species levels.

12. MALVACEAE (Fig. 42)

Disseminule: seed or carpel sectoroid.

External: seed of two types — (*a*) in tribe Hibisceae plump-compressed to ellipsoidal or subspherical and (*b*) in tribes Ureneae and Malveae laterally compressed in a sectorial fashion, in face view circular to irregularly ovate; margin notched. Seed coat smooth to rough or warty, hairy in *Gossypium, Fugosia,* and *Hibiscus.* Hilum flush (in Hibisceae) or depressed within the notch, with a characteristic grill-like structure extending up and adnate to the radicular lobe. Carpel sectoroid containing one seed, frequently reticulate or hairy, sometimes spiny pointed.

Internal: seed coats 2; outer thick commonly impervious to water, inner thin. Embryo peripheral and large folded in campylotropous ovule (anatropous in *Gossypium* and related genera); dominated by large cotyledons; radicle inferior or bent upward. Cotyledons 2; well-developed, usually basally lobed or cordate, frequently nerved, sometimes finely hairy or glandular; folded or convoluted. Endosperm scanty or absent; firm fleshy, hard, or rarely mucilaginous.

Notes. Seeds provide excellent family and generic characteristics. As previously noted, seeds can be used in tribe identification. Far too little use has been made of seed characteristics in this family.

13. VIOLACEAE (Fig. 43)

Disseminule: seed.

External: seed obovoid to subglobose; 1–5 mm long; caruncle present. Seed coat smooth, shiny or dull, occasionally winged or tomentose; bearing a longitudinal raphe; shades of brown, gray, or white. Hilum small, at smaller end of seed.

Internal: seed coat crustaceous, brittle. Embryo axile, straight, and spatulate in anatropous ovule; radicle inferior. Cotyledons 2, flat. Endosperm abundant; soft fleshy, oily.

Notes. Seed characteristics have seldom been used in this family.

FIG. 43. Violaceae. (Magnification × 7.)
FIG. 44. Cucurbitaceae. (Magnification × 1.5.)

14. Cucurbitaceae (Fig. 44)

Disseminule: seed.

External: seed compressed, oblanceolate, pointed at the hilum end, rounded at the opposite end; rarely linear. Seed coat smooth; with or without a distinct thickened margin; winged in tribe Zanonieae; black, brown, yellowish, or whitish. Hilum inconspicuous, seldom noted.

Internal: seed coats 2; outer crustaceous or horny, inner thin. Embryo straight, spatulate or investing in anatropous ovule; cotyledons dominant; radicle short, inferior. Cotyledons 2; foliaceous, veined, fleshy, embracing two distinct leaves. Endosperm absent.

Notes. Seed characteristics are quite strong for most members of this family and may be used at the family, genus, and species levels (Erwin and Haber, 1929; Russell, 1924; Singh, 1953). Barber (1909) discussed seed coat characteristics of twelve species.

15. Cruciferae (Fig. 45)

Disseminule: seed, rarely indehiscent fruit segment (a loment).

External: seed of two types — (*a*) compressed or flattened, notched or cleft usually with a groove or line between the cotyledons and radicle or (*b*) globose. Seed coat minutely reticulate, pitted, tuberculate, or verruculate because outer layer of epidermis is partially dissoluted and the lower portion of their lateral walls persists and becomes thickened, suberized, and sculptured; occasionally winged; brown, black, or white. Hilum inconspicuous and frequently hidden under a peglike funicular remnant which may fill the notch.

Internal: seed coat thin or brittle. Embryo oily; bent in campylotropous ovule; radicle dorsal or lateral; well-developed, usually curved at point of origin and extending along the margin of the seed. Cotyledons 2; in five positions — incumbent, oblique, accumbent, conduplicate, or coiled. Endosperm reduced to 1 or 2 rows of cells, an aleurone layer.

Notes. Development of mucilage which may diffuse or break through the outer cell walls and cuticle of the seed coat while the seed is maturing may cause dissolution of the outer cell walls. When seeds of some species are placed in water, mucilaginous halos develop. Mucilage production is of some value as a specific and sectional or subtribal character (Janchen, 1942).

The classification of the Cruciferae is based on fruit and seed characteristics. The history of the classification of genera in the Cruciferae is summarized by Murley (1951) in her classic study of seeds from northeastern North America.

Although seeds of various genera of Cruciferae have been studied, none has fostered more studies than the economically important genus

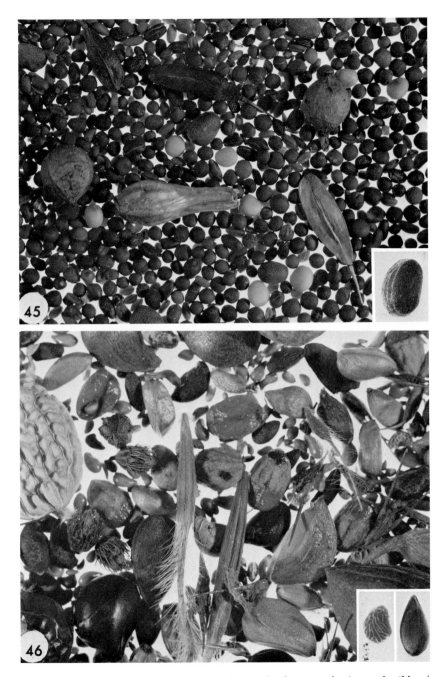

Fig. 45. Cruciferae. (Magnification × 3.) Insert: *Raphanus raphanistrum* L. (Magnification × 4.)

Fig. 46. Rosaceae. (Magnification × 2.) Inserts: *Potentilla recta* L. (Magnification × 10) and *Pyrus malus* L. (Magnification × 1.5).

Brassica (Berggren, 1960, 1962; McGugan, 1948; Musil, 1948; Weinman, 1956).

16. ROSACEAE (Fig. 46)

Disseminule: achene, achene with associated hypanthial parts, seed, or stone (seed within a bony endocarp). Disseminules are quite diverse partly because of the diverse nature of the family and partly because of the varying degree of association of seeds with carpel and hypanthial structures.

External: seed shape variable, generally elongate to rounded. Seed coat smooth to variously marked; usually bearing a longitudinal raphe; variously colored. Hilum distinct; usually subterminal, at or near the end of the raphe. Achene compressed; elongate; occasionally beaked; variously marked. Stones of the *Prunus* and related genera spheroidal to ellipsoidal, smooth to variously marked.

Internal: seed coats 2; outer may be thick and crustaceous or thin; inner thin. Embryo spatulate bent or investing in anatropous to hemianatropous ovule; cotyledons dominating the small inferior or superior radicle. Cotyledons 2; straight; well-developed. Endosperm negligible or absent, except usually present and fleshy in tribe Spiraeae.

Notes. The diverse nature of this family's disseminules makes any circumscription difficult. Whereas this is true at the family level, at the tribe or genus level, seed and fruit characteristics are of value though seldom used. The disseminules for the tribes are described below.

Pomeae—pistil 1; enclosed in a fleshy spurious fruit; several-seeded; seeds within carpel cells which are encased in a succulent hypanthial or appendicular structure.

Potentilleae—pistils numerous; free from each other and hypanthium; one-seeded achenes.

Poterieae—pistils 1 to many; enclosed in a bristly or smooth hypanthium which more or less, permanently encloses the achenes.

Pruneae—pistil solitary developing into a two-layered fruit; a succulent exocarp and a bone-hard endocarp; single seed within the endocarp.

Roseae—pistils 1 to many; enclosed in a fleshy hypanthium (hip) which releases the enclosed achenes.

Rubeae—pistils numerous; free from each other and hypanthium; one-seeded fleshy drupelets which fall as a unit, an aggregate fruit.

Spiraeae—pistils numerous, free from each other and hypanthium; several-seeded; seeds fall free.

Vertes (1913) studied anatomy of fruit and seed coats of selected taxa, and Kelley (1953) described and illustrated seeds of eleven species of *Potenilla.*

17. LEGUMINOSAE (Figs. 47 to 49)

Seed characteristics for the three subfamilies, Caesalpinoideae, Mimosoideae, and Lotoideae are summarized below and then presented separately to insure clarity.

Caesalpinoideae — seed straight, rounded at apex, tapering to basal hilum; seed coat may bear superficial fracture lines, seldom with a closed pleurogram (light line) on each face; embryo straight; cotyledons investing and usually covered by endosperm; ovule anatropous.

Mimosoideae — seed straight, rounded at apex, tapering to basal hilum; seed coat without superficial fracture lines but usually with a horseshoe-shaped pleurogram on each face; embryo straight; cotyledons investing a short radicle and usually covered by endosperm; ovule anatropous.

Lotoideae — seed usually bent, curved or rounded; hilum usually lateral, commonly in a notch; seed coat without fracture lines or pleurograms; embryo bent; cotyledons not investing, radicle well-developed; endosperm present or absent; ovule campylotropous.

For additional information the reader should consult Corner (1951), Isely (1955a,b), Kopooshian (1963), and Pammel (1899).

a. CAESALPINOIDEAE (Fig. 47)

Disseminule: seed.

External: seed symmetrical or nearly so; usually rounded at apex, tapering to base; elongate or rounded; usually compressed, occasionally rounded. Seed coat smooth, rarely rough; usually shiny with a waxy-like cuticle which may, in time, fracture in contiguous circles around the seed; frequently dark brown to black; bearing a spurious closed pleurogram (see Mimosoideae discussion) for some *Cassia* species. Hilum basal, usually inconspicuous.

Internal: seed coat apparently 1; thick and hard. Embryo investing in anatropous or campylotropous ovule; straight, infrequently slightly bent, or rarely convoluted; cotyledons dominating small inferior radicle. Cotyledons 2; usually thin and flat; cordate at base. Endosperm usually present, conspicuous over the lateral faces of the cotyledons and around the base of the embryo; hard and glossy, becoming mucilaginous when moistened.

Notes. Seeds in the subfamily Caesalpinoideae are more uniform than they are in the other two subfamilies. Therefore, their use is somewhat limited at the genus and species levels.

b. MIMOSOIDEAE (Fig. 48)

Disseminule: seed.

External: seed symmetrical or nearly so; usually rounded at apex, tapering to base; elongate or rounded; usually compressed, occasionally

FIG. 47. Leguminosae tribe Caesalpinoideae. (Magnification × 2.5.)
FIG. 48. Leguminosae tribe Mimosoideae. (Magnification × 3.)

rounded. Seed coat smooth, rarely rough; usually shiny; frequently dark brown to black; usually bearing a horseshoe-shaped pleurogram on each face. Hilum basal, usually inconspicuous; occasionally funiculus remains attached to seed perhaps even coiled or folded around seed.

Internal: seed coat apparently 1; thick and hard. Embryo investing in hemianatropous to anatropous ovule; straight; cotyledons dominating small inferior radicle. Cotyledons 2; usually thin and flat; cordate at base. Plumule present or absent. Endosperm present or absent; hard and glossy; most abundant on cotyledonary faces.

Notes. Little has been done with the seeds of this subfamily.

c. LOTOIDEAE (Fig. 49)

Disseminule: seed, or one-seeded fruit or loment segment.

External: seed usually asymmetrical; bent, curved or spherical; frequently notched. Seed coat smooth, rarely rough, usually shiny; color quite variable. Hilum usually appearing apical or lateral; if terminal, at broad end of seed; usually conspicuous; frequently bearing a longitudinal split (a lotoid split); occasionally carunculate.

Internal: seed coat apparently 1; thick and hard. Embryo bent (rarely erect or investing) in anatropous, hemianatropous, amphitropous, or campylotropous ovule; radicle curved; often well-developed. Cotyledons 2; elliptical or ovoid but not cordate at base. Plumule present or absent; usually present in seeds without endosperm. Endosperm present, or absent in tribes Vicieae and Phaseoleae; hard and glossy; most abundant on cotyledonary faces and in lesser amounts around radicle.

Notes. Because of its economic importance, seeds of this subfamily are included in nearly all general seed identification manuals. Unfortunately, seed characteristics are seldom used to their fullest extent in taxonomic studies. Gunn (1970b, 1971a,b) used only seed characteristics in the genus *Vicia* and, to a limited degree, in the tribe Vicieae, for species identification. R. H. Miller (1967) discussed the internal and external characteristics of selected seeds of *Crotalaria*.

18.   ONAGRACEAE (Fig. 50)

Disseminule: seed or nutlet.

External: nutlet ovoid or obpyramidal or compressed and bearing hooked bristles; containing 1 or 2 seeds. Seed angular, compressed angular, or terete. Seed coat smooth or papillose, winged, fringed, or hairy; comose at apex in *Epilobium;* brown or yellowish-brown. Nutlet scar inconspicuous. Hilum quite inconspicuous, seldom noted.

Internal: nutlet coat woody. Seed coat apparently 1; thin to crustaceous. Embryo straight, spatulate in anatropous ovule; radicle near hilum, superior or inferior, rarely centripetal. Cotyledons 2; foliaceous to slightly

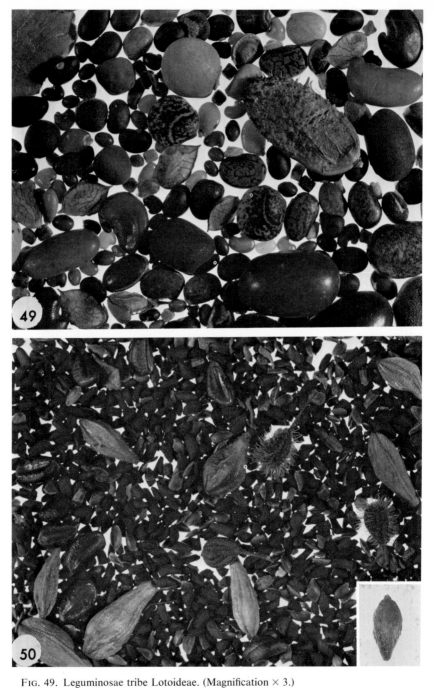

FIG. 49. Leguminosae tribe Lotoideae. (Magnification × 3.)

FIG. 50. Onagraceae. (Magnification × 3.) Insert: *Oenothera missouriensis* Sims. (Magnification × 3.5.)

fleshy; often auricled at base, occasionally slender; rarely curved or con-voluted. Endosperm absent or scant.

Notes. Little information is available about seeds or nutlets of the Onagraceae.

19. EUPHORBIACEAE (Fig. 51)

Disseminule: seed.

External: Seed ovoid to biconvex or planoconvex, occasionally sec-toroid, spheroid, or quadrangular. Seed coat smooth or variously rugose, channeled, or scurfy; dull or shiny; bearing a longitudinal raphe on the ventral surface; variously colored. Hilum on ventral side; subterminal at narrow end of seed; commonly carunculate.

Internal: seed coats 2; outer crustaceous, inner thin. Embryo axile and straight, rarely curved; spatulate, rarely linear, in hemianatropous to anatropous (orthotropous in *Breynia patens*) ovule; radicle well-de-veloped, superior. Cotyledons 2; appressed or separated by a layer of endosperm; thin; usually nerved. Endosperm abundant; soft fleshy to fleshy.

Notes. Seeds of this family are easily recognized by their conspicuous raphe and when present by their caruncles around the subterminal hila. Krochmal (1952) and Murley (1945) have prepared seed keys, descrip-tions, and illustrations for selected species of *Euphorbia*. Pammel (1892) reported on seed coat characteristics of genus *Euphorbia*.

20. UMBELLIFERAE (Fig. 52)

Disseminule: mericarp.

External: mericarp elongate, plump or narrowly oblong, planoconvex to polygonal or strongly flattened. Mericarp coat usually bearing 5, oc-casionally 10, ribs or nerves; ribs corky or thin; occasionally bearing barbs, prickles, warts, or scales; oil ducts between ribs frequently visible as dark lines. Mericarp scar basal; inconspicuous. Stylopodium and calyx frequently persistent at apex of mericarp.

Internal: mericarp coat thick; indehiscent usually not fused with seed coat. Seed coat 1; thin, smooth. Embryo small, linear or rudimentary, occasionally spatulate, in anatropous ovule; radicle inferior. Cotyledons 2 or appearing as 1; closely appressed. Endosperm abundant; firm but watery fleshy.

Notes. The schizocarp (whole fruit) splitting longitudinally at maturity into two segments called mericarps. Therefore, most mericarps have one flattened or occasionally concave side (commissural side) where they joined their twin. In cross section the mericarp is asymmetrical. The mericarps and schizocarps are basic in the classification and identifica-

FIG. 51. Euphorbiaceae. (Magnification × 3.)
FIG. 52. Umbelliferae. (Magnification × 3.5.)

tion of the genera and species. Strong fruit characteristics makes this family one of the easiest to recognize by its "seeds." An excellent pre-Linnaean publication (Morison, 1702) may be one of the first examples of the systematic use of schizocarps. Murley (1946) used schizocarp characteristics in her treatment of Iowa species.

21. ASCLEPIADACEAE (Fig. 53)

Disseminule: seed, often comose.

External: seed ovate, oval, oblong, elliptic, or elongate; strongly dorso-ventrally compressed. Seed coat usually bearing a tuft of silky white hairs (coma) at hilum end; frequently bearing a conspicuous marginal wing; bearing a conspicuous or inconspicuous raphe from hilum to about midpoint or opposite end of ventral face; brown to blackish. Hilum on ventral face; inconspicuous; usually restricted to the wing portion of seed coat immediately below coma; usually a minute, triangular, whitish area.

Internal: seed coat apparently 1; membranous or coriaceous. Embryo straight, spatulate, axile in anatropous ovule; radicle superior, tip just below hilum. Cotyledons 2, appressed, broad, and flat. Endosperm abundant, rarely absent; firm, fleshy to cartilaginous

Notes. The seeds of this family are quite similar. Although seeds contribute a strong family characteristic, their similarity weakens their use at the species and perhaps at the genus levels. Little has been written about these seeds.

Hairs from the coma are often soon deciduous, thus usually absent. They are classified as vegetable silks because of their luster. However, their use in the textile industry is limited because the hairs are brittle and have only slight felting properties. Schorger (1925) analyzed the hairs of *Ascelpias syriaca.*

22. SOLANACEAE (Fig. 54)

Disseminule: seed.

External: seed of two types — (*a*) moderate size, compressed thin or thick, subcircular to elliptical in outline or (*b*) minute, cubical, nearly equidimensional to slightly elongate. Seed coat finely reticulate, punctate, when tops of epidermal cells break down the lateral walls remain as an empty meshwork, or as minute aligned depressions; in *Lycopersicon* bearing spurious hairs which are remnants of lateral epidermal walls; yellowish to brown. Hilum for compressed seeds marginal, occasionally in a submedial notch; for cubical seeds subterminal and flush.

Internal: seed coat 1; coriaceous, or nearly so. Embryo linear and curved (merely bent, annular, or spiralled), or rarely dwarf, in anatropous to hemianatropous ovule. Cotyledons 2; their tips either terminating near base of radicle or spirally incurved; ranging from large to little developed.

FIG. 53. Asclepiadaceae. (Magnification × 3.) Insert: *Asclepias labriformis* M.E. Jones. (Magnification × 1.)

FIG. 54. Solanaceae. (Magnification × 3.) Inserts: *Petunia violacea* Lindl. (Magnification × 12) and *Datura stramonium* L. (Magnification × 7).

Endosperm abundant or nearly so, rarely scant; fleshy and semitransparent.

Notes. Two distinct seed types, compressed with a strongly curved embryo, and minute cuboidal with a bent or dwarf embryo, have tribal significance. The compressed seed with strongly curved embryo is characteristic of tribes Datureae, Nicandreae, and Solaneae. Minute cuboidal seed with the bent or dwarf embryo is characteristic of tribes Cestreae and Salpiglossideae.

Seed characteristics are quite useful in this family at the genus and species level.

23. CONVOLVULACEAE (Fig. 55)

Disseminule: seed.

External: seed irregularly sectoroid to elongate and broader at one end than the other; usually with a strongly developed longitudinal ventral angle. Seed coat smooth to irregularly roughened or tuberculate; glabrous or bearing long or short hairs in definite arrangements. Hilum large; subbasal, at the narrow end; obliquely placed on ventral angle; flush or sunken.

Internal: seed coat apparently 1; thick, bony; occasionally strongly infolded near hilum forming an internal septum. Embryo bent in anatropous ovule; radicle near hilum, inferior. Cotyledons 2; foliaceous; folded or crumpled; sometimes punctate dotted; occasionally bifid. Endosperm hard and clear or semitransparent; mucilaginous when wet.

Note. The genus *Cuscuta* whose seeds have a coiled embryo and no cotyledons or endosperm belongs to the Cuscutaceae.

Seeds of Convolvulaceae and Malvaceae are strikingly similar. From a seed point of view, the two families should be closely related. Convolvulacous seeds have excellent characteristics and should be used at the genus and species levels (Gunn, 1970c,d).

24. BORAGINACEAE (Fig. 56)

Disseminule: nutlet or a one- to rarely four-seeded nut or drupe.

External: nut woody or bony (mainly in tribe Cordieae). Nutlet 1–4 from an ovary; elongate-rounded or obconical, irregularly plump or ovate and compressed, biconvex or planoconvex. Nutlet coat smooth, roughened, warty, or bearing hooked or plain bristles; infrequently winged; dull to shiny; usually black to brown, occasionally white or grayish. Nutlet scar conspicuous or inconspicuous; terminal, subterminal, or lateral. Seed coat smooth. Hilum quite inconspicuous, seldom recorded.

Internal: nutlet coat 1; thick and bony or crustaceous. Seed coat 1; thin. Embryo straight or slightly curved in anatropous or hemianatropous

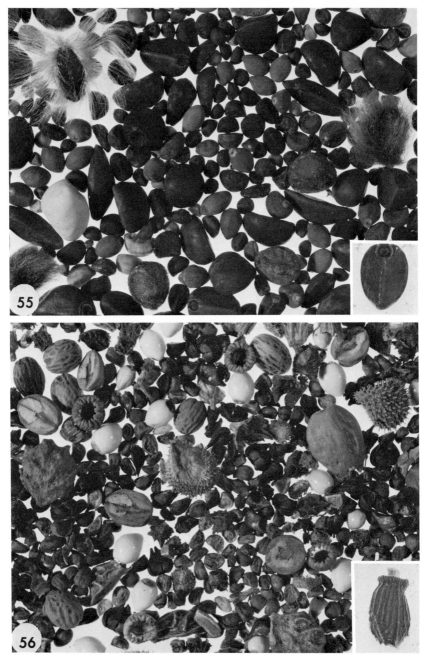

FIG. 55. Convolvulaceae. (Magnification × 1.) Insert: *Ipomoea turbinata* Lagasca y Segura. (Magnification × 2.)

FIG. 56. Boraginaceae. (Magnification × 3.) Insert: *Borago officinalis* L. (Magnification × 3.5.)

ovule; spatulate, linear, or partly investing; radicle prominent, superior or subinferior in some species of the subtribe Cynoglosseae. Cotyledons 2; flat, or plaited and toothed, notched or bifid at apex in tribe Cordieae. Endosperm absent or when present scant to abundant; fleshy to soft fleshy.

Notes. Nutlet or nut characteristics are useful at the genus and perhaps species levels. Little work has been done on seeds of this family.

25. VERBENACEAE (Fig. 57)

Disseminule: nutlet, drupe, or nut.

External: nut within fleshy exocarp usually containing 2–4 seeds; woody or bony. Nutlets 1–4, rarely 8–10; straight or slightly curved; elongate-sectoroid, spherical, roughly elliptical and boat-shaped, or compressed and curved. Nutlet coat smooth, reticulate, or in *Verbena* ventral faces bearing irregularly shaped white "hairs." Nutlet scar basal or sub-basal on ventral side; ventral angle does not continue through scar; inconspicuous. Seed coat smooth. Hilum quite inconspicuous; seldom recorded.

Internal: nutlet coat 1; thick and bony or crustaceous. Seed coat 1; thin. Embryo straight in anatropous ovule; spatulate. Cotyledons 2; flat, or plaited in *Avicennia*. Endosperm absent or when present scant and fleshy to soft fleshy.

Notes. Nutlet or nut characteristics may be useful at the genus and perhaps species levels. Little work has been done on seeds of the Verbenaceae.

26. LABIATAE (Fig. 58)

Disseminule: nutlet.

External: nutlet basically elongate-sectoroid to elongate-rounded, irregularly plump-compressed; ventral side usually with a strongly developed longitudinal angle. Nutlet coat smooth, cellular reticulate or roughened, reticulate, papillose; dull or occasionally shiny; apical hairs present in some genera; rarely winged. Nutlet scar basal or subbasal on ventral side; frequently oblique; ventral angle may or may not continue through scar; inconspicuous to quite conspicuous, sometimes quite large and whitened. Seed coat smooth. Hilum quite inconspicuous; seldom noted.

Internal: nutlet coat 1; quite thick, bony. Seed coat 1; thin. Embryo various, straight and spatulate in endospermous genera, straight and investing in nonendospermous genera, and bent and cylindrical with a long radicle on the back of one cotyledon in *Scutellaria;* ovule hemianatropous to anatropous; radicle inferior, but superior in *Scutellaria.* Cotyledons 2;

FIG. 57. Verbenaceae. (Magnification × 3.5.) Insert: *Verbena* cv. hybrida. (Magnification × 4.)

FIG. 58. Labiatae. (Magnification × 3.) Insert: *Lallemantia iberica* Fisch. & Mey. (Magnification × 4.)

usually large and flattened. Endosperm absent or present; fleshy to soft fleshy.

Notes. Fruit is four-lobed which usually divides into quarters at maturity. Epicarp dry or sometimes fleshy in *Prasium.* Nutlet (one-quarter of a fruit) characteristics are valuable at family, genus, and species levels. Isely (1947) summarizes the seed characteristics of the three tribes represented in the United States.

Ajugeae—nutlet scar quite large, occupying about half the ventral side of nutlet; nutlet coat reticulate; radicle straight, inferior.

Scutellarieae—nutlet scar small; nutlet coat papillose-roughened; radicle curved and superior.

Stachyeae—nutlet scar various but never occupying half ventral surface; nutlet coat smooth; radicle straight, inferior. Most North American species are in this tribe.

Wojciechowska (1966) studied morphology, anatomy, and histochemistry of the nutlets of seventy species which have medicinal value. An English key and illustrations of nutlets with emphasis on fruit and seed coats supplement the text.

27. PLANTAGINACEAE (Fig. 59)

Disseminule: seed, or bony nut.

External: seed elongate, roughly elliptical and boat-shaped to irregularly angular-compressed or lumpy. Seed coat smooth; minutely punctate, striate or rugose; dull or shiny; black, brown, or reddish. Hilum peltate, about medial on ventral face, usually large and conspicuously marked.

Internal: seed coat 1; well-developed; when moistened becoming mucilaginous. Embryo straight (curved around endosperm in *Bougueria*), axile, spatulate to almost linear in a hemianatropous to anatropous ovule, radicle inferior. Cotyledons 2; well-developed, appressed with face wither at right angles or parallel to the dorsoventral surfaces. Endosperm scant or nearly abundant; firm, transparent and waxy in appearance.

Notes. This family is composed of three genera, *Plantago, Bougueria,* and *Littorella* (Hutchinson, 1960). *Plantago* is a ubiquitous weedy genus with circumscissile, two- to several-seeded, thin fruits. The other two genera have indehiscent one-seeded fruits which may be nutlike. These observations and descriptions, like most others, are dominated by the characteristics of *Plantago.* Little has been published on the other two genera.

Peltately attached seeds are not unique to this family. This characteristic is also found in the Menispermaceae, Rubiaceae, and Scrophulariaceae.

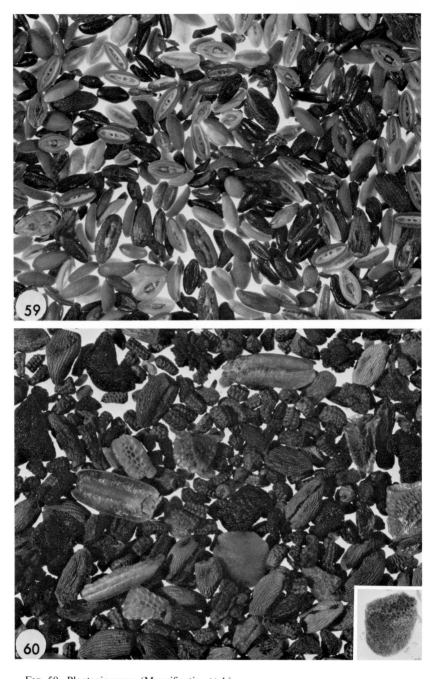

FIG. 59. Plantaginaceae. (Magnification × 4.)

FIG. 60. Scrophulariaceae. (Magnification × 6.) Insert: *Digitalis lanata* Ehrh. (Magnification × 15.)

28. SCROPHULARIACEAE (Fig. 60)

Disseminule: seed.

External: seed generally small, occasionally large and much modified; usually variously angled or oblong, oval, obovoid, globose, rarely subglobular and deeply concave. Seed coat variously roughened, reticulate, foveolate, ridged, furrowed, seldom smooth; occasionally winged; dull; brown to black to yellowish. Hilum basal or lateral in peltate seeds.

Internal: seed coats 2; outer crustaceous or nearly so; inner thin. Embryo axile, dwarf, straight or curved, rarely annular, in anatropous to hemianatropous, rarely campylotropous ovule; linear to spatulate; radicle superior or horizontal and directed toward hilum, sometimes quite large; in relation to seed may be quite large or minute. Cotyledons 2; generally small, occasionally little developed; usually not much broader than radicle. Endosperm abundant to rarely scant or absent; fleshy to cartilaginous.

Notes. Seeds usually formed in a multiseeded capsule. Pressure of the closely packed seeds gives them their typical angular characteristics. Little work has been done on seeds in this family. Seeds of selected species of *Veronica* and related genera have been described and illustrated (Kelley, 1953; Thieret, 1955).

29. RUBIACEAE (Fig. 61)

Disseminule: seed, drupe, or nutlet.

External: drupe or nutlet dry, membranous, and thin to corky; smooth, tuberculate, or hairy; fruit usually splitting longitudinally at maturity into two segments, thus fruit halves have a flattened or concaved face and are asymmetrical in cross section. Seeds vary greatly in size and outline from oval, oblong, to orbicular; compressed, deeply concave, or umbilicate on ventral sides, or planoconvex with a longitudinal ventral slit. Seed coat smooth to reticulate; rarely winged or appendiculate; brown or yellowish-brown. Hilum peltate or basal; conspicuous or inconspicuous.

Internal: drupe coat 1; corky or thin. Seed coat 1 or 2; outer membranous, leathery, crustaceous, or rarely woody; inner thin. Embryo axile, straight or slightly curved in anatropous to hemianatropous ovule; dwarf, linear or spatulate; radicle inferior, less frequently superior, in some embryos strongly developed. Cotyledons 2; flat, foliaceous, rarely involute; not strongly developed, occasionally little developed. Endosperm abundant, fleshy or horny, rarely reduced to a thin layer or absent; ruminate in some genera.

Notes. Seeds and drupes of this family are quite diverse. Little information is available on the seeds and drupes in this family.

FIG. 61. Rubiaceae. (Magnification × 4.) Insert: *Coffea arabica* L. (Magnification × 1.)
FIG. 62. Compositae. (Magnification × 2.) Insert: *Cnidus benedictus* L. (Magnification × 1.)

30. COMPOSITAE (Fig. 62)

Disseminule: achene, with or without pappus; within hardened involucre in tribe Ambrosieae.

External: achene elongate or obconic and truncate at both ends; straight or slightly or strongly curved; basally attached; terete, angled, or compressed. Achene coat smooth, roughened (seldom pitted, laterally ribbed, or reticulate), or strongly or weakly furrowed, longitudinally ribbed, winged; glabrous, hairy, barbed, or bristled; dull or shiny; variously colored. Fruit scar basal; frequently depressed; straight or oblique; well-developed, seldom inconspicuous. Pappus a unique feature of Compositae achenes, present or absent; permanently attached or soon deciduous; arising from a more-or-less raised crown around a depressed style scar; composed of numerous fine bristles or less numerous stiff awns, barbs, or scales, or mixtures of the above. Seed coat, if present, smooth. Hilum quite inconspicuous; seldom noted. Involucre, persistent only in Ambrosieae, enclosing 1–4 seeds; globose to elongate, frequently spiny or bearing hooked bristles, ribs, or hairs.

Internal: involucre coat thick. Achene coat 1; crustaceous, corky ribbed, bony to papery or quite thin. Seed coat absent or 1; thin. Embryo straight, spatulate in anatropous ovule; cotyledons about as long as or twice length of inferior radicle. Cotyledons 2; appressed. Endosperm scant, one or two rows of partially crushed cells, or absent.

Notes. Compositae have strong achene characteristics which usually set it apart from other families. Within the Compositae, achenes are useful at the genus and species levels. Isely (1947) summarized achene characteristics at the tribe level. He noted that an achene key to members of this family would probably have to be artificial. Blake (1928) described and illustrated achenes of thirty-eight species. The importance of achene characteristics was recognized by Babcock and Stebbins (1937, 1938).

31. JUNCACEAE (Fig. 63)

Disseminule: seed.

External: seed minute, ca. 1–1.5 mm; ovoid, subglobose to fusiform; nearly sectoroid in *Luzula*. Seed coat reticulate and roughened; outer coat may be loosened and colorless and drawn out into long or short tails at opposite ends of the seed; in *Luzula* outer coat may form a rounded carunculate appendage over the hilum; inner coat reddish-brown to brown. Hilum quite inconspicuous, terminal.

Internal: seed coats 2; both quite thin. Embryo semilunate in shape, axile, and basal in anatropous ovule; less than half length of seed; radicle inferior. Cotyledon 1; topographically undifferentiated from radicle. Endosperm abundant; fleshy, dense, farinaceous.

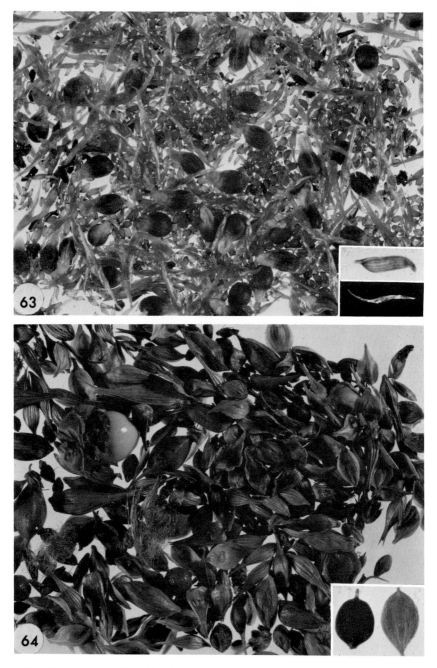

Fig. 63. Juncaceae. (Magnification × 10.) Inserts: *Juncus biflorus* Ell. (Magnification × 20) and *Juncus castaneus* Sm. (Magnification × 5).

Fig. 64. Cyperaceae. (Magnification × 3.) Inserts: *Carex normalis* Mackenzie and *Carex davisii* Schw. & Torr. (Magnification × 5.)

Notes. The foregoing description is primarily based on two widespread genera, *Juncus* (cosmopolitan) and *Luzula* (primarily Northern Hemisphere). Seeds of six other genera from Antarctica, South America, and South Africa have seldom been described.

Juncus—ca. 1 mm long; frequently with two, short- to long-tailed appendages.

Luzula—ca. 1.5 mm; usually sectoroid with distinct raphe on ventral surface; hilum covered by a carunculate appendage.

32. CYPERACEAE (Fig. 64)

Disseminule: achene with or without perigynum, bracts or barbed bristles; a drupe in *Diplasia.*

External: achene naked, or surrounded by a papery perigynum in *Carex,* or subtended by scales or bristles; trigonous or plano- or biconvex; usually ovate to lanceolate in outline, rarely laterally notched or bent; apical beak (stylopodium) like achene or differentiated into a caplike structure. Achene coat smooth to minutely roughened, pitted, irregularly wrinkled or ridged, striate, or rugose. Achene scar basal, irregular, inconspicuous; swollen callus may be present. Seed coat smooth. Hilum basal, quite inconspicuous; seldom noted.

Internal: achene coat 1; crustaceous to thin. Seed coat 1; thin. Embryo capitate to broad in anatropous ovule; with or without a topographically differentiated radicle; small, less than one-half the length of seed; radicle and coleoptile inferior. Cotyledon 1. Endosperm abundant, farinaceous, whitish except for an external yellowish "oil layer" below seed coat.

Notes. Achenes and accessory parts are valuable diagnostic characters at the family, genus, and species levels. The excellent works of Berggren (1969) and Kowal (1958) demonstrate the diagnostic value of achenes. Monographers have also used achene characteristics (Beetle, 1943; Gale, 1944; Hermann, 1970; Mackenzie, 1941; McGivney, 1938; Svenson, 1929, 1932, 1934, 1937, 1939).

33. GRAMINEAE (Figs. 65 and 66)

Seeds of the two subfamilies, Festucoideae and Panicoideae, are different enough to warrant their being treated separately. In fact, seed characteristics are major subfamily characteristics. Van Tieghem (1897) established two basic differences when he reported that the panicoid type had the scutellum free from the coleorhiza, and the coleoptile inserted well above the point of divergence of the scutellum bundle. In the festucoid type, the lower portion of the scutellum is missing or fused to the coleoptile, and the coleoptile is inserted at about the point of divergence of the scutellum bundle. Other differences include the relative size of the

Fig. 65. Gramineae tribe Festucoideae. (Magnification × 5.) Insert: *Triticum aestivum* L. (Magnification × 3.)

Fig. 66. Gramineae tribe Panicoideae. (Magnification × 5.) Insert: *Sorghum bicolor* (L.) Moench. (Magnification × 4.)

embryo to the seed. In the festucoid type, the embryo is relatively small in proportion to the seed, whereas in the panicoid type the embryo is relatively large in proportion to the seed. Reeder (1957, 1962) observed that based on seed characteristics the Panicoideae is a natural and homogeneous subfamily, whereas the Festucoideae is extremely heterogeneous.

The family as a whole produces a caryopsis, a one-seeded fruit which is usually enclosed in a persistent lemma and palea, rarely naked. Therefore, the term "grass seed" usually refers to a caryopsis within an indurate lemma and palea. In some tribes there are associated structures, such as a rachilla, a pair of glumes, or reduced and modified parts of other florets. These accessory parts may also be part of the grass seed. They have identification importance and classification significance. Except for a few genera, identification of isolated caryopses is difficult.

a. FESTUCOIDEAE (Fig. 65)

Disseminule: caryopsis usually within an indurate lemma and palea often with a rachilla and reduced parts of other florets and glumes; or naked caryopsis.

External: caryopsis elongated longitudinally; terete, or compressed and commonly with a ventral longitudinal channel, or furrow and convexly curved dorsal surface; dorsobasal embryo easily located, sometimes a different color. Caryopsis coat (pericarp) smooth or faintly longitudinally striate, rarely variously roughened; glabrous to pubescent. Caryopsis scar basal and opposite embryo on ventral side; usually inconspicuous. Seed coat frequently fused with pericarp; well-developed in some species of *Sporobolus* and *Eleusine* where pericarp is reduced or dehiscent. Hilum when visible, quite inconspicuous.

Internal: caryopsis coat usually thin, rarely loose and membranous, or in some species of the tribe Bambuseae hard and crustaceous, or thick and fleshy. Embryo lateral, small, basal, in anatropous, rarely hemianatropous, or campylotropous ovule; radicle inferior, sheathed by the coleorhiza; plumule superior, sheathed by the coleoptile. Cotyledon 1, scutellate. Endosperm usually farinaceous, varying from fleshy to flinty, firm to hard, rarely liquid or nearly so; outer one or two layers composed of protein-bearing cells, called the aleurone layer.

Notes: Isolated seeds (caryopses and accompanying floral parts) are readily identifiable at family, genus, and perhaps species levels. Seed characteristics are often used in defining the tribes of the subfamily Festucoideae (Chase, 1950), and they have been used at the genus and species levels as well (Colbry, 1957).

b. PANICOIDEAE (Fig. 66)

Disseminule: caryopsis usually with indurate lemma and palea and a sterile lemma and usually a sterile palea; rachilla, if present, 2 in number; naked caryopsis in *Zea*.

External: caryopsis elongated longitudinally or rounded; planoconvex, compressed, or terete; dorsobasal embryo easily located, sometimes a different color. Caryopsis coat (pericarp) smooth; glabrous; occasionally brightly colored. Caryopsis scar basal and opposite embryo on ventral side; usually conspicuous, a reddish to black dot. Seed coat frequently fused with pericarp.

Internal: caryopsis coat thin to coriaceous. Embryo lateral, large (up to half length of seed) in anatropous, rarely hemianatropous or campylotropous ovule; radicle inferior, sheathed in a coleorhiza; plumule superior, sheathed in a coleoptile. Cotyledon 1, scutellate. Endosperm farinaceous, varying from fleshy to flinty, firm to hard; outer one to two layers composed of protein-bearing cells, the aleurone layer.

Notes. Isolated seeds (caryopses and accompanying floral parts) readily identifiable at family, genus, and perhaps species levels. The seed characteristics are often used in defining tribes of the subfamily Panicoideae (Chase, 1950).

34. PALMAE (Fig. 67)

Disseminule: nut, drupe, or seed.

External: nut, fruit, seed varying widely in size from a diameter of a few millimeters to at least 80 cm; generally rounded, occasionally somewhat compressed, oblong or conical; rarely curved. Exocarp dry, smooth, scaly, or fibrous; or fleshy and occasionally oily. Endocarp smooth or wrinkled; in tribe Cocoineae with three adjacent symmetrical depressions and a three-branched raphe. Nut scar inconspicuous. Seed coat often adhering to endocarp; if free smooth to reticulate, bearing one or more raphes. Hilum inconspicuous.

Internal: endocarp thick; fibrous, bony, or woody. Seed coat 1, or fused with endocarp; coriaceous to membranous; forming ingrowths or folds into soft young endosperm which cause the endosperm to be ruminate at maturity. Embryo minute, axile, nearly linear in classification (turbinate, conical or cylindric in shape); ovule anatropous to hemianatropous; straight or bent; sunken in periphery of endosperm usually near the hilum, rarely lateral or apical. Cotyledon 1; topographically little differentiated from radicle or plumule. Endosperm horny or cartilaginous, semitransparent; seldom farinaceous; occasionally oily ruminate or ivorylike; solid or hollow, or indented apically or laterally.

Notes. Palm fruits are usually one-seeded, rarely several-seeded (*Phytelephas, Borassus* and related genera, *Manicaria,* and abnormally in *Cocos* and its related genera).

FIG. 67. Palmae. (Magnification × 2.)

FIG. 68. Liliaceae. (Magnification × 3.) Inserts: *Lilium regale* Wils. and *Camassia scilloides* (Raf.) Cory. (Magnification × 2.)

The largest seed is the double coconut, *Lodoicea maldivica.* The double coconut disseminule includes the bony endocarp. Even when this shell is removed, the double coconut seed is the world's largest seed.

The date palm, *Phoenix,* "seed" is the naked endosperm. The fruit is made up of two layers, fleshy, edible layer and a thin inner membrane around the endosperm.

Palm embryos are seldom dormant and cannot withstand desiccation. Unless an embryo dies, it never ceases to grow within the seed, even when the seed or disseminule is being dispersed.

The internal and external characteristics of disseminules are used in palm classification (Corner, 1966).

35. LILIACEAE (Fig. 68)

Disseminule: seed.

External: seed globose, oblong, angular, or strongly compressed; baccate. Seed coat smooth or variously roughened; occasionally with a wing, a raphe, or with one or two tails; rarely hairy; dull or shiny; usually brown or black. Hilum inconspicuous, seldom noted.

Internal: seed coat 1; varying from thin, suberose, to crustaceous; occasionally quite brittle. Embryo axile, linear to rarely rudimentary in hemianatropous, anatropous, or orthotropous ovule; straight, bent, curved, or curled at upper end; usually less than one-quarter of seed volume. Cotyledon 1; topography little differentiated from radicle. Endosperm abundant; cartilaginous, horny, or fleshy; semitransparent; nonfarinaceous.

Notes. The liliaceae has diverse seed characteristics. Seeds are borne in capsules and one- to several-seeded berries. Some taxonomists have divided the Liliaceae into several closely related families but these families are not formed along seed characteristic lines. Arnott (1962) discussed the seed and seedling characteristics of *Yucca.*

36. IRIDACEAE (Fig. 69)

Disseminule: seed.

External: seed subglobose, to compressed and rounded. Seed coat usually rough, wrinkled or reticulate; occasionally winged; with or without a raphe. Hilum inconspicuous; at small end of seed or at juncture of compressed faces.

Internal: seed coats 2; outer crustaceous, inner thin. Embryo basal, axile, and linear in anatropous ovule; straight or bent; radicle ending near hilum. Endosperm abundant; fleshy to cartilaginous or subhorny; semitransparent; nonfarinaceous.

Notes. Seed characteristics have seldom been used in this family.

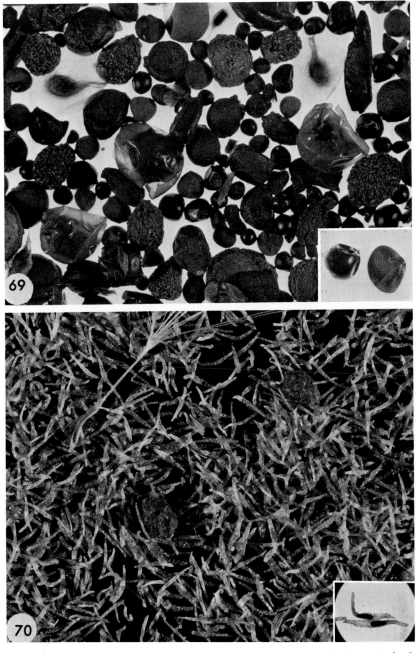

FIG. 69. Iridaceae. (Magnification × 4.) Inserts: *Iris setosa* Pall. and *Iris versicolor* L. (Magnification × 2.)

FIG. 70. Orchidaceae. (Magnification × 10.) Insert: *Cypripedium calceolus* L. var. *pubescens* (Willd.) Correll. (Magnification × 18.)

37. ORCHIDACEAE (Fig. 70)

Disseminule: seed.

External: seed minute, between 0.3 and 2 mm in length; scobicular or fusiform. Seed coat reticulate, loose, usually drawn out into two well-developed opposite tails. Hilum exceedingly inconspicuous; at end of a tail.

Internal: seed coat 1; composed of 1 layer of cells. Embryo micro and undifferentiated in anatropous ovule; in some species seeds have no discernible embryo, or have several to many embryos produced asexually. Endosperm absent.

Notes. Orchid seeds are the smallest of all seeds. The seeds have no endosperm, and there is very little food reserve in the embryo. Usually the embryo is undifferentiated in the capsule. Differentiation takes place after dispersal of the seeds. Each capsule contains many seeds, from about 300 to 3.7 million seeds per capsule. For additional information about orchid seeds, germination, and seedlings the reader is referred to Beer (1863), Clifford and Smith (1969), A. Davis (1946), and Stoutamire (1964, 1965).

REFERENCES

Aesopus. (1850). "Aesopicae fabulae," p. 92. Tauchnitz, Leipzig.

Agricultural Research. (1970). Our shadowed skies put plants under siege. *U.S. Dep. Agr., ARS* May, 8.

Airy-Shaw, H. K. (1966). "Dictionary of Flowering Plants and Ferns," 7th ed. Cambridge Univ. Press, London and New York.

Anghel, G., Bilteanu, G., Bucurescu, N., Burcea, P., Teodorescu, D., and Valiliu, N. (1965). "Indrumator Pentru Determinarea Semintelor de Plante Cultivate." Agro-Silvica, Bucurest.

Anghel, G., Raianu, M., Matei, C., Bucurescu, N., Radulescu, I., Anganu, I., and Velea, C. (1959). "Determinarea Calitatii Semintelor." Academia Republicii Populare Romane.

Arnott, W. J. (1962). Seed, germination, and seedling of Yucca. *Univ. Calif., Berkeley, Publ. Bot.* **35,** 1.

Babcock, E. B., and Stebbins, G. L., Jr. (1937). Genus *Youngia. Carnegie Inst. Wash. Publ.* **484,** 1.

Babcock, E. B., and Stebbins, G. L., Jr. (1938). American species of *Crepis. Carnegie Inst. Wash. Publ.* **504,** 1.

Baranov, A. I. (1969). Species of *Corispermum* (Chenopodiaceae) in northeastern China. *J. Jap. Bot.* **44,** 161.

Barber, K. G. (1909). Comparative histology of fruits and seeds of certain species of Cucurbitaceae. *Bot. Gaz.* **47,** 263.

Barclay, A. S., Gentry, H. S., and Jones, O. (1962). Search for new industrial crops. II. *Lesquerella* (Cruciferae) as a source of new oilseeds. *Econ. Bot.* **16,** 95.

Barton, L. V. (1967). "Bibliography of Seeds." Columbia Univ. Press, New York.

Beer, J. G. (1863). "Beitrage zur Morphologie und Biologie der Familie der Orchideen." Gerold's Son, Vienna.

Beetle, A. A. (1943). Key to North American species of genus *Scirpus* based on achene characters. *Amer. Midl. Natur.* **29**, 533.

Beijerinck, W. (1947). "Zadenatlas der Nederlandsche Flora." Veenman Zonen, Wageningen.

Bell, E. A. (1966). Amino acids and related compounds. *In* "Comparative Phytochemistry" (T. Swain, ed.), pp. 195–209. Academic Press, New York.

Bell, E. A., and O'Donovan, J. P. (1966). Isolation of α- and γ-oxalyl derivatives of α, γ-diaminobutyric acid from seeds of *Lathyrus latifolius,* and detection of α- oxalyl-aminopropionic acid which occurs together with neurotoxin in this and other species. *Phytochemistry* **5**, 1211.

Bennett, E. (1965). Plant introduction and genetic conservation: Genecological aspects of an urgent world problem. *Scot. Plant Breed. Sta. Rec.* p. 27.

Benson, L. (1969). "Cacti of Arizona." Univ. of Arizona Press, Tucson.

Berggren, G. (1960). Beskrivning av vissa odlade och vildvaxande *Brassica-* och *Sinapis-*arters frokaraktarer jamte en harpa grundad bestamningsnyckel. *Medd. Statens Cent. Frokont.* **35**, 28.

Berggren, G. (1962). Reviews on taxonomy of some species of genus *Brassica,* based on their seeds. *Sv. Bot. Tidskv.* **56**, 67.

Berggren, G. (1969). "Atlas of Seeds," Part 2: Cyperaceae. Swed. Nat. Sci. Res. Counc., Stockholm.

Bertsch, K. (1941). Fruchte und samen. *In* "Handbuch der praktischen Vorgeschechtsforchung" (H. Reinerth, ed.), Vol. 1, pp. 1–247. Enke, Stuttgart.

Blake, A. M. (1928). Achenes of some Compositae. *N. Dak. Agr. Exp. Sta., Bull.* **218**, 1.

Bouwer, P. (1927). "Landwirtschaftliche Samenkunde." Newman, Neudamm.

Brezhnev, D. (1970). Mobilization, conservation and utilization of plant resources at N. I. Vavilov All-Union Institute of Plant Industry, Leningrad. *In* "Genetic Resources in Plants – Their Exploration and Conservation" (O. H. Frankel and E. Bennett, eds.), pp. 533–538. Blackwell, Oxford.

Britton, N. L., and Rose, J. N. (1963). "Cactaceae," 2 vols. (reprint). Dover Press, New York.

Brouwer, W., and Stahlin, A. (1955). "Handbuch der Samenkunde fur Landwirtschaft, Gartenbau und Forstwirtschaft." DLG-Verlags-GMBH, Frankfort.

Camus, A. (1948). "Les Chênes: Monographia des genres *Quercus* et *Lithocarpus.*" Lechevalier, Paris.

Chancellor, R. J. (1959). Identification of seedlings of common weeds. *Min. Agr., Fish., Food (London) Bull.* **179**, 1.

Chaney, R. W., Elias, M. K., Dorf, E., Axelrod, D. I., and Condit, C. (1938). "Miocene and Pliocene Floras of Western North America." Carnegie Institution, Washington, D.C.

Chase, A. (1950). Manual of grasses of United States. 2nd ed. *U.S. Dep. Agr., Misc. Publ.* **200**, 1.

Childe, V. G. (1943). "What Happened in History," p. 43. Penguin Books, London.

Clifford, H. T., and Smith, W. K. (1969). Seed morphology and classification of Orchidaceae. *Phytomorphology* **19**, 133.

Colbry, V. L. (1957). Diagnostic characteristics of fruits and florets of economic species of North American *Sporobolus. Contrib. U.S. Nat. Herb.* **34**, 1.

Cole, M. J. (1961). Interspecific relationships and intraspecific variation of *Chenopodium album* in Britain. *Watsonia* **5**, 47.

Corner, E. J. H. (1951). Leguminous seed. *Phytomorphology* **1**, 117.

Corner, E. J. H. (1966). "Natural History of Palms." Univ. of California Press, Berkeley.

Creech, J. (1970). Tactics of exploration and collection. *In* "Genetic Resources in Plants –

their Exploration and Conservation" (O. H. Frankel and E. Bennett, eds.), pp. 221–229. Blackwell, Oxford.

Cronquist, A. (1968). "Evolution and Classification of Flowering Plants." Houghton, Boston, Massachusetts.

Dalla Torre, K. W., and Harms, H. (1900–1907). "Genera Siphonogamarum ad Systema Englerianum Conscripta." G. Engelmann, Leipzig.

Dallimore, W., and Jackson, A. B. (1966). "Handbook of Coniferae and Ginkgoaceae," 4th ed., Arnold, London.

Dasmann, R. F. (1968). "Environmental Conservation," 2nd ed. Wiley, New York.

Davis, A. (1946). Orchid seed and seed germination. *Amer. Orchid Soc., Bull.* **15,** 218.

Davis, G. (1966). "Systematic Embryology of the Angiosperms." Wiley, New York.

de Candolle, A. (1886). "Origine des plantes cultivées." 3rd ed. Ancienne Librarie Germer Bailliere, Paris.

Dobrokhotov, V. N. (1961). "Seeds of Weed Plants." Agricultural Literature, Moscow.

Duke, J. A. (1961). Preliminary revision of genus *Drymaria. Ann. Mo. Bot. Gard.* **48,** 173.

Duke, J. A. (1964a). "Prelude to the Polyclave," Vol. I. Embryo. Roneo, Durham.

Duke, J. A. (1964b). "Prelude," Vol. II. Seed. Roneo, Durham.

Duke, J. A. (1965a). Keys for identification of seedlings of some prominent woody species in eight forest types in Puerto Rico. *Ann. Mo. Bot. Gard.* **52,** 314.

Duke, J. A. (1965b). "Prelude to a Palm Polyclave." Roneo, Durham.

Duke, J. A. (1969a). "Legume Polyclave." Xerox, Columbus.

Duke, J. A. (1969b). "Family Polyclave." Battelle, Columbus.

Duke, J. A. (1969c). On tropical tree seedlings. I. Seeds, seedlings, systems, and systematics. *Ann. Mo. Bot. Gard.* **56,** 125.

Elias, M. K. (1932). Grasses and other plants from tertiary rocks of Kansas and Colorado. *Univ. Kans. Sci. Bull.* **22,** 333.

Elias, M. K. (1935). Tertiary grasses and other prairie vegetation from high plains of North America. *Amer. J. Sci.* **29,** 24.

Elias, M. K. (1942). Tertiary prairie grasses and other herbs from high plains. *Geo. Soc. Amer., Spec. Pap.* **41,** 1.

Elias, M. K. (1946). Taxonomy of tertiary flowers and herbaceous seeds. *Amer. Midl. Natur.* **36,** 373.

Engler, A., and Prantl, K. (1887–1899). "Die Naturlichen Pflanzenfamilien" W. Engelmann, Leipzig.

Erwin, A. T., and Haber, E. S. (1929). Species and varietal crosses in cucurbits. *Iowa, Agr. Exp. Sta., Bull.* **263,** 1.

Fisher, H. H. (1969). Plant introduction in United States. *FAO Plant Intro. Newslett.* No. 22, p. 13.

Fong, C. H. (1969). "Agricultural and Horticultural Seeds in Malaysia." College of Agriculture, Malaya.

Forest Service. (1948). Woody plant seed manual. *U.S. Dep. Agr., Misc. Publ.* **654,** 1.

Fussell, G. E. (1965). "Farming Technique from Prehistoric to Modern Times." Pergamon, Oxford.

Gaertner, J. (1788–1805). "De Fructibus et Seminibus Plantarum," 4 vols. Academiae Carolinae, Stuttgart.

Gale, S. (1944). *Rhynchospora,* section Eurhynchospora, in Canada, United States, and West Indies. *Rhodora* **46,** 89.

Galston, A. W. (1970). Plants, people, and politics. *BioScience* **20,** 405.

Godwin, H. (1956). "History of the British Flora." Cambridge Univ. Press, London and New York.

Grabe, D. F. (1970). "Tetrazolium Testing Handbook," Contrib. 29, Handb. Seed Test. Ass. Off. Seed Anal. Amherst, Massachusetts.

Guinet, P. (1959). Essai d'identification des graines de chenopodes commensaux des cultures ou cultivés en France. *J. Agr. Trop. Bot. Appl.* **6,** 241.

Gunderson, A. (1950). "Families of Dicotyledons." Chronica Botanica, Waltham, Massachusetts.

Gunn, C. R. (1968). Stranded seeds and fruits from southeastern shore of Florida. *Gard. J.* **18**(1), 43.

Gunn, C. R. (1969). *Abrus precatorius:* A deadly gift. *Gard. J.* **19**(1), 2.

Gunn, C. R. (1970a). *Vicia menziesii* Sprengel. *In* "Flora Hawaiiensis" (O. Degener and I. Degener, eds.), Suppl. Authors, Waialua.

Gunn, C. R. (1970b). Key and diagrams for seeds of one hundred species of *Vicia* (Leguminosae). *Proc. Int. Seed. Test. Ass.* **35,** 773.

Gunn, C. R. (1970c). Seeds of United States noxious and common weeds in Convolvulaceae, excluding genus *Cuscuta. Proc. Ass. Off. Seed Anal.* **59,** 101.

Gunn, C. R. (1970d). History and taxonomy of purple moonflower, *Ipomoea turbinata* Lagasca y Segura. *Proc. Ass. Off. Seed Anal.* **59,** 116.

Gunn, C. R. (1971a). Seeds of native and naturalized vetches of North America. *U.S. Dep. Agr., Agr. Handb.* **392,** 1.

Gunn, C. R. (1971b). Seeds of the tribe Vicieae (Leguminosae) in North American Agriculture. *Proc. Ass. Off. Seed Anal.* **60,** 48.

Guppy, H. B. (1917). "Plants, Seeds, and Currents in the West Indies and Azores." Williams & Norgate, London.

Gustafsson, A., and Simak, M. (1963). X-ray photography and seed sterility in *Phragmites communis* Trin. *Hereditas* **49,** 442.

Haftorn, S. (1956). Synzoic seed dispersal by birds in Norway. *Blyttia* **14,** 103.

Hanelt, P., and Tschiersch, B. (1967). Blausaureglykosiduntersuchungen am gaterslebner Wickensortiment. *Kulturpflanze* **15,** 85.

Harada, K. (1934). Diagnosis of minute seeds by means of impressions upon celluloid-membrane ('sump'-figures). *J. Jap. Bot.* **10,** 238.

Hardin, C. M. (1969). "Overcoming World Hunger." Prentice-Hall, Englewood Cliffs, New Jersey.

Harris, D. R. (1969). Agricultural systems, ecosystems and the origins of agriculture. *In* "Domestication and Exploitation of Plants and Animals." (P. J. Ucko and G. W. Dimblby, eds.), pp. 3–15. Duckworth, London.

Harz, C. D. (1885). "Landwirtschaftliche Samenkunde." Parey, Berlin.

Heinisch, O. (1955). "Samenatlas." Deut. Akad. Landwirt., Berlin.

Helbaek, H. (1953). Early crops in southern England. *Proc. Prehist. Soc.* **18,** 194.

Helbaek, H. (1954). Prehistoric food plants and weeds in Denmark. *Dan. Geol. Unders.* [Ath.], *Raekke 4* **2,** 250.

Hermann, F. J. (1970). Manual of Carices of Rocky Mountains and Colorado Basin. *U.S. Dep. Agr., Forest Serv., Agr. Handb.* **374,** 1.

Herron, J. W. (1953). Study of seed production, seed identification, and seed germination of *Chenopodium* spp. *Cornell Univ., Agr. Exp. Sta., Memo.* **320,** 1.

Horanic, G. E., and Gardner, F. E. (1967). Improved method of making epidermal imprints. *Bot. Gaz.* **128,** 144.

Hubner, R. (1955). "Der Same in der Landwirtschaft." Neumann, Radebeul.

Humphrey, W. (1970). Ditches are quicker. *Life* **69** (Aug. 7), 58.

Hutchins, R. E. (1965). "Amazing Seeds." Dodd, Mead, New York.

Hutchinson, J. (1960). "Families of Flowering Plants," 2 vols. Oxford Univ. Press, London and New York.

Hutchinson, J. (1969). "Evolution and Phylogeny of Flowering Plants." Academic Press, New York.

Hyland, H. L. (1970). Description and evaluation of wild and primative introduced plants. *In* "Genetic Resources in Plants—Their Exploration and Conservation" (O. H. Frankel and E. Bennett, eds.), pp. 413–419. Blackwell, Oxford.

Isely, D. (1947). Investigations in seed classification by family characteristics. *Iowa, Agr. Exp. Sta., Res. Bull.* **351,** 1.

Isely, D. (1955a). Observations on seeds of Leguminosae: Mimosoideae and Caesalpinioideae. *Iowa Acad. Sci.* **62,** 142.

Isely, D. (1955b). Key to seeds of Caesalpinioideae and Mimosoideae of north-central states. *Iowa Acad. Sci.* **62,** 146.

Jackson, H. H. T. (1961). "Mammals of Wisconsin." Univ. of Wisconsin Press, Madison.

Janchen, E. (1942). Das system der cruciferen. *Oester. Bot. Z.* **91,** 1.

Jensen, H. A. (1969). Content of buried seeds in arable soil in Denmark and its relation to weed population. *Dan. Bot. Ark.* **27,** 1.

Katz, N. J., Katz, S. V., and Kipiani, M. G. (1965). "Atlas and Keys of Fruits and Seeds occurring in the Quaternary Deposits of the USSR." Nauka, Moscow.

Kelley, W. R. (1953). Study of seed identification and seed germination of *Potenilla* spp. and *Veronica* Spp. *Cornell Univ., Agr. Exp. Sta., Memo.* **317,** 1.

Kiffmann, R. (1955–1960). "Bestimmungsatlas fur Samereien der Wiesen- und Weidepflanzen des mitteleuropaischen Flachlandes," Parts A–G. Freising, Weihenstephan.

Kopooshian, H. A. (1963). Seed Character Relationships in Leguminosae," Microfilm 63-7257. University Microfilms, Ann Arbor, Michigan.

Korsmo, E. (1935). "Weed Seeds." Gyldendal, Oslo.

Kowal, T. (1953). Key for determination of genera *Chenopodium* L. and *Atriplex* L. *Monogr. Bot.* **1,** 87.

Kowal, T. (1958). Study on morphology of fruits of European genera from subfamilies Scirpoideae Pax, Rhynchosporoideae A. & G. and some genera of Caricoideae Pax. *Monogr. Bot.* **6,** 97.

Kowal. T. (1966). Systematic studies on seeds of genera *Delia* Dum., *Spergula* L., and *Spergularia* Presl. *Monogr. Bot.* **21,** 245.

Kowal, T., and Wojterska, H. (1966). Systematic studies on seeds of genus *Dianthus* L. *Monogr. Bot.* **21,** 271.

Krochmal, A. (1952). Seeds of weedy *Euphorbia* species and their identification. *Weeds* **1,** 243.

Krummer, A. P. (1951). "Weed Seedlings." Univ. of Chicago Press, Chicago.

Lawrence, G. H. M. (1951). "Taxonomy of Vascular Plants." Macmillan, New York.

Le Maout, E., and Decaisne, J. (1876). "General System of Botany" (English edition by Mrs. Hooker). Longmans, Green, New York.

Lhotska, M. (1957). "Urcovani Semen a Plodu Vzemedelske Praxi." Ministerstvo Potravinarskeho, Prague.

Lockey, S. D., Jr., and Dunkleberger, L. (1968). Anaphylaxis from an Indian necklace. *J. Amer. Med. Ass.* **206,** 2900.

Lubbock, J. (1892). "Contribution to our Knowledge of Seedlings," 2 vols. Kegan Paul, Trench, Trubner, London.

McClure, D. S. (1957). Seed characters of selected plant families. *Iowa J. Sci.* **31,** 649.

MacGinitie, H. D. (1941). "Middle Eocene Flora from the Central Sierra Nevada." Carnegie Institution, Washington, D.C.

McGivney, Sister M. V. de P. (1938). Revision of subgenus Eucyperus found in United States. *Catholic Univ. Amer., Biol. Ser.* **26.**

McGugan, J. M. (1948). Seeds and seedlings of genus *Brassica*. *Can. J. Res. Sect. C* **26,** 520.

Mackenzie, K. K. (1941). "North American Cariceae," 2 vols. N. Y. Bot. Gard., New York.

MacMillan, C. (1902). Suggestions on the classification of seeds. *Bot. Gaz.* **34,** 224.

Mangelsdorf, P. C. (1965). Evolution of maize. *In* "Essays on Crop Plant Evolution" (J. B. Hutchinson, ed.), pp. 23–49. Cambridge Univ. Press, London and New York.

Manning, W. E. (1949). Genus *Alfaroa. Bull. Torrey Bot. Club* **76,** 196.

Marek, S. (1954). Morphology and anatomical features of fruits of genera *Polygonum* L., *Rumex* L. and keys for their determination. *Monogr. Bot.* **2,** 77.

Marek, S. (1958). European genera of Polygonaceae in light of anatomical and morphological investigations on their fruits and seeds. *Monogr. Bot.* **6,** 57.

Martin, A. C. (1946). Comparative internal morphology of seeds. *Amer. Midl. Natur.* **36,** 513.

Martin, A. C. (1954). Identifying *Polygonum* seeds. *J. Wildl. Manage.* **18,** 514.

Martin, A. C., and Barkley, W. D. (1961). "Seed Identification Manual." Univ. of California Press, Berkeley.

Miller, R. C. (1950). Oldest bird nest? *Pac. Discovery* **3**(4), 29.

Miller, R. H. (1967). *Crotalaria* seed morphology, anatomy, and identification. *U.S. Dep. Agr. Tech. Bull.* **1373,** 1.

Milner, M., Lee, M. R., and Katz, R. (1952). Radiography applied to grain and seeds. *Food Technol.* **6,** 44.

Milner, M., Katz, R., Lee, M. R., and Pyle, W. B. (1953). Application of Polaroid-Land process to radiographic inspection of wheat. *Cereal Chem.* **30,** 169.

Moggridge, J. T. (1873). "Harvesting Ants and Trapdoor Spiders, with Observations on their Habits and Dwellings." L. Reeve, London.

Morison, R. (1702). "Plantarum Umbelliferarum." Sheldon, Oxford.

Muir, J. (1937). Seed-drift of South Africa. *Repub. S. Afr. Dep. Agr., Bot. Surv. Mem.* No. 16, 1.

Murley, M. (1945). Distribution of Euphorbiaceae in Iowa, with seed keys. *Iowa J. Sci.* **19,** 415.

Murley, M. (1946). Fruit key to Umbelliferae in Iowa, with plant distribution records. *Iowa J. Sci.* **20,** 349.

Murley, M. (1951). Seeds of Cruciferae of northeastern North America. *Amer. Midl. Natur.* **46,** 1.

Musil, A. F. (1948). Distinguishing species of *Brassica* by their seeds. *U.S., Dep. Agr., Misc. Publ.* **643,** 1.

Musil, A. F. (1963). Identification of crop and weed seeds. *U.S., Dep. Agr., Agr. Handb.* **219,** 1.

Netolitsky, F. (1926). Anatomie der angiospermensamen. *In* "Handbuch der Pflanzenanatomie (K. Linsbauer, ed.), vol. 10, No. 14, Borntraeger, Berlin.

Nicholson, J. F., Milner, M., Munday, W. H., Kurtz, O. L., and Harris, K. L. (1953). Evaluation of five procedures for determining of internal insect infestation of wheat. Use of X-rays. *Ass. Off. Agr. Chem.* **36,** 150.

Pammel, L. H. (1892). On seed-coats of genus *Euphorbia. Trans. Acad. Sci. St. Louis* **5,** 543.

Pammel, L. H. (1899). Anatomical characters of seeds of Leguminosae, chiefly genera of Gray's manual. *Trans. Acad. Sci. St. Louis* **9,** 90.

Panton, G. A., and D. Donaldson, eds. (1869–1874). "'Gest hystoriale' of destruction of Troy." Trubner, London.

Parker, B. M. (1952). "Seeds and Seed Travels." Harper, New York.

President's Science Advisory Committee (1967). "The World Food Problem," Vol. 1. The White House, Washington, D.C.

Quinn, V. (1936). "Seeds: Their Place in Life and Legend." Stokes, New York.

142   CHARLES R. GUNN

Reeder, J. R. (1957). Embryo in grass systematics. *Amer. J. Bot.* **44,** 756.
Reeder, J. R. (1962). Bambusoid embryo: A reappraisal. *Amer. J. Bot.* **49,** 639.
Reid, E. M., and Chandler, M. E. J. (1926). "Bembridge Flora," Vol. 1. British Museum, London.
Reid, E. M., and Chandler, M. E. J. (1933). "London Clay Flora." British Museum, London.
Ricker, P. L. (1961). Seeds of wild flowers. *Yearb. Agr. (U.S. Dep. Agr.)* pp. 288–294.
Ritter, W. E. (1929). Nutritial activities of the California woodpecker. *Quart. Rev. Biol.* **4,** 455.
Ritter, W. E. (1938). "California Woodpecker and I." Univ. of California Press, Berkeley.
Rockcastle, V. N. (1961). Seeds. *Cornell Univ. Sci. Leafl.* **54,** 1.
Russell, P. (1924). Identification of commonly cultivated species of *Cucurbita* by means of seed characters. *J. Wash. Acad. Sci.* **14,** 265.
Russell, P. (1961). Seed collection, United States Department of Agriculture. *Amer. Hort. Mag.* **40,** 325.
Sauer, C. O. (1952). "Agricultural Origins and Dispersal." Amer. Geogr. Soc., New York.
Schorger, A. W. (1925). Seed hairs of milkweed. *Ind Eng. Chem.* **17,** 642.
Scurti, J. (1948). Chiave analitica per il reconoscimento della piante infestanti attraverso i semi. *Ann. Sper. Agr.* [n.s.] **2,** Suppl., 1.
Selsam, M. E. (1957). "Play with Seeds." Morrow, New York.
Shorland, F. B. (1963). Distribution of fatty acids in plant lipids. *In* "Chemical Plant Taxonomy" (T. Swain, ed.), pp. 253–303. Academic Press, New York.
Singh, B. (1953). Studies on structure and development of seeds of Cucurbitaceae. *Phytomorphology* **3,** 224.
Singh, B. (1964). Development and structure of angiosperm seed. I. *Bull. Nat. Bot. Gard.* **89,** 1.
Smirnova, E. S. (1965). Types of seed structures in phanerogamous plants in the phylogenetic aspect. *J. Gen. Biol.* **26,** 310.
Smith, C. E., Jr. (1969). From Vavilov to present—a review. *Econ. Bot.* **23,** 2.
Smith, C. E., Jr. (1971). Preparing herbarium specimens of vascular plants. *U.S., Dep. Agr., Agr. Inform. Bull.* **348.**
Stearn, W. T. (1965). Origin and later development of cultivated plants. *J. Roy. Hort. Soc.* **90,** 279.
Stoddard, E. M. (1965). Identifying plants by leaf epidermal characters. *Conn., Agr. Exp. Sta., New Haven, Circ.* **227.**
Stoutamire, W. P. (1964). Seeds and seedlings of native orchids. *Mich. Bot.* **3,** 107.
Stoutamire, W. P. (1965). Strange seeds. *Cranbrook Inst. Sci. Newslett.* **34,** 94.
Svenson, H. K. (1929). Monographic studies in genus *Eleocharis.* I. *Rhodora* **31,** 121.
Svenson, H. K. (1932). Monographic studies in genus *Eleocharis.* II. *Rhodora* **34,** 193.
Svenson, H. K. (1934). Monographic studies in genus *Eleocharis.* III. *Rhodora* **36,** 377.
Svenson, H. K. (1937). Monographic studies in genus *Eleocharis.* IV. *Rhodora* **39,** 210.
Svenson, H. K. (1939). Monographic studies in genus *Eleocharis.* V. *Rhodora* **41,** 1.
Swaminathan, M. S., and Kamra, S. K. (1961). X-ray analysis of anatomy and viability of seeds of some economic plants. *Indian J. Gen. Plant Breed.* **21,** 129.
Sykes, W. H. (1835). Descriptions of new species of Indian Ants. *Trans. Roy. Entomol. Soc. London* **1,** 99.
Terrell, E. E. (1968). Biometric and taxonomic uses of cellulose acetate plastic. *Rhodora* **70,** 552.
Thieret, J. W. (1955). Seeds of *Veronica* and allied genera. *Lloydia* **18,** 37.
Ucko, P. J., and Dimbleby, G. W. (1969). "Domestication and Exploitation of Plants and Animals," pp. xvii–xxi. Duckworth, London.

Uyeki, H. (1927). Seeds of genus *Pinus,* as an aid to identification of species. *Agr. Forest. Coll. (Saigon) Bull.* **2,** 1.

van Tieghem, P. (1897). Morphologie de l'embryon et de la plantule chez les graminees et les cyperacees. *Ann. Sci. Natur. Bot. Biol. Veg.* [7] **8,** 259.

Vasil'chenko, I. T. (1965). "Identification of Weed Seedlings." Kolos, Leningrad.

Vavilov, N. I. (1949–1950). "Origin, Variation, Immunity and Breeding of Cultivated Plants" (English edition by K. S. Chester) *Chron. Bot.* **13,** 1.

Vertes, K. (1913). "Anatomisch-entwicklungsgeschichtliche Untersuchungen uber einige nutzbare Fruchte und der Samen." Selbstverlage, Bern.

Vinal, W. G. (1919). Mainly pedagogy of seeds with some seeds of pedagogy." *Nature Study Rev.* **15,** 213.

Walters, S. M. (1961). Shaping of angiosperm taxonomy. *New Phytol.* **60,** 74.

Weinmann, I. (1956). "Untersuchungen zur samendiagnostik von *Brassica*-arten und -sorten unter besonderer Berucksichtigung chemisch-physikalischer methoden. *Z. Pflanzenzuecht.* **36,** 1.

Wheeler, W. M. (1926). "Ants," p. 267. Columbia Univ. Press, New York.

Wiegand, K. M. (1895). Structure of fruit in order Ranunculaceae. *Amer. Microsc. Soc.* **16,** 69.

Wojciechowska, B. (1966). Morphology and anatomy of fruits and seeds in the family Labiatae with particular respect to medicinal species. *Monogr. Bot.* **21,** 3.

3

SEED STORAGE AND LONGEVITY

James F. Harrington

I. Historical Background

A. *Purposes of Seed Storage*

Essential to man's change from hunter or herder to cultivator was storage of seeds from harvest to the following planting. This required forethought and forbearance, safeguarding seeds in times of famine from those who hungered for them. Such seeds also needed safeguarding from birds, rodents, insects, and microorganisms. Even when man succeeded in this effort, he might sometimes find that his carefully guarded seeds had lost their germinability and would not sprout when planted. Thus, man learned also to guard against high temperature and high humidity in the stored seeds. Not until man had solved such storage problems could he settle down to an agrarian life.

As civilization became more complex, man found that some areas produced better seed crops than other areas, so trade in seeds developed and needs increased. The producer, the seedsman, and the user were all concerned with storage and with the added problems involved in moving seeds from production areas to areas of use. Seeds then required special packaging and protection of germinability during transit. Since grain and seeds produced in years of high yield must be saved against possible years of poor harvests, storage conditions must maintain seeds with high germinability from harvest to several plantings over a period of years. Minimal storage conditions cannot meet these longer storage needs.

The dry areas of Egypt and the Near East, being very favorable for seed storage, may be a major reason why they were among the first to develop agararian civilizations. In other areas of the world, high humidity, alone or with high temperatures, makes seed storage much more difficult, and even today severe famines can occur in these areas when little or no holdover seeds are viable.

As civilization encroaches into areas where species of domestic plants originated, there is a tendency for races and related species with different genetic makeup to disappear. These, even though not economically worthwhile by themselves, may contain genes invaluable for future generations. Consequently, a desire to preserve these possibly useful genes developed recently, and gene storage facilities have been created. Two of these (Fort Collins, Colorado, U.S.A., and Hiratsuka, Japan) have been built to preserve many different genes of cultivated and related plants for as long as possible, using the latest research knowledge on seed storage. This is best done by long-term storage of seeds (hundreds of years are hoped for some species) since repeated growing out increases the generations and leads to loss of many of these genes by crop failure, outcrossing, genetic drift, and human error.

Seed storage requirements range in complexity from those of a farmer saving his seeds to plant a few months later, through longer storage of carry-over seeds and breeding material, to the longest possible storage of seeds in plant gene banks. Storage problems increase in complexity as severity of ambient climate and length of storage increase.

B. Primitive Means of Seed Storage

The ancient farmer depended on facilities he had at hand to store his seeds. The containers could be clay pots, woven baskets, or even holes dug in the ground. The need to protect against pests was obvious. Other early observations, however, were that viability was quickly lost in undried seed or seed stored under high-humidity high-temperature conditions. Hence, care was taken to dry the seeds in the sun and then store them dry. A common practice was to store the seeds in a crock or basket hung from the ceiling near the cooking fire. Such an arrangement kept the seeds dry and reasonably free of attacks by pests. In areas of little or no rainfall, the pit method was frequently used to store seeds after they were thoroughly dried.

These means of farm storage are used even today in developing countries. Loss from insects and rodents is frequently high, however, and by planting time the seeds have lost much of their vigor and are low in germination capacity. The need is recognized for small farm storages that are cheap, easy to construct by the farmer, and effective in keeping out pests and maintaining germinability. This problem is receiving much research.

C. Early Attempts to Measure Longevity

One of the first problems of seed storage studied was the question of how long seeds can be stored and still remain viable. Ewart (1908) has an excellent review of this early work. These workers did not consider the role of environmental or physiological factors in determining longevity in a given sample of seeds. Nevertheless, they were able to obtain comparative data among kinds or species under similar storage conditions. They found that seeds of some species cannot be dried to a moisture content in equilibrium with ambient relative humidities and that these seeds will not survive storage for more than a few weeks or months at most. However, they found that certain hard-seeded species, particularly in the Leguminosae, could survive on herbarium sheets, in a variety of climates for many years, even more than a century. They found that almost all crop seeds fell in an intermediate class under storage conditions considered favorable. Crop seeds could be dried and stored for one to several years. Seed longevity records among species and crops are elaborated more fully in Section III, A.

II. Collection of Seeds for Storage

A. *Physiological Maturity*

At the moment of fertilization a zygote is formed. If all goes well, the zygote will become an embryo in a mature seed. From the moment of its creation the zygote possesses all the genetic information necessary to produce a new plant. Much development must take place, however, before a mature seed containing an embryo developed from the zygote is produced. This development has been discussed in Chapters 2 and 3, Volume I of this treatise. It is probably true that under proper nutrient-culture techniques the single-cell zygote could be developed into a normal plant, as has been done by Steward *et al.* (1966) from free cells from various carrot tissues or by Guha and Maheshwari (1966) from a pollen grain. Without such cultural techniques, however, the zygote, the early stages of the developing embryo, and even the early stages of the entire ovule cannot survive if removed from the mother plant. The culture medium must include all the nutrients and hormones that would have been provided by the mother plant for a normal seed to develop.

As the ovule develops into an immature seed, a point is reached with many species at which the seeds can be removed from the plant and, if sown immediately without drying out, will germinate and produce a normal plant. Seeds of many species in this immature state will germinate immediately but develop dormancy before they mature (see, for example, Khan and Laude, 1969). If dried in this immature state, however, these seeds will die and, therefore, cannot be stored for more than a few hours or days at most. Some species have the ability to germinate without embryo-culture techniques after the seed reaches one-fourth to one-half of its mature size and is in the premilk stage. Tomato is a good example of a plant of which the immature seed can germinate and produce a normal plant when removed from the fruit when only one-fourth in size. Such fragile seeds cannot withstand the extreme stresses that mature seed can and still produce a seedling. The conditions of germination must approach the ideal for the species. Harlan and Pope (1922) found that some barley seeds were viable 6 days after pollination. McAlister (1943) harvested seed of several grasses in the premilk stage and obtained fair germination.

Finally the seed reaches maturity. For seeds of most species, though not all, mature seeds can be dried to equilibrium with ambient relative humidities and then stored.

In recent years several workers have done considerable research on the biochemistry of the development of the seed following fertilization.

They have found that there are two waves of growth in seeds: (*1*) structural development—cell division, cell enlargement, increase in concentration of many enzymes and hormones, and production of additional deoxyribonucleic acid (DNA) and ribonucleic acid (RNA) needed to create all these substances and control metabolic processes; and (*2*) a large increase in dry weight near maturity as nutrients flow into the seed and are converted into reserve foods. Some of the reserve foods for storage are starch, hemicelluloses, lipids, proteins, and phytin. The relative amounts vary among species.

McAlister (1943) studied the effect of maturity on the longevity of seeds of eight grasses. Seeds harvested at the premilk stage (13–16 days after bloom) germinated well shortly after harvest, but, in storage at a moisture content of 7 to 9%, they declined in germination more rapidly than seeds harvested in the milk stage. Seeds harvested in the milk stage declined in germination slightly faster than seeds harvested in the dough and mature stages. These latter two stages did not differ in seed longevity. Even more important, because of low vigor of seeds in the premilk stage, almost no plant stand was obtained from these seeds in field plantings even the first year. The plant stand from milk-stage seeds averaged around one-half that from mature seeds. Dough-stage seeds equaled mature seed in stands in some species and were slightly less in others. These data show very well the adverse effect on longevity of seed immaturity.

1. EFFECT OF STRESSES BEFORE PHYSIOLOGICAL MATURITY ON LONGEVITY

From the time of fertilization to physiological maturity, stresses can occur that will influence the longevity of the mature seed. These stresses may lead to maturity of a seed without its being fully developed or while it still lacks some essential. The result is impaired longevity.

Growing the mother plant under conditions in which one mineral or another is deficient may influence the longevity of the mature seed. The seed is a plant sink, however, with mineral nutrients flowing to the seed rather than to other parts of the plant, which can then reduce the plant growth and, therefore, the yield of seeds. The seeds that do mature may, nevertheless, have an adequate supply of the deficient nutrient. Harrington (1960a) studied the effect of severe deficiencies of nitrogen, phosphorus, potassium, and calcium during plant growth. Seeds of pepper, carrot, and lettuce plants severely deficient in N, K, or Ca declined more rapidly in germination capacity during 8 years of storage than did seeds from plants given a balanced nutrition. A severe phosphorus deficiency in the mother plant did not affect the storage life of the resulting seed. Abnormal pea seedlings result from deficiencies in the mother plant of boron (Leggatt, 1948) and manganese (Glasscock and Wain, 1940). It is also entirely possible that storage life would be affected by such deficiencies. Thus, a mother plant grown under certain mineral-deficient conditions will produce seeds with deficiencies that shorten their storage life.

A wealth of published data substantiate that other stresses which a mother plant may undergo during seed development may also lead to reduced yield and abnormal seedlings. Records of seed companies indicate that these other stresses may also reduce the longevity of seeds produced under such stress conditions. These other stresses include those from water, temperature, high soil salt, plant diseases, insect damage, and frost damage.

Internal water stress in plants can result not only from a lack of rainfall but also from hot winds of low humidity, such as sometimes occur in the Central Valley of California. Transpiration may be more rapid than the capacity of roots to absorb water from the soil, even though there may be considerable water in the soil. If internal water deficits result from lack of water in the soil or from excessive transpiration when the seed is in the rapidly developing stage, development of the embryo may be irreversibly damaged. On subsequent maturity and storage, seeds deteriorate more rapidly than other seeds of the same lot that were more mature at the time of the moisture stress or had not yet reached the stage of rapid development. When too many seeds in a lot are thus harmed, the lot declines from acceptable germination to below-standard germination much faster than normally.

In many seed-producing areas, temperature is a frequent serious stress; either too high or too low temperatures during maturation of a seed crop can inhibit seed development. High-temperature periods cause blasting of flowers or are lethal to pollen, thus preventing seed set. Such results are readily seen and easily related to poor seed yields. High temperatures that literally cook the seeds produce nonviable seeds. Temperatures of 60° to 65°C have been recorded in the onion seed umbel when air temperature reached about 40° to 45°C. Short of reduced seed yield and seed not viable at harvest, heat of lower intensity can cause damage that leads to a rapid loss of viability in storage.

Freezing injury as the crop matures often adversely affects seed quality. Corn grown in upper latitudes may be hit by early frosts, damaging the seed and lowering the keeping quality (Rossman, 1949). The extent of the freezing injury is a function of duration of the freeze, minimum temperature, seed moisture content at the time of the freeze, physiological maturity of the seed, and the species and cultivar. Seed damage is generally slight if at the time of the freeze the seed is below 20% in moisture content.

In summary, environmental stresses to the mother plant during seed development can reduce the storage life of the mature seed. Much research has been done on the effect of these stresses on yield and germination immediately after harvest, but investigations of effects on seed longevity are meager. This problem should be studied more thoroughly.

2. DETERMINATION OF PHYSIOLOGICAL MATURITY

When is a seed mature? This is an important point since storage starts only from the moment the seed is mature. The most generally accepted measure of maturity is the time when the seed has reached its maximum dry weight, a point called physiological maturity. This means that nutrients are no longer flowing into the seed from the mother plant. An abscission layer is also probably forming at the hilum (or has formed). The seed is still high in moisture (30% or higher), however, so that continued drying is necessary for minimum damage in harvest or for success in storage. At physiological maturity the seeds of most species can be dried to a low moisture content without loss in viability. Earlier in seed development this is not true. Therefore, a significant transformation has occurred in the components of the seed cell, particularly the proteins, in that they can now be dehydrated and rehydrated without losing their function as enzymes, membranes, genes, and so on. At the time that seed dry weight is maximum, however, maturation is not yet complete. Reduction of moisture content occurs subsequent to physiological maturity. Seeds of many species have dormancies which disappear during storage, as in the phytochrome dormancy in lettuce, which disappears in dry storage, and in the vernalization required by some seeds. Also, in some species such as carrot, many embryos are immature at physiological maturity and continue to develop after the seed is removed from the plant. Thus, maximum dry weight may provide an index of physiological maturity, but not always.

Maximum fresh weight does not indicate physiological maturity, because the maturing seed begins losing water while nutrients are still moving into the seed and important biochemical processes are occurring, such as an increase in ribonuclease activity and a decrease in RNA.

Determining maximum dry weight is not easy, for it requires that a seed field be sampled and dry weights determined until a relatively steady weight among successive samples is reached. Accompanying physiological maturity are changes in the fruit or the seed coat which show visual signs that can be used to determine maturity. For example, when a tomato fruit turns red, its seeds are physiologically mature; when the seed coat of an onion seed grates on a fingernail, the onion seed is physiologically mature; and in many species when the fruit dehisces, as with many legumes, or abscises, as with maple or elm, the seed can be considered physiologically mature.

Physiological maturity, even if not a precise point in the life of a seed or not precisely determinable, is still of extreme importance since it marks the moment when the seed begins to age. At this moment the seed has its highest vigor, thereafter declining to senescence and eventually no longer able to germinate. Research has not yet discovered any way to prevent senescence, but many ways have been found to slow the decline. Thus, proper storage of seed becomes a problem of utilizing some or all of the many ways of prolonging the life of the mature seed consistent with the storage life desired and the costs involved. As we shall see, it is possible to destroy germinability in a few seconds or in some species maintain it for at least a thousand years, depending on the storage environment.

B. Harvesting Injuries

As stated, seed storage begins the moment the seed is physiologically mature. The seed is still on the plant, however, and may be about to shatter to the ground. Since that may not be a good storage place, or at least not convenient for man, the seed must be harvested — a process that may injure it and shorten its storage life.

A first difficulty is that in a field or forest not all of the seeds of a given crop mature at the same time. Thus, the harvester must pick a time when he will obtain maximum yields with a minimum of immature seed. In crops that do not shatter, the early maturing seeds can be left until all the seeds have matured. An example is carrot. If shatter is a problem, the harvester may harvest whole plants or the fruiting parts and place them on a canvas, where the early seeds will shatter and the later seeds will mature. This happens in onion, for example. Of course, it is possible, though usually not economical, to harvest each seed or fruit as it matures, by repeatedly going over the crop by hand. This is sometimes done with expensive flower seed crops, such as pansies and petunias. Usually the harvester combines the whole crop, losing the earliest seeds by shatter and harvesting immature seeds along with mature ones. With most crops, therefore, the seed harvest contains immature seeds as well as seeds that have been mature for some time. As noted, immature seeds have a shorter longevity than mature seeds and, thus, are less desirable for storage. However, seeds that mature some time before harvest have been stored in the field since physiological maturity, which often does not constitute good storage. Such seeds may age considerably even before they are harvested. Storage on the plant, usually called weathering, is subject to all the environmental influences that affect longevity. High humidity, rain, high temperature, fungal infection, insect damage, ultraviolet light from the sun all damage the longevity of these seeds before they are harvested.

The harvesting operation is frequently a further factor in reducing seed longevity in subsequent storage. The amount of damage done in harvesting varies greatly among species. Among the crops least injured by normal harvesting procedures are wheat and barley, but even the seeds of these can be injured if the harvesting machine is run too fast or is adjusted improperly. However, snap beans, particularly some of the most desirable cultivars, are extremely subject to injury in threshing. Bean seeds with splits, cracked cotyledons, snake heads (in which the cotyledons are broken from the embryo axis), bald heads (in which the epicotyl is damaged or broken) have injuries that are immediately apparent at the first germination test. In contrast, bruising, crushing, and internal cracks may not affect germination immediately after harvest but, nevertheless, can hasten aging in storage. Damage to the seed crops can be lessened by reducing cylinder speeds, coating beater bars with rubber, and harvesting only when seed moisture is in the right range for minimum damage. All these practices, of course, slow the harvesting operation, requiring more machines or a longer harvest time and increasing harvest expense. So the tendency is always to push the harvest a little faster, increasing seed injury and, therefore, reducing its longevity in storage.

C. *Preparation of Seeds for Storage*

1. SEED CLEANING

When a seed lot is harvested it contains trash, broken seeds, and light seeds and may have a high moisture content. In this condition storage life will not be maximum. The seed must be cleaned to remove everything except sound seed. In fact, more and more evidence indicates that the seeds of greatest longevity in a given lot are those with the greatest density. Only seeds of high density should be kept for long storage. Separation by density is included in the seed-cleaning operation. Seed cleaning can, however, be damaging to seeds, especially dry seeds. Thus, care must be taken to minimize damage during cleaning, and seed moisture during cleaning needs to be higher than desirable for long-term storage. Proper cleaning and handling procedures for seeds are well covered in the handbook by Harmond *et al.* (1968).

Inert matter, such as dried sticks, leaves, clods of soil, and broken seeds, interferes with the flow of air through seeds in storage and may be a breeding place for insects and fungi in storage. Light, shriveled, and cracked seeds are low in vigor, lose viability rapidly, or may already be dead. It is usually desirable to remove these poor-quality seeds prior to storage. Since this must be done eventually if the seeds are to be salable, it is better to discard these seeds before storage.

2. WAYS OF DRYING SEEDS

One of two moisture situations is encountered at seed harvest. In one, the seed crop is dry at harvest and no further drying is necessary; in the other, the seeds have too high a moisture content for storage. The latter seeds may be in moist fruits (such as tomato) or they may have a high moisture content because the season is so short that harvest must begin before seed moisture has reached equilibrium with ambient relative humidity (as is the case with maize in Iowa) or the harvest may occur during the rainy season (as is the case with spinach in the Netherlands or with rice in some harvest periods in Southeast Asia). For all of these situations, the seed must be dried after extraction from the fruit or after harvest. This must be done as quickly as possible to reduce seed moisture to equilibrium with a relative humidity (RH) below 70%. If such seeds are not dried, respiration heating and storage fungi may cause loss in germination and a short storage life.

Drying can be done in several ways, as exemplified below.

Seeds can be dried by spreading them on the ground, on a paved surface, or on a canvas for sun drying. This method fails during rainy periods or those of high humidity. Capacity for seed germination can be decreased if the days are extremely hot. Temperatures of 45°C are not uncommon in the Imperial Valley of California when onion seeds are being dried; in the direct sun the seed temperature may exceed 60°C. Shading or drying on screen-bottom trays will reduce the drying temperature.

In many farm drying operations in the Midwest of the United States, maize is placed in a bin or storage and ambient air is blown through the ears or seed. As will be explained below, seed moisture is a function of the relative humidity of the ambient air. Such drying is inexpensive. It is effective as long as the relative humidity of the air is lower than the equilibrium moisture content of the seeds. If the air relative humidity exceeds the seed equilibrium moisture, the seeds will gain moisture. If the seeds are not dried fast enough, storage fungi will invade the seeds, thereby reducing longevity.

The most usual drying operation uses heated air. The temperature of the drying air is heated to increase the moisture gradient between the moist seeds and the air, thus drying the seeds faster. Rapid drying can cause injury, however, by checking of the seed coat or cracking of the endosperm or cotyledons as a result of rapid shrinking of outer parts of the seed while the inner parts are still undried. High temperature can also seriously injure the seed or immediately kill it. If the temperature is high but not quite so high as to kill, seed vigor may be reduced and seed longevity decreased. As the seed dries, the maximum safe drying temperature increases. The literature is conflicting as to the maximum safe drying temperature for different species of seed. However, for grains, beets, and many grasses, 45°C is a safe maximum temperature. For most vegetable seed, 35°C is considered safe.

Since there is an upper temperature limit for safe drying of seeds, a problem arises in trying to dry seeds down to moisture contents low enough for packaging in moistureproof containers or for long-term storage where equilibrium with 25 to 20% RH may be desired. If the ambient temperature is 25°C and the RH of the air is 70%, heating the air to 35°C reduces its RH to only 38%, and even heating the air to 45°C reduces the RH only to 24%. Thus, using heated air alone will not safely dry seeds down to these desired moisture contents. By using a combination of air dehumidification and heat in a closed-air circuit system, an RH as low as 10% can be achieved and the seeds can be dried without injury in 4 to 8 hours from an air-dry moisture content to desired low moisture contents.

III.   Seed Physiology and Storage

A.   *Seed Longevity*

1.   REVIEWS CONCERNED WITH SEED LONGEVITY

As already stated, Ewart (1908) published the first comprehensive review of seed longevity along with much new data of his own. Since then the subject has been reviewed with increasing frequency. In chronological order the most significant reviews of seed longevity are Crocker (1909, 1938, 1948), Crocker and Barton (1953), Anderson and Alcock (1954), Owen (1956), Barton (1961), United States Department of Agriculture (1961), and Harrington (1970). In addition, James (1961, 1963) and Barton (1967) have published bibliographies which contain comprehensive references to research on seed storage and seed longevity.

2.   CURVE OF SEED LIFE

As an individual seed ages, it moves imperceptibly from one stage of deterioration to the next. The symptoms of age appear in a set order. The first symptoms of aging of a seed are slower growth, inability to germinate at the extremes of its environmental range, and greater susceptibility to attacks by microorganisms at its environmental extremes. (This loss of vigor is discussed in Chapter 6 of Volume I of this treatise.) Next, abnormalities appear. (These are discussed in Chapter 4, Volume II of this treatise.) Such an abnormal seed has deteriorated appreciably from its original vigorous state but is still very much alive. Deterioration continues. The radicle becomes shorter and shorter and the cotyledons fail to break out of the seed coat. A point is reached where there may be no visible growth outside of the seed coat, and the seed is classified as dead. At this stage it may be possible to find enzymic activity and even live cells in isolated parts of the seed. But, finally, even these symptoms of life disappear. At no time is there a sharp change in behavior as the seed ages.

In practice, our concern is not the longevity of an individual seed but that of a quantity or lot of seeds of one species or cultivar. A lot may consist of many tons of seed in an elevator or be only a few grams of seed of a breeding line. A lot may be a blend of several harvests of a variable cultivar, even those of different years, or it may consist of seeds from a genetically uniform inbred ear of maize. The lot may contain only uninjured fully mature seeds, or seeds of variable maturity, variously injured and containing inert matter and weed seed from improper cleaning. The term "lot" is used extensively in the seed trade to identify a group of seeds to which an identifying number is given. It is normally possible to retrace the history of a particular lot of seed through records of seed companies that have handled seeds of a specific lot.

Thus, the aging of a quantity or lot of seeds is not uniform but is a function of the past history of each individual seed in the lot. Figure 1 illustrates, in a generalized manner, the loss of vigor and germination

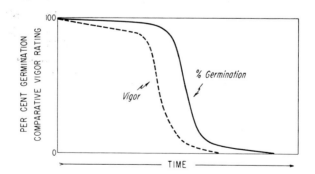

Fig. 1. The decline in vigor and germination of a seed lot with time. The absolute value for time is primarily a function of plant species, temperature of storage, and seed moisture. The steepness of the sharp slope is primarily a function of the genetic purity, of uniformity of maturity and handling of the lot.

with time of a seed lot or the curve of seed life in that lot. If the lot is homogeneous, the drop in germination may be precipitous; if the lot is very heterogeneous, the decline may be very gradual. The time when the rapid drop begins will depend on both the previous history of the seeds and the present storage conditions. A loss in vigor, however it is measured, precedes the loss in germination, though the decline in vigor normally parallels a decrease in germination over most of the curve, because an equilibrium is reached. As more seeds begin to decline in vigor, the first seeds to show a loss in vigor die; thus, the number of less vigorous seeds tends to remain constant over a considerable portion of the curve.

3. INEVITABILITY OF DETERIORATION

This decline in vigor, and finally death, of an individual seed, and the loss in vigor and germination of a seed lot, have not yet been stopped by man. However, we now know much about the environmental factors that influence aging and thus are able to slow the aging process with proper storage conditions. Further, we are beginning to learn something of the biochemistry of aging in seeds. As we learn more, we may be able to apply such knowledge by chemically or physically treating seeds to prolong their life even in environments adverse to seed longevity.

The two most important environmental factors that influence the speed of seed aging are the relative humidity of the air, which controls seed moisture content, and temperature, which affects the rates of biochemical processes in seeds. Harrington (1959) gives two rules of thumb which are of general validity and are useful guides to the effects of seed moisture and ambient temperature on the rates of seed aging.

1. For each 1% increase in seed moisture the life of the seed is halved. This rule applies when seed moisture content is between about 5 and 14%. Below 5% the speed of aging may increase because of autoxidation of seed lipids. Above about 14%, storage fungi destroy the capacity for seed germination. These factors are discussed more fully below.

2. For each 5°C increase in seed temperature the life of the seed is halved.

This rule applies between at least 0° and 50°C. Below freezing temperatures and above 50°C the available data are insufficient to test the rule, although seeds with a moisture content of less than 14% can be stored safely at below freezing, and very dry seeds of some species can withstand temperatures up to 100°C for short periods.

These two rules apply independently. Thus, seeds having 10% moisture and stored at 20°C will survive about as long as those with 8% moisture stored at 30°C. A good example to illustrate these two rules is the aging of seeds of onion, which deteriorate rapidly. Rocha (1959) found that the germination of onion seeds was reduced more than half in 10 weeks at 13% seed moisture and 25°C. However, Barton (1966a) maintained onion seeds with almost no loss in germination for 18 years (936 weeks) at below 6% moisture content and 5°C. Using the data of Rocha and the two rules of thumb, under the storage condition of Barton's experiment the germination of onion seeds should have declined 50% in 20,480 weeks, well beyond the 936 weeks of storage (13% H_2O less 6% H_2O is 7% H_2O or 2^7; and 25° less 5°C is 20°C or 2^4; and $2^7 + 2^4$ is 2^{11}, or 2048 times the 10 weeks or 20,480 weeks.) Other investigators have killed onion seeds at 50°C in a few hours or maintained the germination capacity of onion seeds for 22 years, but the moisture contents of these seeds were not reported. Much data available on many other species show the adverse effects on seed longevity of high seed moisture and high temperature.

Therefore, the major requirements for storage of most seeds are to create storage conditions of low relative humidity and low temperature. Seed longevity is also affected by the composition of the storage atmosphere. High oxygen tends to hasten loss in viability, particularly for seeds of high moisture content. High CO_2, N_2 or a vacuum may retard deterioration under some circumstances, but these effects are minor compared with the temperature and moisture effects. Ultraviolet light reaching seeds will hasten deterioration. Radiation can damage seeds in storage.

4. Variation among Species in Seed Longevity

Among species, the range in longevity is greater than one might suspect, even under identical storage conditions. On the one hand, seeds of chayote cannot survive much longer than a month even under ideal storage conditions, and, on the other hand, Ødum (1965) presents archaeological evidence that seeds of lamb's-quarters can survive in the cool moist soil of Denmark for 1700 years and subsequently germinate. Seeds of these two species are exceptional. There may be many species in these two classes, including several of economic importance. Unfortunately records of seed longevity exist for only a few of the many species of plants. Most, if not all, of these records were obtained from seeds stored under less than ideal conditions. Even for these records, most tests do not indicate the maximum seed longevity because they were concluded before that point was reached or are still in progress. However, it has been deemed advisable to prepare tables that show the present knowledge of seed longevity. Table I lists species with seeds found to be short lived, and Table II lists species with records of seed longevity of 10 years of more.

The species with short-lived seeds listed in Table I include several aquatic species, nut tree species, such as oaks, walnuts, and chestnuts, and some tropical species. Since the aquatic species listed are all native of northeastern United States (a very small area), it is reasonable to assume that there exist many other aquatic species of which the seeds cannot be dried and, therefore, are short-lived. A study of seed longevity of aquatic plants in other areas of the world should be made in order to obtain a more complete picture of aquatic species which are short-lived as well as the aquatic species of which the seeds can withstand drying and, therefore, are probably long-lived.

The information on longevity of seeds of tropical species is equally meager. Almost all the tropical species listed in Table I are economic species, probably because only their seeds have been of concern. Many other species with short-lived seeds exist in the tropics but no one has yet studied them.

Even among species with short-lived seeds there is a wide variation in seed longevity. Chayote and cacao seeds not only cannot be dried but also low temperatures are lethal to these seeds. Thus the two most important factors for improving longevity of most seeds—low temperature and low relative humidity—cannot be used to prolong the life of these seeds; therefore the life of these seeds is measured in days, not years. However, several *Citrus* species have seeds which also cannot be dried and yet, as can be seen from Table I, may have high germination after 4 years when stored under refrigeration in high humidity and using a surface fungicide to control fungi.

A group of trees including elms, poplars, and birches has very short-lived seeds in their normal habitat, but it appears that the seeds of at least some species of these trees can be carefully dried and then refrigerated, maintaining the ability to germinate for several years. Possibly these species should not be classified in the group with short-lived seeds.

Table II lists species of which the seeds have been found to have longevities of 10 years or over and are, therefore, classified as species with long-lived seeds. There are two types of long-lived seeds, those that survive best under low moisture and cool conditions and those that survive in the soil under dormancy-inducing conditions. Table II has columns for each type of environment. It can be seen that the seeds of several species are long-lived under both environments. (In Table II, the storage temperature and seed moisture are listed if the author reported this information.)

Many of the species with long-lived seeds have hard seeds. These seeds dry down to as low as 4% seed moisture and do not regain moisture until the seed coat is pierced by abrasion or microbial action. These seeds thus have a natural low-moisture content advantageous for long

TABLE I

PLANT SPECIES WITH SHORT-LIVED SEEDS[a]

Plant species	Longevity (and germination)	Environment	Ref.
Taxaceae			
Taxus (±8 spp.)—yews	>5 yr	2°C, moist	Forest Service, 1948
T. baccata L.—English yew	—	Cannot be dried	Forest Service, 1948
T. brevifolia Nutt.—Pacific yew	—	Cannot be dried	Forest Service, 1948
T. canadensis Marsh.—Canada yew	—	Cannot be dried	Forest Service, 1948
Najadaceae—naiads and pondweeds			
Naja marina L.	7 mo(15%)	3°C, in water	Muenscher, 1936
N. flexilis (Willd.) R. & S.	7 mo(87%)	3°C, in water	Muenscher, 1936
N. gracillima (A.Br.) Magnus	2 mo(13%)	3°C, in water	Muenscher, 1936
N. minor Allione	7 mo(2%)	3°C, in water	Muenscher, 1936
Potamogeton amplifolius Tuckerm.	7 mo(30%)	3°C, in water	Muenscher, 1936
P. foliosus Raf.	7 mo(64%)	3°C, in water	Muenscher, 1936
P. praelongus Wulf.	7 mo(11%)	3°C, in water	Muenscher, 1936
Alismaceae			
Sagittaria latifolia Willd.—arrowhead	7 mo(42%)	3°C, in water	Muenscher, 1936
Butomaceae			
Butomus umbellatus L.	7 mo(64%)	3°C, in water	Muenscher, 1936
Hydrocharitaceae			
Vallisneria sp.	Viviparous	—	Hemsley, 1895
V. americana Michx.—eelgrass	5 mo(87%)	3°C, in water	Muenscher, 1936
Gramineae			
Saccharum officinarum L.	—	0°C, air-dry, CO_2 atm.	Verret, 1928
Zizania aquatica L.—wild rice	14 mo(88%)	1°C, in water	Duvel, 1906
Cyperaceae			
Eleocharis calva Torr.—spike rush	7 mo(6%)	3°C, in water	Muenscher, 1936
Scirpus validus Vahl.—great bulrush	5 mo(6%)	3°C, in water	Muenscher, 1936

Species	Duration	Storage condition	Reference
Palmaceae			
Acrocomia sclerocarpa Mart. – grugru palm	—	Cannot be dried	Guppy, 1912
Areca cathecu L. – areca nut	—	Cannot be dried	—
Attalea excelsa Mart. – Pallia palm	—	Cannot be dried	Guppy, 1912
Cocos nucifera L. – coconut	16 mo	Ambient	Child, 1964
Mauritia spp.	2 wk	Cannot be dried	Guppy, 1912
Oredoxa spp.	—	Cannot be dried	Guppy, 1912
Sabal spp.	—	Cannot be dried	Guppy, 1912
Thrinax spp.	—	Cannot be dried	Guppy, 1912
Araceae			
Acorus calamus L. – sweet flag	7 mo(54%)	3°C, in water	Muenscher, 1936
Peltandra virginica (L.) Kunth. – arum	7 mo(96%)	3°C, in water	Muenscher, 1936
Pontederieceae			
Orontium aquaticum L. – golden club	5 mo(92%)	3°C, in water	Muenscher, 1936
Amaryllidaceae			
Crinum capense Auth.	Germinates while drying		Guppy, 1912
Salicaceae			
Populus tremuloides Michx. – quaking aspen	2 yr(70%)	−5°C, 10% RH	Moss, 1938
Salix discolor Muhl. – pussy willow	6 wk max.	Room temp., >50% RH	Forest Service, 1948
S. nigra Marsh. – black willow	6 wk max.	Room temp., >50% RH	Forest Service, 1948
Juglandaceae			
Carya (±20 spp.)	5 yr	5°C, 90% RH	Forest Service, 1948
C. glabra (Mill.) Sweet – pignut hickory	—	Cannot be dried	Forest Service, 1948
C. illinoensis K. Koch. – pecan	—	Cannot be dried	Forest Service, 1948
C. ovata (Mill.) K. Koch. – shagbark hickory	—	Cannot be dried	Forest Service, 1948
Juglans (±15 spp.)	5 yr	5°C, >85% RH	Forest Service, 1948
J. californica S. Wats. – Calif. black walnut	—	Cannot be dried	Forest Service, 1948
J. cinerea L. – butternut	—	Cannot be dried	Forest Service, 1948
J. nigra L. – black walnut	—	Cannot be dried	Forest Service, 1948
J. regia L. – English walnut	1 yr	0°C, 85% RH	Woodroof, 1967

TABLE I (Continued)

Plant species	Longevity (and germination)	Environment	Ref.
Corylaceae			
Betula alleghaniensis Britt.–yellow birch	4 yr (good %)	–	Clausen, 1965
B. lenta L.–cherry birch	1½ yr (77% n.s.l.[b])	Room temp., 7.4% H₂O[c]	Joseph, 1929
B. lutea Michx.–silver birch	1½ yr (84% n.s.l.[b])	5°C, 5.9% H₂O	Joseph, 1929
B. papyrifera Marsh.–paper birch	4 yr (good %)	5°C, 7.6% H₂O	Clausen, 1965
B. populifolia Marsh.–gray birch	1½ yr (71% n.s.l.[b])	5°C, 7.6% H₂O	Joseph, 1929
Carpinus caroliniana Walt.–American hornbeam	–	Cannot be dried	Jones, 1920
Corylus (±15 spp)	2 yr	5°C, >85% RH	Koopman, 1963
C. americana Marsh.–hazelnut	–	Cannot be dried	Forest Service, 1948
C. avellana L.–European filbert	–	Cannot be dried	Forest Service, 1948
Fagaceae			
Castanea spp.	9 mo max.	Cannot be dried	Jones, 1920
C. dentata (Marsh) Borkh.–American chestnut	–	Cannot be dried	Forest Service, 1948
Castanopsis chrysophylla (Dougl.) A.DC.–golden chinquapin	<5 yr	5°C, moist	Forest Service, 1948
C. sempervirens (Kell.) Dudley–Sierra chinquapin	<5 yr	5°C, moist	Forest Service, 1948
Fagus (±10 sp.)	<2 yr	Cannot be dried	Forest Service, 1948
F. grandifolia Ehrh.–American beech	–	Cannot be dried	Forest Service, 1948
F. sylvatica L.–European beech	–	Cannot be dried	Forest Service, 1948
Lithocarpus densiflorus (Hook. & Arn.) Rehd.–tanoak	–	Cannot be dried	Forest Service, 1948
Quercus (±300 spp.)	3 yr max.	2°C, moist, ventilated	Johannsen, 1921
Q. suber L.–cork oak	8 mo(86%)	0°C, 35% H₂O (<23% H₂O lethal)	Mirov, 1943
Ulmaceae			
Ulmus americana L.–American elm	16 mo (n.s.l.[b])	5°C, 6.5% H₂O	Barton, 1939

Species	Longevity	Storage conditions	Reference
Moraceae			
Artocarpus heterophyllus Lam. –jackfruit	—	Cannot be dried	Campbell, 1970
A. incisa L. –breadfruit	—	Cannot be dried	Guppy, 1912
Proteaceae			
Macadamia integrefolia Maiden & Betche – smooth-shelled macadamia nut	6 mo(<50%)	Ambient	Storey, 1969
M. ternifolia F. Muell. –macadamia nut	11 mo(8%)	Ambient	Hamilton, 1957
M. tetraphylla L. Johnson –rough-shelled- macadamia nut	6 mo(<50%)	Ambient	Storey, 1969
Nymphaeaceae			
Nymphaea tuberosa Paine –white water-lily	7 mo(56%)	3°C, in water	Muenscher, 1936
Nymphozanthus variegatus Fern. –yellow pond-lily	7 mo(43%)	3°C, in water	Muenscher, 1936
Ceratophyllaceae			
Ceratophyllum demersum L. –hornwort	7 mo(73%)	3°C, in water	Muenscher, 1936
Eriocaulon septangulare With. –pipewort	7 mo(94%)	3°C, in water	Muenscher, 1936
Ranunculaceae			
Caltha palustris L. –marsh marigold	—	Cannot be dried	Jones, 1920
Magnoliaceae			
Liriodendron tulipifera L. –tulip tree	4 yr	Cannot be dried	Paton, 1945
Lauraceae			
Cinnamomum zeylenicum Nees. –cinnamon	1 mo max.	Ambient	Kannan and Balakrishnan, 1967
Persea americana Mill. –avocado	15 mo(+%)	5°C, >90% RH	Halma and Frolich, 1949
Sassafras albidum (Nutt.) Nees –sassafras		3–5°C, seal	Forest Service, 1948
Rosaceae			
Eriobotrya japonica Lindl. –loquat	6 mo(92% n.s.l.[b])	5°C, high RH, semisealed	Zink and Ojirma, 1965
Oxalidaceae			
Averrhoa carambola L. –carambola	—	Cannot be dried	Campbell, 1970
Oxalis (±500 sp.)	—	Cannot be dried	Jones, 1920
Erythroxylaceae			
Erythroxylum coca Lamk. –cocaine	30 days (2%)	3°C, high RH	McClelland, 1944

TABLE I (*Continued*)

Plant species	Longevity (and germination)	Environment	Ref.
Rutaceae			
Citrus aurantifolia Swingle—lime	6 mo(n.s.l.[b])	2°C, 88% RH, fungicide	Childs and Hrnciar. 1949
C. aurantium L.—sour orange	11 mo(22%)	5°C, 3.8% H_2O	Barton, 1943
C. grandis Osbeck—pummelo	8 mo(n.s.l.[b])	2°C, 88% RH, fungicide	Childs and Hrnciar, 1949
C. karna Raf.	6 mo(> 80%)	8–15°C, 61% H_2O, Fungicide	Chacko and Singh, 1958
C. limon Burm. f.—lemon	8 mo(n.s.l.[b])	2°C, 88% RH, fungicide	Childs and Hrnciar. 1949
C. limonia Osbeck—rough lemon	16 mo(46%)	5°C, 3.8% H_2O	Barton, 1943
C. paradisi Macf.—grapefruit	12 mo(n.s.l.[b])	5°C, 37% H_2O	Barton, 1943
C. reticulata Blanco—tangerine, mandarin	4 yr(88%)	7°C, > 58% H_2O, fungicide	Bitters, 1970
C. sinensis Osbeck—sweet orange	4 yr(71%)	7°C, > 58% H_2O, fungicide	Bitters, 1970
Fortunella margarita Swingle—kumquat	8 mo(n.s.l.[b])	2°C, 88% RH, fungicide	Childs and Hrnciar, 1949
Poncirus trifoliata Raf.—trifoliate orange	4 yr(91%)	7°C, > 58% H_2O, fungicide	Bitters, 1970
Meliaceae			
Swietenia sp.—large-leaf mahogany	—	Cannot be dried	Lopez, 1938
Malpighiaceae			
Malpighia glabra L.—Barbados cherry	—	Cannot be dried	Campbell, 1970
Euphorbiaceae			
Aleurites fordii Hemsl.—tung	4.5 yr(35%)	1°–3°C, high RH, in sand	Large *et al.*, 1947
Hevea brasiliensis Muell.—rubber	3 mo(51%)	5°–10°C, in water, sealed	Cardosa *et al.*,1966
Anacardiaceae			
Mangifera indica L.—mango	80 days max.	Room temp., 50% RH (3°C lethal)	Bajpai and Trivedi, 1961

Species	Storage	Conditions	Reference
Aceraceae			
Acer saccharinum L.—silver maple	6 mo(100%)	10°C over water (<30% H_2O lethal)	Jones, 1920
Hippocastanaceae			
Aesculus californica (Spach) Nutt.—California buckeye	—	Cannot be dried	Forest Service, 1948
A. glabra Willd.—Ohio buckeye	—	Cannot be dried	Forest Service, 1948
A. hippocastanum L.—horsechestnut	15 mo(+%)	−1°C, sealed, high RH	Widmoyer and Moore, 1968
Sapindaceae			
Blighia sapida Koenig.—akee	3 wk(8%)	Cannot be dried	Campbell, 1970
Litchi chinensis Sonn.—litchifruit		3°C, high RH	McClelland, 1944
Sterculiaceae			
Cola nitida (Vent.) Schott & Endl.—kola	—	Cannot be dried	Clay, 1964
Theobroma cacao L.—cocoa	4 mo(52%)	25–30°C, 31–33% H_2O (10°C or <24% H_2O lethal)	Barton, 1965
Theaceae			
Camellia sinensis (L.) O. Ktze.—tea	10 mo(50%)	0°C, 100% RH	Visser and Tillerkeratne, 1958
Dipterocarpaceae			
Dryobalanops aromatica Gaertn.	24 days max.	12°C, moist (5°C or drying lethal)	Jensen, 1971
Guttiferae			
Garcinia mangostana L.—mangosteen	8 wk max.	21°–29°C, moist, ventilated	Winters and Rodrigues-Colon, 1953
Flacourtiaceae			
Dovyalis hebecarpa Warb.—kitembilla	—	Cannot be dried	Campbell, 1970
Flacourtia indica Merr.—govenor's plum	—	Cannot be dried	Campbell, 1970
Nyssaceae			
Nyssa aquatica L.—water tupelo	—	Cannot be dried	Forest Service, 1948
N. sylvatica Marsh.—black tupelo	—	Cannot be dried	Forest Service, 1948
Myrtaceae			
Eugenia uniflora L.—Surinam cherry	—	Cannot be dried	Campbell, 1970
Myrciaria cauliflora Berg.—jaboticaba	—	Cannot be dried	Campbell, 1970

TABLE I (*Continued*)

Plant species	Longevity (and germination)	Environment	Ref.
Rhizophoraceae			
Bruguiera sp.	Viviparous	—	Guppy, 1912
Rhizophora mangle L. — mangrove	Viviparous	—	La Rue and Muzik, 1951
R. mucronata	Viviparous	—	Hemsley, 1895
Trapaceae			
Trapa natans L. — water chestnut	7 mo(92%)	3 °C, in water	Muenscher, 1936
Sapotaceae			
Achras zapota L. — sapodilla	—	Cannot be dried	Campbell, 1970
Calocarpum sapota Merr. — sapote	—	Cannot be dried	Campbell, 1970
Chrysophyllum cainito L. — star apple	—	Cannot be dried	Campbell, 1970
Ebenaceae			
Diospyros virginiana L. — persimmon	—	Cannot be dried	Kester, 1970
Rubiaceae			
Coffea spp. — coffee	22 mo max.	25°C, 52% H_2O	de Fluiter, 1939
C. arabica L. — Arabica coffee	47 wk(22%)	Warm, humid (0°C or <10% H_2O lethal)	Haarer, 1962
C. canephora Froehner — Robusta coffee	16 mo(41%)	4°–7°C, 19% H_2O (0°C or <10% H_2O lethal)	Huxley, 1964
Cucurbitaceae			
Sechium edule — chayote	—	Cannot be dried	MacGillivray, 1953
Lobeliaceae			
Lobelia dortmanna L. — water lobelia	7 mo(87%)	3 °C, in water	Muenscher, 1936

[a] The seeds are killed by drying, by cold temperatures, or by some not yet understood mechanism.
[b] n.s.l. = no significant loss in germination.
[c] % H_2O = percent water in seed, fresh weight basis.

life. They survive even when no special care is taken to provide the ideal environment. However, in future experiments using the knowledge of the need for low seed moisture and low-temperature storage, seeds of many other species will be found to survive for equally long periods. This is partially illustrated by the data of James *et al.* (1964) for vegetable seeds and Madsen (1962) for agronomic seeds. Their data show that the germination of dry, cool stored seeds of cultivated crops can still be high after 20 to 30 years of storage, although seeds of most cultivated crops are considered to be fairly short-lived. Future tests of these seeds will no doubt show still greater longevity.

The data for longevity of seeds in soil is exact only for the buried seed experiments where the date of placing in the soil is known. The Duvel experiment (Toole and Brown, 1946), which had to be terminated after 39 years, and the Beal experiment (Darlington and Steinbauer, 1961), now over 80 years old, are the two most famous of these tests. However, there is strong evidence that many weed seeds survive in soil for long periods. Therefore, data from experiments where soil is sampled for seedling emergence from pastures, meadows, or forests when the date of last cultivation is known are included in Table II. It is possible that seeds reached soil depths after the last cultivation through the action of earthworms, cattle tromping in the seeds, or seeds falling down cracks in the soil, so some reservations must be made about the exact age of such seeds. However, the fact that certain weeds are repeatedly found in such soils gives validity to the assumption that some of these seeds have survived for many years in the soil. Even more reservations must be made about the claims for seed longevity based on archeological evidence or carbon dating of associated material as Godwin (1968) has pointed out. In Table II these longevities are listed as reported but with a question mark next to the date.

Two types of seed which must remain viable in nature for long periods are seeds of desert plants that may wait years for rain and seeds of parasitic plants that must survive until the proper host plant grows near them. There is little data on longevity of desert seeds although Went and Munz (1949) have started a storage study on the seeds of several desert species. The data on the longevity of seeds of parasitic plants are even more meager. That seeds of many species remain viable for many years in soil without germinating is surprising. There the seeds are subject to high moisture (and many become fully imbibed), high temperature, and fluctuating temperature. These are all environmental factors known to hasten aging of stored seeds. However, in the soil, seeds either maintain the dormant condition present at harvest or have an induced secondary dormancy. Hard seeds are only one dormancy involved in long-term survival of

TABLE II

SPECIES WITH SEEDS SHOWN TO HAVE LONGEVITIES OF 10 YEARS OR MORE (EITHER UNDER DRY CONDITIONS OR IN SOIL)

Plant species	Dry storage			Soil storage		References
	Environment[a]	Age (yr.)	Germ.[b]	Environment[c]	Age[d] (yr.)	
Pinaceae						
Abies concolor (Gord) Hoopes – white fir	5°C, < 10% H_2O	21	6	–	–	Schubert, 1952
A. grandis (Dougl.) Lindl. – lowland fir	<8°C, 9% H_2O	16	+	–	–	Barton, 1953
A. magnifica A. Murr. – red fir	5°C, < 10% H_2O	16	8	–	–	Schubert, 1952
A. nobilis (Dougl.) Lindl. – noble fir	<8°C, 9% H_2O	16	50	–	–	Barton, 1953
Larix dahuria Turez. – Dahurian larch	5°C, < 10% H_2O	>11	2	–	–	Schubert, 1952
L. occidentalis – Nutt. Western Larch	5°C, < 10% H_2O	16	5	–	–	Schubert, 1952
L. sibirica Ledef. – Siberian larch	4°C, 8% H_2O	25	Good	–	–	Heit, 1967b
Picea abies (L.) Karst. – Norway spruce	−4°C, sealed	15	40	–	–	Barton, 1961
P. glauca (Moench) Voss – white spruce	−4°C, sealed	15	40	–	–	Barton, 1961
P. rubens Sarg. – red spruce	<10°C, sealed	10	+	–	–	Holmes and Buszewicz, 1958
P. sitchensis Carr – Sitka spruce	5°C, sealed	>11	6	–	–	Schubert, 1952
Pinus albicaulis Engelm. – white-bark pine	5°C, sealed	12	1	–	–	Mirov, 1946
P. aristata Engelm. – bristlecone pine	5°C, sealed	10	64	–	–	Mirov, 1946
P. attenuata Lemm. – knobcone pine	Cone on tree	±40	+	–	–	Coker, 1909

Species	Storage conditions					Reference
P. balfouriana Jeffrey – foxtail pine	5°C, sealed	20	+	—	—	Holmes and Buszewicz, 1958
P. banksiana Lamb. – Jack pine	Cone on tree	±20	+	—	—	Coker, 1909
P. canariensis C. Smith – Canary pine	5°C, sealed	12	13	—	—	Mirov, 1946
P. chihuahuana Engelm. – Chihuahua pine	Cones on tree	±20	+	—	—	Coker, 1909
P. clausa (Chapm.) Vasey – sand pine	Cones on tree	±20	+	—	—	Coker, 1909
P. contorta Dougl. – lodgepole pine	Cone in tree	±150	+	—	—	Mills, 1915
P. echinata Mill. – shortleaf pine	2°C, 10% H_2O	17	+	—	—	Holmes and Buszewicz, 1958
P. elliotti Engelm. – slash pine	3°C, 10% H_2O	15	95	—	—	Uebersezig, 1947
P. jeffreyi A. Murr. – Jeffrey pine	5°C, < 10% H_2O	18	62	—	—	Schubert, 1952
P. lambertiana Dougl. – sugar pine	5°C, < 10% H_2O	15	50	—	—	Schubert, 1952
P. monticola Dougl. – western white pine	5°C, < 10% H_2O	15	40	—	—	Schubert, 1952
P. muricata D. Don – bishop pine	Cone on tree	±25	+	—	—	Coker, 1909
P. nigra Arnold – Austrian pine	4°C, 7% H_2O	10	99	—	—	Heit, 1967b
P. patula Schl. & Cham. – Jelecote pine	5°C, sealed	21	+	—	—	Holmes and Buszewicz, 1958
P. pinaster Ait. – cluster pine	5°C, sealed	12	35	—	—	Mirov, 1946
P. pinea L. – Italian stone pine	In cone	±40	+	—	—	Fancourt, 1856
P. ponderosa Laws. – ponderosa	5°C, < 10% H_2O	17	94	—	—	Schubert, 1952
P. pungens Lamb. – Table-Mountain pine	Cones on tree	±20	+	—	—	Coker, 1909
P. radiata D. Don – Monterey pine	5°C, < 10% H_2O	21	86	—	—	Schubert, 1952
P. resinosa Ait. – red pine	4°C, 8% H_2O	30	80	—	—	Heit, 1967b
P. serotina Michx. – pond pine	Cones on tree	14	51	—	—	Coker, 1909
P. sylvestris L. – Scots pine	5°C, dry	15	85	—	—	Heit, 1967b

TABLE II (*Continued*)

Plant species	Dry storage			Soil storage		References
	Environment[a]	Age (yr.)	Germ[b]	Environment[c]	Age[d] (yr.)	
P. strobus L. – eastern white pine	3°C, 8% H_2O	10	93	–	–	Roe, 1948
P. taeda L. – loblolly pine	–4°C, sealed	15	45	–	–	Barton, 1961
P. virginiana Mill. – scrub pine	–	–	–	Buried	10	Toole and Brown, 1946
Pseudotsuga macrocarpa (Vasey) Mayr. – big cone spruce	5°C, < 10% H_2O	19	2	–	–	Schubert, 1952
P. menziesii (Mirb.) – Franco – Douglas fir	5°C, < 10% H_2O	16	31	–	–	Schubert, 1952
Tsuga heterophylla (Raf) Sarg. – western hemlock	5°C, sealed	15	13	–	–	Schubert, 1952
Taxodiaceae						
Sequoia gigantea (Lindl.) Decme. – giant sequoia	Dry lab	18	16	–	–	Tourney, 1930
S. sempervirens (D. Don) Endl. – redwood	Dry lab	11	7	–	–	Holmes and Buszewicz, 1958
Cupressaceae						
Chamaecyparis lawsoniana (A. Murr.) Parl. – Port Orford cedar	5°C, < 10% H_2O	16	13	–	–	Schubert, 1952
Cupressus arizonica Greene – smooth Arizona cypress	5°C, < 10%H_2O	> 11	35	–	–	Schubert, 1952
C. forbesii Jepson – Tecate cypress	5°C, < 10%H_2O	12	12	–	–	Schubert, 1952
C. goveniana Gard. – Gowen cypress	5°C, < 10%H_2O	> 11	53	–	–	Schubert, 1952
C. lusitanica Mill. – Portuguese cypress	5°C, < 10%H_2O	21	10	–	–	Schubert, 1952
C. macnabiana A. Murr.	5°C, < 10%H_2O	> 11	18	–	–	Schubert, 1952
C. macrocarpa Hartw. – Monterey cypress	Storage	11	14	–	–	Holmes and Buszewicz, 1958

Species	Storage conditions			Habitat		Reference
C. nevadensis Abrams	5°C, < 10% H_2O	> 11	18	—	—	Schubert, 1952
C. pygmaea Sarg.—Mendocino cypress	5°C, < 10% H_2O	> 11	66	—	—	Schubert, 1952
C. sargentii Jepson	5°C, < 10% H_2O	< 11	15	—	—	Schubert, 1952
Juniperus spp.	5°C, sealed	10	+	—	—	Heit, 1967b
Gramineae						
Agrostis spp.	Dry lab	11	> 50	Pasture	±68	Mackay and Tonkin, 1967; Chippendale and Milton, 1934
A. alba L.—red top	Dry lab	11	19	Meadow	±58	Filter, 1932; Brenchley, 1918
A. hyemalis (Walt.) BSP.—hairgrass	—	—	—	Forest	±42	Livingston and Allesio, 1968
A. perennans (Walt.) Tuckerm.—thin grass	—	—	—	Forest	±42	Livingston and Allesio, 1968
A. vulgaris With.	—	—	—	Forest	±100	Peter, 1893
Agropyron repens (L.) Beauv.—quackgrass	—	—	—	Buried	10	Toole and Brown, 1946
Aira caespitosa L.	—	—	—	Forest	±100?	Peter, 1893
Andropogon spp.	—	—	—	Forest	±112	Oosting and Humphreys, 1940
	—	—	—	Pasture	±50	Carruthers, 1911; Chippendale and Milton, 1934
Anthoxanthum odoratum L.—sweet vernal grass	Dry lab	10	1	—	—	Filter, 1932
Avena elatior L.—tall oat grass	Dry lab	11	8	—	—	Brenchley, 1918
A. pubescens Huds.—downy oat	—	—	—	Meadow	±42?	Haferkamp et al., 1953
A. sativa L.—oats	Storage	32	84	—	—	Went, 1957
Bouteloua aristidoides Griseb.—needle grama grass	—	—	—	Desert	±20	Went, 1957
B. barbata Lag.—six weeks grama grass	—	—	—	Desert	±20	Peter, 1893
Bromus hordeaceus L.—soft chess	—	—	—	Forest	±18	Carruthers, 1911
Cynosurus cristatus L.—dogstail grass	Dry lab	12	2	—	—	—

TABLE II *(Continued)*

Plant species	Dry storage			Soil storage		References
	Environment[a]	Age (yr.)	Germ.[b]	Environment[c]	Age[d] (yr.)	
Dactylis glomerata L.—orchard grass	Dry lab	10	3	Pasture	±60	Carruthers, 1911 / Chippendale and Milton, 1934
Danthonia spicata (L.) Beauv.—wild oat grass	—	—	—	Forest	±75	Livingston and Allesio, 1968
Digitaria ischaemum Schreb.—small crabgrass	—	—	—	Pasture	±50	Prince and Hodgdon, 1946
D. sanguinalis (L.) Scop.—large crabgrass	—	—	—	Pasture	±50	Prince and Hodgdon, 1946
Festuca arundinacea Schreb.—tall fescue	Dry lab	10	1	—	—	Carruthers, 1911
F. elatior L.—meadow fescue	Dry lab	11	1	—	—	Filter, 1932
F. octoflora Walt.	—	—	—	Forest	±112?	Oostung and Humphreys, 1940
F. rubra L.—red fescue	—	—	—	Pasture	±50	Chippendale and Milton, 1934
Glyceria fluitans (L.) R.Br—Manna grass	—	—	—	Pasture	±22	Chippendale and Milton, 1934
Holcus lanatus L.—velvet grass	—	—	—	Pasture	±68	Chippendale and Milton, 1934
Hordeum vulgare L.—barley	Storage	32	96	—	—	Haferkamp *et al.*, 1953
Lolium multiflorum Lam.—Italian ryegrass	Dry lab	12	8	—	—	Carruthers, 1911
L. perenne L.—perennial ryegrass	Dry lab	11	6	Meadow	±58	Carruthers, 1911 / Brenchley, 1918
Melica nutans L.	—	—	—	Forest	±20?	Peter, 1893
Oryza sativa L.—rice	Lab, sealed	10	62	—	—	Kondo and Kasahara, 1940
Oryzopsis asperifolia Michx.—mountain rice	—	—	—	Forest	±80?	Livingston and Allesio, 1968
Panicum antidotale Retz.-Blue panic grass	Dry lab	12	3	—	—	Myers, 1940

Species	Storage					Reference
P. capillare L. – old witchgrass	—	—	—	Forest	±46	Livingston and Allesio, 1968
P. clandestimum L.	—	—	—	Forest	±80?	Livingston and Allesio, 1968
P. decompositum R. Br.	Dry lab	10	3	—	—	Myers, 1940
P. lanuginosum Ell.	—	—	—	Forest	±46	Livingston and Allesio, 1968
Paspalidium jubiflorum D. K. Hughes	Dry lab	10	1	—	—	Myers, 1940
Phalaris arundinacea L. – reed canary grass	Dry lab	16	+	Buried	30	Toole and Brown, 1946
Phleum pratense L. -timothy	Dry lab	16	+	Buried	21	Sifton, 1920 / Madsen, 1962
Poa annua L. Annual bluegrass		—	—	Pasture	±68	Chippendale and Milton, 1934
P. compressa L. – Canada bluegrass		—	—	Forest	±20	Peter, 1893
P. nemoralis L. – wood		—	—	Forest	±36	Peter, 1893
P. pratensis L. – Kentucky bluegrass	Dry lab	11	23	Buried	39	Filter, 1932 / Toole and Brown, 1946
P. trivialis L. rough bluegrass		—	—	Pasture	±68	Chippendale and Milton, 1934
Secale cereale L. – rye	Storage	32	1	—	—	Haferkamp et al., 1953
Setaria glauca (L.) Beauv. – yellow foxtail		—	—	Buried	30	Darlington and Steinbauer, 1961
S. verticillata (L.) Beauv.		—	—	Buried	39	Toole and Brown, 1946
S. viridis (L.) Beauv. – green foxtail		—	—	Buried	39	Toole and Brown, 1946
Sorghum vulgare Pers. – sorghum	Lab, <11% H_2O	17	98	—	—	Robertson et al., 1943
Sporobolus airoides Torr. – hairgrass dropseed		—	—	Buried	21	Toole and Brown, 1946
S. cryptandrus (Torr.) A. Gray sand dropseed		—	—	Buried	39	Toole and Brown, 1946
Triticum aestivum L. – wheat	Storage	32	85		—	Haferkamp et al., 1953
T. dicoccon Schrank. – emmer	Storage	31	87		—	Haferkamp et al., 1953
T. durum Desf. – durum wheat	Storage	31	87		—	Haferkamp et al., 1953

TABLE II (*Continued*)

Plant species	Dry storage			Soil storage		References
	Environment[a]	Age (yr.)	Germ.[b]	Environment[c]	Age[d] (yr.)	
T. repens L.—couch grass	—	—	—	Meadow	±42?	Brenchley, 1918
T. spelta L.—spelt	Storage	31	87	—	—	Haferkamp et al., 1953
Zea mays L.—maize	Storage	37	70	—	—	Haferkamp et al., 1953
Cyperaceae						
Carex spp.	—	—	—	Pasture	±68	Chippendale and Milton, 1934
C. glauca Scop.	—	—	—	Forest	±45	Peter, 1893
C. muricata L.	—	—	—	Forest	±100	Peter, 1893
C. pallescens L.	—	—	—	Forest	±100?	Peter, 1893
C. remota L.	—	—	—	Forest	±100?	Peter, 1893
C. sylvatica Huds.	—	—	—	Forest	±100	Peter, 1893
Cyperus spp.	—	—	—	Forest	±76	Livingston and Allesio, 1968
C. compressus L.	—	—	—	Forest	±33	Oosting and Humphreys, 1940
C. esculentus L.—yellow nutgrass	—	—	—	Buried	21	Toole and Brown, 1946
C. flavescens L.	—	—	—	Forest	±85	Oosting and Humphreys, 1940
C. globulosus Aubl.	—	—	—	Forest	±58	Oosting and Humphreys, 1940
Scirpus spp.	—	—	—	Pasture	±50	Prince and Hodgdon, 1946
S. setaceus L.	—	—	—	Pasture	±50	Chippendale and Milton, 1934
Stenophyllus capillaris (L.) Britt.	—	—	—	Forest	±85	Oosting and Humphreys, 1940
Juncaceae						
Juncus articulatus L.	—	—	—	Pasture	±60	Chippendale and Milton, 1934
J. bufonius L.	—	—	—	Pasture	±68	Chippendale and Milton, 1934
J. bulbosus L.	—	—	—	Pasture	±50	Chippendale and Milton, 1934
J. communis May.	—	—	—	Pasture	±68	Chippendale and Milton, 1934
J. effusus L.—common bogrush	—	—	—	Forest	±75	Livingston and Allesio, 1968
J. filiformis L.	—	—	—	Forest	±100?	Peter, 1893
J. glaucus Sebth.	—	—	—	Forest	±100	Peter, 1893

Species	Storage conditions				±	Reference
J. squarrosus L.	—	—	—	Pasture	±22	Chippendale and Milton, 1934
J. tenuis Willd. – path rush	—	—	—	Forest	±73	Livingston and Allesio, 1968
J. triglumus L.	—	—	—	Pasture	±50	Chippendale and Milton, 1934
Luzula campestris Br. – field wood rush	—	—	—	Pasture	±68	Chippendale and Milton, 1934
L. forsteri DC.	Dry lab	—	—	Forest	±100	Peter, 1893
Liliaceae						
Allium cepa L. – onion	Dry lab	22	33	—	—	James et al., 1964
Iridaceae						
Sisyrinchium angustifolium Mill. – blue-eyed grass	—	—	—	Forest	±36?	Livingston and Allesio, 1968
Watsonia meriana (L.) Mill.	Dry lab	51	13	—	—	Ewart, 1908
Cannaceae						
Canna indica L. – Achira	—	—	—	Grave	±530	Sivori et al., 1968
C. paniculata Ruiz & Pav.	Dry lab	69	7	—	—	Schjelderup-Ebbe, 1936
Casuarinaceae						
Casuarina suberosa Otto & Dietr.	Dry lab	12	62	—	—	Ewart, 1908
Myricaceae						
Myrica pensylvanica Loisel – bayberry	3°C, sealed	15	56	—	—	Heit, 1967c
Corylaceae						
Alnus rugosa (DuRoi) Spreng – speckled alder	3°C, sealed	10	n.s.l.[b]	—	—	Heit, 1967c
Betulaceae						
Betula alba L. – white birch	—	—	—	Forest	±100	Peter, 1893
B. lutea Michx. – yellow birch	5°C, sealed	12	44	—	—	Holmes and Buszewicz, 1958
B. papyrifera Marsh. – paper birch	—	—	—	Forest	±47	Livingston and Allesio, 1968
Ulmaceae						
Ulmus americana L. – American elm	−4°C, 3% H_2O	15	75	—	—	Barton, 1961
Moraceae						
Cannabis sativa L. – hemp	Lab, sealed	19	+	—	—	Kondo et al., 1950
Morus multicaulis Perr. – mulberry	Lab, 4% H_2O	14	High	—	—	Takagi, 1939

TABLE II (*Continued*)

Plant species	Dry storage			Soil storage		References
	Environment[a]	Age (yr.)	Germ.[b]	Environment[c]	Age[d] (yr.)	
Urticaceae						
Boehmeria nivea (L.) Gaud. – ramie	—	—	—	Buried	39	Toole and Brown, 1946
Urtica dioica L. – stinging nettle	—	—	—	Under church	±600?	Ødum, 1965
Proteaceae						
Hakea cycloptera R.Br.	Lab	10	60	—	—	Ewart, 1908
Polygonaceae						
Polygonum aviculare L. – knotweed	—	—	—	Under church	±400?	Ødum, 1965
P. convolvulus L. – black bindweed	—	—	—	Forest	±36	Peter, 1893
P. hydropiper L. – smartweed	—	—	—	Buried	50	Darlington and Steinbauer, 1961
P. pennsylvanicum L. – smartweed	—	—	—	Buried	30	Toole and Brown, 1946
P. persicaria L. – lady's thumb	—	—	—	Buried	30	Toole and Brown, 1946
P. scandens L. – false buckwheat	—	—	—	Buried	39	Toole and Brown, 1946
P. tomentosum Schrank. – willoweed	—	—	—	Buried	16	Madsen, 1962
Rumex acetosella L. – sheep sorrel	—	—	—	Buried	26	Madsen, 1962
R. acetosa L. – garden sorrel	—	—	—	Pasture	±68	Chippendale and Milton, 1934
R. crispus L. – curley dock	—	—	—	Buried	80	Darlington and Steinbauer, 1961
R. nemorosus Schrad.	—	—	—	Forest	±35	Peter, 1893
R. obtusifolius L. – broad-leaf dock	—	—	—	Buried	39	Toole and Brown, 1946
R. salicifolius Weinm. – willow-leaf dock	—	—	—	Buried	39	Toole and Brown, 1946
Chenopodiaceae						
Atriplex patula L.	—	—	—	Meadow	±58	Brenchley, 1918

Species	Storage				Condition	Years	Reference
Beta vulgaris L.—beet	Lab	—	30	58	Buried	21	James et al., 1964 / Toole and Brown, 1946
Blitum virgatum L. (Amb.)	Herbarium	—	35	+	—	—	Heinis, 1943
Chenopodium album L.—lamb's quarters	—	—	—	—	Buried / Under church	39 / ±1700?	Toole and Brown, 1946 / Ødum, 1965
C. hybridum L.—maple-leaved goosefoot	—	—	—	—	Buried	39	Toole and Brown, 1946
C. polyspermum L.	—	—	—	—	Forest	±35	Peter, 1893
Spinacia oleracea L.—spinach	Lab, 7% H_2O	—	11	n.s.l.[b]	—	—	Madsen, 1957
Amaranthaceae							
Amaranthus fimbriatus (Gray) Wats.	—	—	—	—	Desert	±20	Went, 1957
A. retroflexus L.—pigweed	—	—	—	—	Buried	40	Darlington and Steinbauer, 1961
Thymelaeaceae							
Daphne mezereum L.	Lab	35	1	—	—	—	Schjelderup-Ebbe, 1936
Nyctaginaceae							
Boerhaavia annulata Cov.	—	—	—	—	Desert	±20	Went, 1957
Aizoaceae							
Mollugo cerviana (L.) seringe	—	—	—	—	Desert	±20	Went, 1957
M. verticillata L.—carpetweed	—	—	—	—	Forest	±85	Oosting and Humphreys, 1940
Portulacaceae							
Portulaca oleracea L. purslane	—	—	—	—	Buried	40	Darlington and Steinbauer, 1961
Basellaceae							
Basella rubra L.—red basella	Lab	—	10	10	—	—	Ewart, 1908
Caryophyllaceae							
Arenaria serphyllifolia L.—sandwort	—	—	—	—	Meadow	±58	Brenchley, 1918
Cerastium arvense L.—field chickweed	—	—	—	—	Forest	±100?	Peter, 1893
C. caespitosum	—	—	—	—	Under church	±600?	Ødum, 1965
C. triviale Link	—	—	—	—	Forest	±45	Peter, 1893

TABLE II *(Continued)*

Plant species	Dry storage			Soil storage		References
	Environment[a]	Age (yr.)	Germ.[b]	Environment[c]	Age[d] (yr.)	
C. vulgatum L.—mouse-ear chickweed	—	—	—	Pasture	68	Chippendale and Milton, 1934
Dianthus caryophyllus L.—carnation	Lab	10	47	—	—	Goss, 1937
D. chinensis L.—pink	Lab	10	24	—	—	Goss, 1937
Holosteum umbellatum L.—jagged chickweed	—	—	—	Forest	±36	Peter, 1893
Sagina decumbeus (Ell.) T & G.	—	—	—	Forest	±33?	Oosting and Humphreys, 1940
S. procumbens L.—pearlwort	—	—	—	Pasture	±60	Chippendale and Milton, 1934
Saponaria vaccaria L.—cowherb	—	—	—	Buried	10	Toole and Brown, 1946
Scleranthus annuus L.—annual knawel	—	—	—	Buried	13	Madsen, 1962
Spergula arvensis L.—corn spurry	—	—	—	Under church	±1700	Ødum, 1965
Spergularia rubra (L.) Presl.	Lab	12	37	—	—	Ewart, 1908
Stellaria graminea L.	—	—	—	Forest	±36	Peter, 1893
S. media (L.) Cyrill—chickweed	—	—	—	{ Buried	30	Darlington and Steinbauer, 1961
				{ Under church	±600	Ødum, 1965
S. uliginosa Murr.	—	—	—	Pasture	±50	Chippendale and Milton, 1934
Nymphaceaceae						
Nelumbo asperifolium ?	Herbarium	47	+	—	—	Becquerel, 1906
N. nucifera Gaertn.—East Indian lotus	Herbarium	237	+	Peatbog	±1040	{ Ramsbottom, 1942 { Libby, 1951
N. pentapetala Fern.—American lotus	Herbarium	55	+	—	—	Becquerel, 1906

Species	Storage				Location	Longevity (years)	Reference
Nymphaea gigantea Hook.—Australian water lily	Lab	—	16	16	—	—	Ewart, 1908
Ranunculaceae							
Aquilegia canadensis L.—American columbine		—	—	—	Forest	±47?	Livingston and Allesio, 1968
Delphinium ajacis L.—larkspur	8°C, 6% H_2O	—	16	43	—	—	Crocker, 1945
D. chinense Fisch.—delphinium	−15°C, 6% H_2O	—	16	33	—	—	Crocker, 1945
Nigella damascena L.—love-in-a-mist	Lab	—	10	18	—	—	Goss, 1937
Ranunculus acris L.—tall buttercup		—	—	—	Meadow	±31?	Brenchley, 1918
R. bulbosus L.—bulbous buttercup		—	—	—	Meadow	±51	Brenchley, 1918
R. flammula L.—smaller spearwort		—	—	—	Pasture	±50	Chippendale and Milton, 1934
R. repens L.—creeping buttercup		—	—	—	Under church	±600	Ødum, 1965
Papaveraceae							
Papaver sp.		—	—	—	Meadow	±31	Brenchley, 1918
P. heterophyllum (Benth.) Greene—wind poppy	Lab	—	10	53	Chaparral	±40	Rowntree, 1930
P. rhoeas L.—corn poppy	Lab	—	—	—	—	—	Goss, 1937
P. rhoeas L.—corn poppy (wild)		—	10	22	Buried	26	Madsen, 1962
Eschscholtzia californica Cham.—California poppy	Lab	—	—	—	—	—	Goss, 1937
Fumariaceae							
Fumaria officinalis L.—common fumatory		—	—	—	Under church	±600?	Ødum,
Capparidaceae							
Capparis nobilis F. Muell.	Lab	—	10	7	—	—	Ewart, 1908
Cruciferae							
Barberea vulgaris R.Br.—yellow rocket		—	—	—	Field	±10	Muenscher, 1955

TABLE II (*Continued*)

Plant species	Dry storage			Soil storage		References
	Environment[a]	Age (yr.)	Germ[b]	Environment[c]	Age[d] (yr.)	
Brassica arvensis (L.) Ktze.—charlock	—	—	—	Forest	±20	Peter, 1893
B. campestris L.—bird rape	—	—	—	Meadow	±58	Brenchley, 1918
B. campestris L. var. rapifera Metzg.—turnip	Lab	15	+	Under church	±600?	Ødum, 1965
B. hirta moench—white mustard	Lab	10	21	Buried	23	James *et al.*, 1964 Madsen, 1962
B. kaber (DC) L.C. Wheeler—wild mustard	Lab	10	12	Buried	26	Kjaer, 1940 Madsen, 1962
B. napus L.—rape	Lab	12	3	—	—	Carruthers, 1911
B. napus v. napobrassica (L.) Reichen—rutabaga	Lab	—	—	Buried	16	Carruthers, 1911 Madsen, 1962
B. nigra (L.) Koch—black mustard	—	—	—	Buried	50	Darlington and Steinbauer, 1961
B. oleracea L.—cabbage	Lab	19	19	—	—	James *et al.*, 1964
B. sinapsis Visiani	—	—	—	Pasture	±60	Chippendale and Milton, 1934
Capsella bursa-pastoris (L.) Medic.—shepherd's purse	—	—	—	Buried	35	Darlington and Steinbauer, 1961
Corydallis sempervirens (L.) Pers	—	—	—	Forest	±80	Livingston and Allesio, 1968
Lepidium virginicum L.—peppergrass	—	—	—	Buried	40	Darlington and Steinbauer, 1961
Lobulari martima (L.) Desu Lam.—sweet alyssum	Lab	10	20	—	—	Goss 1937
Matthiola incana R. Br.—stock	Lab	10	54	—	—	Goss, 1937
Nasturtium palustra DC—marsh cress	—	—	—	Forest	±45?	Peter, 1893
Neslia paniculata (L.) Resv.—bull mustard	—	—	—	Buried	10	Toole and Brown, 1946

Species	Storage			Condition	±	Reference
Raphanus raphanistrum L.—wild radish	—	—	—	Forest	±36	Peter, 1893
R. sativus L.—radish	Lab, 4% H$_2$O	16	n.s.l.[b]	—	—	Madsen, 1957
Sinapis thalianum J. Gray	—	—	—	Forest	±18	Peter, 1893
Sisymbrium altissimum L.	—	—	—	Buried	10	Toole and Brown, 1946
Thlaspi arvense L.—field penny cress	—	—	—	Buried	30	Toole and Brown, 1946
Saxifragaceae						
Ribes rotundifolium Michx.—wild gooseberry	—	—	—	Under tree	±70	Fivaz, 1931
Rosaceae						
Alchemilla alpina L.	—	—	—	Forest	±30	Peter, 1893
A. arvensis Scop.—lady's mantle	—	—	—	Pasture	±68	Chippendale and Milton, 1934
Fragaria spp.—strawberry	<5°C, dry	23	89	—	—	Scott and Draper, 1970
F. vesca L.—European strawberry	—	—	—	Forest	±100	Peter, 1893
Potentilla argentea L.—silver cinquefoil	—	—	—	Forest	±20	Livingston and Allesio, 1968
P. erecta Hampi	—	—	—	Pasture	±100	Chippendale and Milton, 1934
P. norvegica L.—rough cinquefoil	—	—	—	Buried	39	Toole and Brown, 1946
P. reptans L.	—	—	—	Pasture	±68	Chippendale and Milton, 1934
P. tormentilla Neck	—	—	—	Forest	±100	Peter, 1893
Prunus pensylvanica L.F—pin cherry	3°C, sealed	10	76	—	—	Heit, 1967c
Rubus spp.—black-and raspberries	<5°C, dry	12	+	Forest	±65	Heit, 1967c / Livingston and Allesio, 1968
R. fruticosus L.—blackberry	—	—	—	Pasture	±50	Chippendale and Milton, 1934
R. idaeus L.—European red raspberry	—	—	—	Forest	±100	Peter, 1893
Spiraea latifolia—meadowsweet	—	—	—	Forest	±40	Livingston and Allesio, 1968
S. tomentosa L.—steeplebush	—	—	—	Pasture	±50	Prince and Hodgdon, 1946
Leguminosae						
Acacia acinacea Lindl.	Lab	51	4	—	—	Ewart, 1908
A. alata R.Br.	Lab	30	12	—	—	Ewart, 1908

TABLE II *(Continued)*

Plant species	Dry storage Environment[a]	Age (yr.)	Germ.[b]	Soil storage Environment[c]	Age[d] (yr.)	References
A. aneura F. Muell	Lab	20	100	—	—	Ewart, 1908
A. armata R.Br.	Lab	51	11	—	—	Ewart, 1908
A. bossiaeoides A. Cunn.	Lab	57	6	—	—	Ewart, 1908
A. brachybotrya Benth.	Lab	57	21	—	—	Ewart, 1908
A. calamifolia Sweet.	Lab	18	80	—	—	Ewart, 1908
A. cornigera Willd.	Herbarium	36	+	—	—	Becquerel, 1906
A. dealbata Link. silver wattle	Herbarium	23	14	—	—	Holmes and Buszewicz, 1958
A. decurrens Willd.–green wattle	Lab	17	63	—	—	Ewart, 1908
A. diffusa Lind.	Lab	59	10	—	—	Ewart, 1908
A. doratoxylon Cunn.	Lab	20	9	—	—	Ewart, 1908
A. farnesiana (L.) Willd.–sweet acacia	Lab	>35	1	—	—	Schjelderup-Ebbe, 1936
A. glaucescens Willd.	Lab	20	93	—	—	Ewart, 1908
A. lanigera A. Cunn.	Lab	20	26	—	—	Ewart, 1908
A. leprosa Andrews	Lab	51	28	—	—	Ewart, 1908
A. longifolia Willd.–Sydney wattle	Lab	68	5	—	—	Ewart, 1908
A. lunata (Mill.) Britt.	Lab	48	25	—	—	Ewart, 1908
A. melanoxylon Ait.–blackwood acacia	Lab	51	12	—	—	Ewart, 1908
A. merralli F. Muell.	Lab	10	6	—	—	Ewart, 1908
A. montana Benth.	Lab	58	2	—	—	Ewart, 1908
A. myrtifolia (Sm.) Willd.	Lab	55	5	—	—	Ewart, 1908
A. neriifolia Benth.–bald acacia	Lab	17	4	—	—	Ewart, 1908
A. nervosa DC.	Lab	30	4	—	—	Ewart, 1908
A. oswaldi F. Muell.	Lab	10	66	—	—	Ewart, 1908

Species	Storage					Reference
A. penninervis DC.	Lab	67	2	—	—	Ewart, 1908
A. pentadenia Lindl.	Lab	30	15	—	—	Ewart, 1908
A. pycnantha Benth. – golden wattle	Lab	16	80	—	—	Rees, 1911
A. senegal (L.) Willd.	Lab	51	5	—	—	Ewart, 1908
A. sinisii A. Cunn.	Lab	31	23	—	—	Ewart, 1908
A. suaveolens (Sm.) Willd.	Lab	51	4	—	—	Ewart, 1908
A. terminalis Macbride	Lab	30	16	—	—	Ewart, 1908
A. verniciflua A. Cunn.	Lab	41	5	—	—	Ewart, 1908
Albizia decurrens?	Lab	31	+	—	—	Dent, 1942
A. julibrissin Durazz. – silk tree	Herbarium	147	+	—	—	Ramsbottom, 1942
A. lebbek Benth. – lebbek tree	Lab	31	+	—	±23	Dent, 1942
A. lophantha (Willd.) Benth.	Lab	50	33	Forest	—	{ Ewart, 1908 / Rees, 1911
A. odoratissima (L.f.) Benth.	Lab	31	+	—	—	Dent, 1942
Anthyllis vulneraria L. – kidney vetch	Lab	90	4	—	—	Turner, 1933
Astragalus alpinus L.	Lab	40	9	—	—	Schjelderup-Ebbe, 1936
A. antiselli A. Gray	Lab	18	3	—	—	Ewart, 1908
A. brachyceras Ledeb.	Herbarium	37	+	—	—	Becquerel, 1906
A. edulis Coso.	Lab	43	1	—	—	Schjelderup-Ebbe, 1936
A. glycyphylloides DC.	Lab	51	8	—	—	Ewart, 1908
A. glycyphyllos L.	Lab	30	25	—	—	Filter, 1932
A. macrocephalus Willd.	Lab	64	6	—	—	Schjelderup-Ebbe, 1936
A. massiliensis (Mill) Lam.	Herbarium	86	10	—	—	Becquerel, 1934
A. utriger Pall	Lab	82	6	—	—	Schjelderup-Ebbe, 1936
Baptisia alba (L.) R.Br.	Lab	61	1	—	—	Schjelderup-Ebbe, 1936
B. commutata ?	Lab	43	1	—	—	Schjelderup-Ebbe, 1936
Bauhinia sp.	Lab	10	100	—	—	Ewart, 1908
Bossiaea ensata Sieb.	Lab	10	23	—	—	Ewart, 1908
B. heterophylla Vent	Lab	15	80	—	—	Ewart, 1908

TABLE II (*Continued*)

Plant species	Dry storage			Soil storage		References
	Environment[a]	Age (yr.)	Germ.[b]	Environment[c]	Age[d] (yr.)	
B. stephensoni F. Muell	Lab	12	44	—	—	Ewart, 1908
Caesalpinia bonducella Fleming	Lab	15	20	—	—	Ewart, 1908
C. digyna Roppl. in Geo.	Lab	41	+	—	—	Dent, 1942
Canavalia ensiformis DC.–jack bean	Lab	10	100	—	—	Ewart, 1908
C. obtusifolia (Lam.) DC.	Lab	60	31	—	—	Ewart, 1908
Cassia australis Sims.	Lab	10	45	—	—	Ewart, 1908
C. bicapsularis L.	Herbarium	115	40	—	—	Becquerel, 1934
C. brewsterii F. Muell.	Lab	19	90	—	—	Ewart, 1908
C. eremophila Vogel	Lab	10	25	—	—	Ewart, 1908
C. fistula L.–golden shower	Lab	31	+	—	—	Dent, 1942
C. laevigata Willd.	Lab	16	26	—	—	Ewart, 1908
C. marilandica L.–American senna	—	—	—	Buried	30	Toole and Brown, 1946
C. multijuga (L.)C. Rich.	Herbarium	158	100	—	—	Becquerel, 1934
C. pleurocarpa F. Muell.	Lab	15	46	—	—	Ewart, 1908
C. sophera L.	Lab	27	4	—	—	Ewart, 1908
C. suratensis Burm. f.	Lab	43	100	—	—	Rees, 1911
Cercis sp.–redbud	Lab, <4% H_2O	50	+	—	—	Heit, 1967c
Cicer arietinum L.–garbanzo	Lab	17	25	—	—	Ewart, 1908
Coronilla montana Scop.	Lab	52	14	—	—	Schjelderup-Ebbe, 1936
Crotalaria laburnifolia L.	Lab	10	74	—	—	Ewart, 1908
C. mitchellii Benth.	Lab	13	78	—	—	Ewart, 1908
C. ramossima Roxb.	Herbarium	38	+	—	—	Becquerel, 1906
C. vitellina Ker. Gawl.	Lab	13	92	—	—	Ewart, 1908

Species	Condition					Reference
Cytisus albus Hacquet	Lab	51	78	—	—	Ewart, 1908
C. austriacus L. Austrian broom	Herbarium	63	10	—	—	Becquerel, 1934
C. biflorus L' Her.	Herbarium	84	20	—	—	Becquerel, 1934
C. candicans Lam.	Lab	80	62	—	—	Ewart, 1908
C. caucasicus ?	Lab	62	23	—	—	Schjelderup-Ebbe, 1936
C. scoparius Link. – Scotch broom	Herbarium	81	+	—	—	Turner, 1933
C. villosus Pourret	Lab	50	18	—	—	Ewart, 1908
Cytisus spachiana (Webb) Kuntze	Lab	51	2	—	—	Ewart, 1908
Daviesia cordata J. E. Smith	Lab	67	1	—	—	Ewart, 1908
D. crenulata Turcz.	Lab	18	59	—	—	Ewart, 1908
Dichrostachys glomerata (Forsk.) Chiov.	Herbarium	81	50	—	—	Becquerel, 1934
Dillwynia ericifolia J. E. Smith	Lab	25	22	—	—	Ewart, 1908
D. floribunda J. E. Smith – parrot pea	Lab	32	19	—	—	Ewart, 1908
Dioclea pauciflora Rusby	Herbarium	93	20	—	—	Becquerel, 1934
Dolichos funarius Molina	Herbarium	37	+	—	—	Becquerel, 1906
D. lablab L. – hyacinth bean	Lab	10	8	—	—	Ewart, 1908
Ervum lens L. – lentil	Herbarium	65	10	—	—	Becquerel, 1934
Erythrina variegata Stickm.	Lab	32	7	—	—	Ewart, 1908
E. vespertilio Mitchell – coral tree	Lab	15	25	—	—	Ewart, 1908
Galega orientalis Lam. caucauses goatrue	Lab	51	2	—	—	Ewart, 1908
Genista anglica L.	Lab	42	20	—	—	Ewart, 1908
Gleditsia triacanthos L. – honey locust	Lab, < 4% H_2O	50	+	—	—	Heit, 1967c
Glycine max (L.) Merr. – soybean	Storage	13	+	—	—	Robertson *et al.*, 1943
Glycyrrhiza echinata L. licorice	Lab	43	1	—	—	Schjelderup-Ebbe, 1936
Goodia lotifolia Salisb.	Lab	105	8	—	—	Ewart, 1908

TABLE II (*Continued*)

Plant species	Dry storage			Soil storage		References
	Environment[a]	Age (yr.)	Germ.[b]	Environment[c]	Age[d] (yr.)	
Gompholobium latifolium Sm.	Lab	64	2	—	—	Ewart, 1908
Gymnocladus dioica (L.) Koch. – Kentucky coffee tree	Lab, 4% H$_2$O	50	+	—	—	Heit, 1967c
Hardenbergia monophylla Benth.	Lab	51	15	—	—	Ewart, 1908
Hovea heterophylla A. Cunn.	Lab	45	33	—	—	Ewart, 1908
H. linearis (J. E. Smith) R.Br.	Lab	105	17	—	—	Ewart, 1908
H. longifolia R.Br.	Lab	50	9	—	—	Ewart, 1908
Indigofera australis Willd.	Lab	57	6	—	—	Ewart, 1908
I. cytisoides L. Indigo	Lab	81	5	—	—	Ewart, 1908
Jacksonia spinosa R.Br.	Lab	72	4	—	—	Ewart, 1908
J. thesioides A. Cunn.	Lab	29	36	—	—	Ewart, 1908
Kennedia apetala ?	Lab	77	13	—	—	Schjelderup-Ebbe, 1936
K. monophylla Vent.	Lab	50	15	—	—	Ewart, 1908
K. prostrata R. Ait.	Lab	50	8	—	—	Ewart, 1908
K. rubicunda (Schneev) Vent.	Lab	65	10	—	—	Ewart, 1908
Labichea lanceolata Benth.	Lab	12	2	—	—	Ewart, 1908
Laburnum anagyroides alschingeri(Vis.) C. Schneid.	Lab	62	1	—	—	Schjelderup-Ebbe, 1936
Lathyrus silvestris L.–flat pea	Lab	40	1	—	—	Schjelderup-Ebbe, 1936
Lespedeza intermedia (S. Wats) Britt.–bushclover	—	—	—	Buried	39	Toole and Brown, 1946
Leucaena leucocephala (Lam.) deWit–Leadtree	Herbarium	99	30	—	—	Becquerel, 1934
L. pulverulenta (Schlecht.) Benth.	Lab	18	42	—	—	Ewart, 1908
Lotus corniculatus L.–birdsfoot trefoil	Lab	18	22	Forest	±100?	Ewart, 1908; Peter, 1893

Species	Storage				Condition	Value	Reference
L. tetragonolobus purpureus Moench—winged pea	Lab		16	4	—	—	Ewart, 1908
Lotus pedunculatus Cav.	Lab	—	100	1	—	—	Youngman, 1952
Lupinus arcticus Wats.			—	—	Artic	±10,000?	Porsild et al., 1967
L. perennis L.—sun-dial lupine	Lab		30	72	—	—	Filter, 1932
L. polyphyllus Lindl.—Washington lupine	Lab		49	75	—	—	Schjelderup-Ebbe, 1936
L. rivularis Dougl.	Lab		45	2	—	—	Schjelderup-Ebbe, 1936
Macuna gigantea (Willd.)DC.	Lab		11	60	—	—	Ewart, 1908
Medicago intertexta var. ciliaris (L.) Heyn.	Lab		40	7	—	—	Schjelderup-Ebbe, 1936
M. polymorpha var. vulgaris (Benth.) Shinners	Lab		17	8	—	—	Ewart, 1908
M. lupulina L.—black medic	Lab		50	68	Buried	26	Ewart, 1908 / Madsen, 1962
M. orbicularis (L.) Bart.	Lab		>78	22	—	—	Turner 1933
M. sativa L.—alfalfa	Lab		78	22	—	—	Turner, 1933
M. scutellata (L.) Mill.—snail medic	Lab		37	2	—	—	Ewart, 1908
M. truncatula Gaertn.	Lab		54	3	—	—	Ewart, 1908
Melilotus alba Desr.—white sweet clover	Lab		77	18	Field	±10	Ewart, 1908 / Muenscher. 1955
M. altissima Thuill.	Lab		40	32	—	—	Schjelderup-Ebbe, 1936
M. bonplandi Ten.	Lab		40	10	—	—	Ewart, 1908
M. indica (L.) All.	Lab		82	22	—	—	Ewart, 1908
M. neopolitana Ten.	Lab		45	1	—	—	Ewart, 1908
M. officinalis (L.) Lam.—yellow sweet clover	Herbarium		55	30	—	—	Becquerel. 1934
M. siculus (tuna) B. D. Jacks.	Lab		51	6	—	—	Ewart, 1908
Mimosa sp.	Lab		67	3	—	—	Ewart, 1908
M. asperata L.	Lab		26	36	—	—	Ewart, 1908
M. distachya Cav.	Herbarium		52	+	—	—	Becquerel, 1906

TABLE II (*Continued*)

Plant species	Dry storage			Soil storage		References
	Environment[a]	Age (yr.)	Germ.[b]	Environment[c]	Age[d] (yr.)	
M. pudica L. – sensitive plant	Lab	44	21	—	—	Ewart, 1908
Mirbelia oxyloboides F. Muell.	Lab	16	12	—	—	Ewart, 1908
M. reticulata J. E. Smith	Lab	47	3	—	—	Ewart, 1908
Onobrychis viciifolia Scop. – sainfoin	Storage	14	36	—	—	Haferkamp et al., 1953
Ononis rotundifolia L.	Lab	44	2	—	—	Schjelderup-Ebbe, 1936
Ornithopus sativus Brot-serrodella	Lab	11	59	—	—	Filter, 1932
Oxylobium callistachys Benth.	Lab	25	82	—	—	Ewart, 1908
O. cuneatum Benth.	Lab	50	5	—	—	Ewart, 1908
O. epipticum R.Br.	Lab	63	19	—	—	Ewart, 1908
O. lineare Benth.	Lab	51	20	—	—	Ewart, 1908
O. parviflorum Benth.	Lab	32	21	—	—	Ewart, 1908
O. trilobatum Benth.	Lab	49	4	—	—	Ewart, 1908
Oxytropis campestris DC.	Lab	40	6	—	—	Schjelderup-Ebbe, 1936
O. nigrescens DC.	Lab	61	1	—	—	Schjelderup-Ebbe, 1936
Phaseolus lunatus L. – lima bean	Lab	>15	1	—	—	James et al., 1964
P. mungo L. – black gram	Lab	20	44	—	—	Ewart, 1908
P. pilosus H.B.K.	Lab	18	12	—	—	Ewart, 1908
P. vulgaris L. – bean	Lab	22	30	—	—	James et al., 1964
Phylacium bracteosum Bennet.	Lab	17	33	—	—	Ewart, 1906
Pisum sativum L. – pea	Storage	31	78	—	—	Haferkamp et al., 1953
Pithecellobium pruninosum Benth.	Lab	16	22	—	—	Ewart, 1908
Podalyria calyptrata Willd.	Lab	51	4	—	—	Ewart, 1908
P. sericea R.Br.	Lab	51	1	—	—	Ewart, 1908
Prosopis julif velvet m	Herbarium	50	60	Buried	10	{Glendening and Paulsen, 1955 / Tschirley and Martin. 1960}

Species	Storage					Reference
Psoralea pinnata L.	Lab	51	10		—	Ewart, 1908
Pultenaea baeckeoides A. Cunn.	Lab	57	2		—	Ewart, 1908
P. daphnoides Wendl.	Lab	19	7		—	Ewart, 1908
P. retusa J. E. Smith	Lab	22	4		—	Ewart, 1908
P. stipularis J. E. Smith	Lab	16	8		—	Ewart, 1908
P. villosa Willd.	Lab	16	25		—	Ewart, 1908
Robinia pseudoacacia L. – black locust	Lab, < 4% H$_2$O	50	+	Buried	39	Heit, 1967c; Toole and Brown, 1946
Rhynchosia minima (L.) D.C.	Lab	18	8		—	Ewart, 1908
Sebania bispinosa (Jacq.) W.F. Wight	Lab	41	24		—	Schjelderup-Ebbe, 1936
Strophostyles helvola (L.) Elliot	Lab	61	11		—	Schjelderup-Ebbe, 1936
Swainsona galegifolia (Andrews) Darling pea R. Br.	Lab	25	11		—	Ewart, 1908
Trifolium agrarium L.	Lab	22	48		—	Ewart, 1908
T. arvense L. – rabbit-foot clover	Herbarium	68	20		—	Becquerel, 1934
T. campestre Schreb.	Lab	30	87	Forest	±18	Filter, 1932; Peter, 1893
T. filiforme L.	Lab	19	32		—	Ewart, 1908
T. glomeratum L. – cluster clover	Lab	53	10		—	Ewart, 1908
T. hybridum L. – alsike clover	Lab	17	19	Buried	30	Sifton, 1920; Toole and Brown, 1946
T. hybridum var. *pratense* Rabenk.	Herbarium	27	+		—	Becquerel, 1906
T. incarnatum L. – crimson clover	Lab	30	16		—	Filter, 1932
T. medium Huds. – zigzag clover	Lab	40	3	Pasture	±68	Schjelderup-Ebbe, 1936; Chippendale and Milton, 1934
T. minus J. E. Smith	Storage	—	—		—	Filter, 1932
T. pannonicum Jacq.	—	30	39		—	Youngman, 1952
T. pratense L. – red clover	Lab	100	1	Buried	39	Toole and Brown, 1946

TABLE II (*Continued*)

Plant species	Dry storage Environment[a]	Age (yr)	Germ.[b]	Soil storage Environment[c]	Age[d] (yr.)	References
T. repens L. – white clover	Lab	26	3	Buried	30	Madsen, 1962
				Under church	±600	Toole and Brown, 1946; Ødum, 1965
T. striatum L. – striate clover	Lab	90	14	—	—	Turner, 1933
T. strictum L.	Lab	47	11	—	—	Ewart, 1908
Thermopsis montana Nutt.	Lab	43	20	—	—	Schjelderup-Ebbe, 1936
Trigonella foenumgraecum L. – fenugreek	Lab	12	5	—	—	Ewart, 1908
Vicia angustifolia L.	Lab	—	—	Forest	±18	Peter, 1893
V. hirsuta (L.) S.F. Gray	Lab	30	12	Buried	25	Filter, 1932; Madsen, 1962
V. orobus DC. – upright vetch	Lab	44	1	Forest	±45	Schjelderup-Ebbe, 1936
V. tenuifolia Roth	Lab	—	—	—	—	Peter, 1893
V. villosa Roth – hairy vetch	Lab	30	6	—	—	Filter, 1932
V. unguiculata (L.)Walp.–Cowpea	Lab	10	66	—	—	Ewart, 1908
Vigna glabra Savi.	Lab	48	33	—	—	Ewart, 1908
Viminaria denudata J. E. Smith – rush Broom	Lab	18	38	—	—	Ewart, 1908
Wisteria maideniana F.M. Bailey	Lab	10	43	—	—	Ewart, 1908
Geraniaceae						
Geranium dissectum L. – cut-leaved crane's bill	Lab	15	11	—	—	Madsen, 1962
G. pusillum Burm. – small-flowered crane's bill	—	—	—	Buried	21	Madsen, 1962
G. robertianum L. – herb Robert	—	—	—	Forest	±37?	Peter, 1893
Oxalidaceae						
Oxalis spp.	—	—	—	Pasture	±50	Prince and Hodgdon, 1946

Species	Storage				Habitat	±	Reference
O. corniculata L.—Creeping woodsorrel	—	—	—	—	Forest	±85	Oosting and Humphreys, 1940
O. florida Salisb.							
O. stricta L.—yellow wood-sorrel	—	—	—	—	Forest	±85	Oosting and Humphreys, 1940
					Forest	±15	Oosting and Humphreys, 1940
Tropaeolaceae							
Tropaeolum majus L.—common nasturtium	Lab	—	10	61	—	—	Goss, 1937
Linaceae							
Linum catharticum L.	—	—	—	—	Forest	±100	Peter, 1893
L. usitatissimum L.—flax	Lab, 8% H$_2$O	—	18	58	—	—	Dillman and Toole, 1937
Meliaceae							
Flindersia australis R.Br.	Lab	—	12	3	—	—	Ewart, 1908
Euphorbiaceae							
A. moluccana (L.) Wills.—candlenut tree	Lab	—	79	74	—	—	Ewart, 1908
Acalypha rhomboidea Raf.—three-seeded mercury	—	—	—	—	Field	±10	Muenscher, 1955
A. virginica L.	—	—	—	—	Forest	±25?	Livingston and Allesio, 1968
Euphorbia exigua L.—dwarf spurge	—	—	—	—	Forest	±22	Peter, 1893
E. helioscopia L.—sun spurge	Lab	—	10	12	Pasture	±68	{Ewart, 1908 / Chippendale and Milton, 1934}
E. maculata L.—spotted spurge	—	—	—	—	Pasture	±50	Prince and Hodgdon, 1946
E. micromera Boiss.	—	—	—	—	Desert	±20	Went, 1957
E. peplus L.—petty spurge	Lab	—	57	3	Forest	±36	{Ewart, 1908 / Peter, 1893}
E. setiloba Engelm.	—	—	—	—	Desert	±20	Went, 1957
Mercurialis annua L.—mercury	Lab	—	40	1	—	—	Schjelderup-Ebbe, 1936
Petalostigma quadriloculare F. Muell.	Lab	—	12	13	—	—	Ewart, 1908
Pseudanthus ovalifolius F. Muell.	Lab	—	57	2	—	—	Ewart, 1908

TABLE II (*Continued*)

Plant species	Dry storage			Soil storage		References
	Environment[a]	Age (yr.)	Germ.[b]	Environment[c]	Age[d] (yr.)	
Anacardiaceae						
Cotinus sp.	Lab, <4% H_2O	50	+	—	—	Heit, 1967c
Rhus sp. — sumac	Lab, <4% H_2O	50	+	Grave	±200?	Heit, 1967c / Heit, 1967a
R. glabra L. — smooth sumac	Lab	20	67	Buried	39	Heit, 1967a / Toole and Brown, 1946
R. hirta (L.) Sudsworth — staghorn sumac	Lab	20	57	—	—	Heit, 1967a
Aceraceae						
Acer rubrum L. — red maple	Lab	23	100	—	—	Rees, 1911
Rhamnaceae						
Alphitonia excelsa (Fenzl) Reissek	Lab	12	74	—	—	Ewart, 1908
Ceanothus spp.	Lab, <4% H_2O	50	+	—	—	Heit, 1967c
Tiliaceae						
Entelea arborescens R.Br.	Lab	51	47	—	—	Ewart, 1908
Corchorus capsularis L. — roundpod jute	Lab	10	5	—	—	Ewart, 1908
C. olitorius L. — potherb-jute	Lab	10	36	—	—	Ewart, 1908
Malvaceae						
Abutilon avicennae Gaertn.	Lab	57	2	—	—	Ewart, 1908
A. canariense ?	Lab	71	1	—	—	Ewart, 1908
A. indicum Sweet	Lab	17	+	—	—	Dent, 1942
A. mitchelli Benth.	Lab	10	44	—	—	Ewart, 1908
A. oxycarpum F. Muell.	Lab	14	52	—	—	Ewart, 1908
A. reflexum Sweet	Lab	40	23	—	—	Schjelderup-Ebbe, 1936
A. theophrasti Medic. — velvet leaf	—	—	—	Buried	39	Toole and Brown, 1946

Species	Storage					Reference
Althea rosea (L.) Cav.–hollyhock	Lab	10	37	—	—	Goss, 1937
Anoda crenatiflora Orteg.	Lab	41	27	—	—	Schjelderup-Ebbe, 1936
A. wrightii A. Gray	Lab	42	20	—	—	Schjelderup-Ebbe, 1936
Gossypium hirsutum L.–upland cotton	Lab	25	6	—	—	Simpson, 1946
G. sturtianum J. H. Willis	Lab	22	14	—	—	Ewart, 1908
Hibiscus californicus Kell.	Lab	41	1	—	—	Schjelderup-Ebbe, 1936
H. esculentus L.–okra	Lab	10	20	—	—	Ewart, 1908
H. heterophyllus Vent.	Lab	10	39	—	—	Ewart, 1908
H. macrophyllus-Hornemann rosemallow	Lab	24	+	—	—	Dent, 1942
H. militaris Cav.–halberd leaved rosemallow	—	—	—	Buried	39	Toole and Brown, 1946
H. panduraeformis Burm. f.	Lab	16	2	—	—	Ewart, 1908
H. tiliaceus L.–linden hibiscus	Lab	14	5	—	—	Ewart, 1908
H. trionum L.–flower-of-an-hour	Lab	70	33	—	—	Ewart, 1908
Kitaibelia vitifolia Willd.	Lab	43	2	—	—	Schjelderup-Ebbe, 1936
Kydia calycina Roxb.	Lab	10	29	—	—	Ewart, 1908
Lavatera cretica L.–crete tree-mallow	Lab	65	1	—	—	Schjelderup-Ebbe, 1936
L. plebeia Sims.	Lab	18	7	—	—	Ewart, 1908
L. pseudo-olbia Desf.	Herbarium	64	20	—	—	Becquerel, 1934
L. thuringiaca L.	Lab	54	1	—	—	Schjelderup-Ebbe, 1936
Malva moschata L.–musk mallow	Lab	42	12	—	—	Schjelderup-Ebbe, 1936
M. rotundifolia L.–common mallow	Lab	16	4	Buried	20	Ewart, 1908 / Darlington and Steinbauer, 1961
Modiola caroliniana (L.) Don	Lab	50	3	—	—	Ewart, 1908
Ochroma lagopus Sw.	Lab	24	+	—	—	Dent, 1942
Sida cordifolia L.	Lab	41	4	—	—	Schjelderup-Ebbe, 1936
S. hederacea (Dougl.) Torr.–alkali mallow	Lab	16	4	—	—	Bruch, 1961

TABLE II (*Continued*)

Plant species	Dry storage			Soil storage		References
	Environment[a]	Age (yr.)	Germ.[b]	Environment[c]	Age[d] (yr.)	
S. intricata F. Muell.	Lab	57	3	—	—	Ewart, 1908
Sphaeralcea abutiloides Endl.	Lab	74	43	—	—	Schjelderup-Ebbe, 1936
Thespesia macrophylla Blume	Lab	35	2	—	—	Ewart, 1908
Bombacaceae						
Adansonia gregorii F. Muell.	Lab	12	4	—	—	Ewart, 1908
Sterculiaceae						
Hermannia angularis Jacq.	Lab	51	23	—	—	Ewart, 1908
H. sandersonii Harv.	Lab	12	2	—	—	Ewart, 1908
Hypericaceae						
Ascyrum hypericoides L.	—	—	—	Buried	39	Toole and Brown, 1946
Hypericum gentianoides (L.) BSP — pineweed	—	—	—	Forest	±85	Oosting and Humphreys, 1940
H. hirsutum L.	—	—	—	Forest	±36	Peter, 1893
H. humifusum L.	—	—	—	Pasture	±68	Chippendale and Milton, 1934
H. perforatum L. — Klamath weed	—	—	—	Forest	±40	Livingston and Allesio, 1968
Cistaceae						
Helianthemum canadense Michx. — frostweed	—	—	—	Forest	±47?	Livingston and Allesio, 1968
Violaceae						
Viola arvensis Murr.	—	—	—	Under church	±400?	Ødum, 1965
V. tricolor L. — wild pansy	—	—	—	Meadow	±31?	Brenchley, 1918
Myrtaceae						
Callistemon lanceolatus Sweet	Lab	16	75	—	—	Ewart, 1908
C. rigidus R.Br.	Lab	22	3	—	—	Ewart, 1908
Eucalyptus amygdalina Labill.	Lab	11	4	—	—	Ewart, 1908
E. botryoides J. E. Smith	Lab	16	30	—	—	Rees, 1911

Species						Reference
E. calophylla Lindl.	Lab	32	5	–	–	Ewart, 1908
E. coccifera Hook. f.	Lab	11	3	–	–	Ewart, 1908
E. cornuta Labill.	Lab	22	9	–	–	Ewart, 1908
E. corynocalyx F. Muell. – sugar gum	Lab	16	68	–	–	Rees, 1911
E. diversicolor F. Muell.	Lab	24	11	–	–	Ewart, 1908
E. foecunda Schauer	Lab	10	4	–	–	Ewart, 1908
E. globulus Labill. – Tasmanian bluegum	Lab	20	2	–	–	Ewart, 1908
E. gomphocephala DC.	Lab	30	1	–	–	Ewart, 1908
E. goniocalyx F. Muell.	Lab	15	4	–	–	Ewart, 1908
E. gunnii Hook. f.	Lab	11	2	–	–	Ewart, 1908
E. leptopoda Benth.	l ab	30	1	–	–	Ewart, 1908
E. miniata Schauer	Lab	20	2	–	–	Ewart, 1908
E. obcordata Turcz.	Lab	15	8	–	–	Ewart, 1908
E. obliqua L'. Her.	Lab	25	1	–	–	Ewart, 1908
E. odorata Behr.	Lab	10	10	–	–	Ewart, 1908
E. paniculata J. E. Smith	Lab	10	4	–	–	Ewart, 1908
E. patens Benth.	Lab	10	8	–	–	Ewart, 1908
E. pauciflora Spreng.	Lab	12	6	–	–	Ewart, 1908
E. pilularis J. E. Smith	Lab	11	26	–	–	Ewart, 1908
E. punctata DC.	Lab	15	8	–	–	Ewart, 1908
E. rostrata Schlecht.	Lab	37	6	–	–	Ewart, 1908
E. rudis Endl. – desert gum	Lab	10	12	–	–	Ewart, 1908
E. siderophloia Benth.	Lab	10	2	–	–	Ewart, 1908
E. tereticornis Smith – forest red gum	Lab	15	2	–	–	Ewart, 1908
E. urnigera Hook. f.	Lab	11	6	–	–	Ewart, 1908
E. viminalis Labill. – ribbon gum	Lab	12	15	–	–	Ewart, 1908
Leptospermum scoparium Forst. – manuka tea tree	Lab	16	2	–	–	Ewart, 1908
Syncarpia laurifolia Ten.	Lab	10	6	–	–	Ewart, 1908

TABLE II *(Continued)*

Plant species	Dry storage			Soil storage		References
	Environment[a]	Age (yr.)	Germ.[b]	Environment[c]	Age[d] (yr.)	
Onagraceae						
Epilobium montanum L.	—	—	—	Forest	±100	Peter, 1893
Oenothera biennis L.—evening primrose	—	—	—	Buried	80	Darlington and Steinbauer, 1961
Araliaceae						
Aralia hispida Vent.—bristly sarsaparilla	—	—	—	Forest	±80?	Livingston and Allesio, 1968
Umbelliferae						
Aethusa cynapium L.—fools parsley	—	—	—	Forest	±20	Peter, 1893
Anthriscus cerefolium (L.) Hoffm —chervil	—	—	—	Forest	±20	Peter, 1893
Apium graveolens L.—celery	—	—	—	Buried	39	Toole and Brown, 1946
Caucalis sp.	—	—	—	Meadow	±31	Brenchley, 1918
Conopodium denudatum Koch.	—	—	—	Pasture	±68	Chippendale and Milton, 1934
Daucus carota L.—carrot	Lab	31	7	—	—	James *et al.*, 1964
D. carota L.—wild carrot	—	—	—	Buried	20	Madsen, 1962
Pastinaca sativa L.—parsnip	Lab	28	30	—	—	James *et al.*, 1964
P. sativa L.—wild parsnip	—	—	—	Buried	16	Toole and Brown, 1946
Ericaceae						
Vaccinium spp.—blueberries	<5°C, dry	12	+	—	—	Heit, 1967c
Primulaceae						
Anagallis arvensis L.—poor man's weatherglass	—	—	—	Pasture	±68	Chippendale and Milton, 1934
A. foemina Mill.—blue pimpernal	—	—	—	Pasture	±60	Turner, 1933
A. tenella L.	—	—	—	Pasture	±22	Chippendale and Milton, 1934

Centunculus minimus L. — chaffweed	—	—	—	Forest	±32	Peter, 1893
Loganiaceae						
Polypremum procumbens L.	—	—	—	Forest	±85	Oosting and Humphreys, 1940
Gentianaceae						
Centaurium umbellatum Gilib. — centaury	—	—	—	Forest	±100?	Peter, 1893
Apocynaceae						
Alyxia ruscifolia R. Br.	Lab	10	12	—	—	Ewart, 1908
Ochrosia poweri Bailey	Lab	11	5	—	—	Ewart, 1908
Convolvulaceae						
Convolvulus arvensis L. — field bindweed	Herbarium	50	62	Buried	16	Brown and Porter, 1942 / Bruch, 1961
C. flavus Willd.	Lab	64	14	—	—	Schjelderup-Ebbe, 1936
C. mauritanicus Boiss.	Lab	10	38	—	—	Ewart, 1908
C. sepium L. — hedge bindweed	—	—	—	Buried	39	Toole and Brown, 1946
Cuscuta epilinum Weihe — flax dodder	—	—	—	Buried	10	Toole and Brown, 1946
C. europaea L.	Lab	10	6	—	—	Ewart, 1908
C. polygonorum Englem. — smartweed dodder	—	—	—	Buried	39	Toole and Brown, 1946
Ipomoea sp.	Lab	43	6	—	—	Turner, 1933
I. coccinea L.	Lab	40	60	—	—	Schjelderup-Ebbe, 1936
I. eriocarpa R. Br.	Lab	10	33	—	—	Ewart, 1908
I. lacunosa L. — small-flower morning glory	—	—	—	Buried	39	Toole and Brown, 1946
I. pilosa Sweet	Lab	40	1	—	—	Schjelderup-Ebbe, 1936
Polemoniaceae						
Gilia capitata Dougl. — gilia	Lab	10	58	—	—	Goss, 1937
Hydrophyllaceae						
Emmenanthe penduliflora Benth. — whispering bells	—	—	—	Chaparral	±40	Rowntree, 1930

TABLE II (*Continued*)

Plant species	Dry storage			Soil storage		References
	Environment[a]	Age (yr.)	Germ.[b]	Environment[c]	Age[d] (yr.)	
Boraginaceae						
Myosotis arvensis (L.) Hill. – field forget-me-not	—	—	—	Meadow	±42	Brenchley, 1918
M. stricta Link	—	—	—	Forest	±20	Peter, 1893
Verbenaceae						
Verbena sp. – verbena	Lab	10	1	—	—	Goss, 1937
V. hastata L. – blue vervain	—	—	—	Buried	39	Toole and Brown, 1946
V. urticifolia L. – nettle-leaved vervain	—	—	—	Buried	39	Toole and Brown, 1946
Labiatae						
Calamintha acinos Man.	—	—	—	Forest	±39	Peter, 1893
Galeopsis tetrahit L. – common hemp nettle	—	—	—	Pasture	±68	Chippendale and Milton, 1934
Glecoma hederacea L. – ground ivy	—	—	—	Under church	±400?	Ødum, 1965
Lamium album L.	—	—	—	Under church	±600?	Ødum, 1965
L. galeobdolon Crantz	—	—	—	Forest	±100?	Peter, 1893
L. purpureum L.	—	—	—	Under church	±600?	Ødum, 1965
Mentha aquatica L. – water mint	—	—	—	Pasture	±68	Chippendale and Milton, 1934
M. arvensis L.	—	—	—	Forest	±45	Peter, 1893
Nepeta cataria L. – catnip	Lab	15	4	—	—	Ewart, 1908
Prunella vulgaris L. – self-heal	—	—	—	Pasture	±68	Chippendale and Milton, 1934
Stachys arvensis L.	—	—	—	Forest	±45	Peter, 1893
S. nepetifolia Desf.	Herbarium	77	1	—	—	Becquerel, 1934
S. silvatica L.	—	—	—	Forest	±40	Peter, 1893
Solanaceae						
Atropa belladonna L.	—	—	—	Forest	±100	Peter, 1893
Capsicum frutescens L. – red pepper	Lab	28	69	—	—	James et al., 1964

Datura inermis Jacq.	Lab	—	40	—	—	—	Schjelderup-Ebbe, 1936
D. stramonium L.—Jimson weed			—	—	Buried	39	Toole and Brown, 1946
Duboisia myoporoides R. Br.	Lab	—	10	—	—	—	Ewart, 1908
Hyoscyamus niger L.—black henbane			—	—	Under church	±600?	Ødum, 1965
Lycopersicum esculentum Mill.—tomato	Lab		33	84	Buried	10	James *et al.*, 1964 / Toole and Brown, 1946
Nicotiana tabacum L.—tobacco	Lab, sealed		20	92	Buried	39	Schloesing and Leroux, 1943 / Toole and Brown, 1946
Petunia hybrida Vilm.	Storage		11	80	—	—	Kempf, 1970
Salpiglosis sinuata Ruis & Pav.—salpiglossis	Lab		10	70	—	—	Goss, 1937
Schizanthus wisetonensis Low.—butterfly flower	Lab		10	60	—	—	Goss, 1937
Solanum carolinense L.—horse nettle					Forest	±112?	Oosting and Humphreys, 1940
Solanum elaeagnifolium Cav.—white horse nettle	—		—	—	Buried	11	Bruch, 1961
S. melongena L.—eggplant	−4°C, 7%H_2O		20	86	—	—	Barton, 1961
S. nigrum L.—black nightshade	—		—	—	Buried	>39	Toole and Brown, 1946
S. sodomaeum L.	Lab		51	1	—	—	Ewart, 1908
S. tuberosum L.—white potato	Lab		20	17	—	—	Wollenweber, 1942
Scrophulariaceae							
Digitalis purpurea L.—foxglove			—	—	Pasture	±68	Chippendale and Milton, 1934
Linaria canadensis (L.) Dumont—toadflax			—	—	Forest	±58	Oosting and Humphresy, 1940
L. elatine Mill.			—	—	Forest	±22	Peter, 1893
L. minor (L.) Desf.—small toadflax			—	—	Meadow	±42?	Brenchley, 1918
L. vulgaris Hill—butter-and-eggs			—	—	Forest	±20	Peter, 1893
Rhinanthus crista-galli L.			—	—	Pasture	±68	Chippendale and Milton, 1934

TABLE II (*Continued*)

Plant species	Dry storage			Soil storage		References
	Environment[a]	Age (yr.)	Germ.[b]	Environment[c]	Age[d] (yr.)	
Scrophularia nodosa L.—figwort	—	—	—	Forest	±100?	Peter, 1893
Verbascum blattaria L.—moth mullein	—	—	—	Buried	80	Darlington and Steinbauer, 1961
V. thapsus L.—common mullein	—	—	—	Buried	39	Toole and Brown, 1946
Veronica agrestis L.—field speedwell	—	—	—	Under church	±600?	Ødum, 1965
				Forest	±20	Peter, 1893
V. arvensis L.—corn speedwell	—	—	—	Forest	±35	Peter, 1893
V. didyna Ten.	—	—	—	Forest	±45	Peter, 1893
V. officinalis L.—common speedwell	—	—	—	Forest	±100	Peter, 1893
Veronica peregrina L.—purslane speedwell	—	—	—	Forest	±33	Oosting and Humphreys, 1940
V. serpyllifolia L.—thyme-leaved speedwell	—	—	—	Pasture	±68	Chippendale and Milton, 1934
V. tournefortii C.C.Gmel.—large-field speedwell	—	—	—	Meadow	±58?	Brenchley, 1918
Plantaginaceae						
Plantago lanceolata L.—buckhorn plantain	—	—	—	Buried	16	Darlington and Steinbauer, 1961
P. major L.—broad-leaved plantain	—	—	—	Buried	40	Darlington and Steinbauer, 1961
P. rugelii DC.—Rugel's plantain	—	—	—	Buried	21	Toole and Brown, 1946
P. virginica L.	—	—	—	Forest	±15	Oosting and Humphreys, 1940
Rubiaceae						
Galium saxatile L.	—	—	—	Forest	±100?	Peter, 1893
G. tricorne Stokes	—	—	—	Forest	±20	Peter, 1893

Species	Storage			Habitat	Longevity	Reference
Houstonia spp.	—	—	—	Forest	±85	Oosting and Humphreys, 1940
H. caerulea L.—bluets	—	—	—	Pasture	±50	Prince and Hodgdon, 1946
Caprifoliaceae						
Diervilla lonicera Mill.—bush honeysuckle	—	—	—	Forest	±42?	Livingston and Allesio, 1968
Lonicera tatarica L.—Tartarian honeysuckle	−3°C, sealed	15	84	—	—	Heit, 1967c
Sambucus nigra L.—European alder	—	—	—	Under church	±500?	Ødum, 1965
S. racemosa L.—red alder	—	—	—	Forest	±100?	Peter, 1893
Viburnum dentatum L.—arrowwood	0°C, sealed	10	6	—	—	Heit, 1967c
V. lentago L.—sheepberry	0°C, sealed	10	69	—	—	Heit, 1967c
Valerianaceae						
Valerianella dentata Polich	—	—	—	Forest	±36	Peter, 1893
Cucurbitaceae						
Citrullus vulgaris Schrad.—watermelon	Lab	30	92	—	—	James *et al.*, 1964
Cucumis melo L.—muskmelon	Lab	30	96	—	—	James *et al.*, 1964
C. sativus L.—cucumber	Lab	30	77	—	—	James *et al.*, 1964
Cucurbita pepo L.—squash, marrow	Lab, sealed	10	55	—	—	Kondo and Kasahara, 1940
Campanulaceae						
Campanula rotundifolia L.—bluebell	—	—	—	Forest	±20	Peter, 1893
C. trachelium L.—nettle-leaved bellflower	—	—	—	Forest	±100?	Peter, 1893
Specularia perfoliata (L.) A.DC.—Venus' looking glass	—	—	—	Forest	±85?	Oosting and Humphreys, 1940
Lobeliaceae						
Lobelia cardinalis L.—cardinal flower	5°C, 7% H_2O	25	62	—	—	Barton, 1960
L. inflata L.—Indian tobacco	—	—	—	Pasture	±50	Prince and Hodgdon, 1946

TABLE II (*Continued*)

Plant species	Dry storage			Soil storage		References
	Environment[a]	Age (yr.)	Germ.[b]	Environment[c]	Age[d] (yr.)	
Goodeniaceae						
Scaevola hookeri F. Muell.	Lab	15	5	—	—	Ewart, 1908
Compositae						
Achillea millefolium L.—common yarrow	Lab	10	3	Forest	±18	Carruthers, 1911; Peter, 1893
Ambrosia artemisiifolia L.—ragweed	—	—	—	Buried	>39	Toole and Brown, 1946
A. trifida L.—great ragweed	—	—	—	Buried	21	Toole and Brown, 1946
Anthemus cotula L.—dog fennel	—	—	—	Buried	30	Darlington and Steinbauer, 1961
Arctium lappa L.—great burdock	—	—	—	Buried	39	Toole and Brown, 1946
Aster spp.	—	—	—	Forest	±85	Oosting and Humphreys, 1940
A. acuminatus Michx.—wild aster	—	—	—	Forest	±47?	Livingston and Allesio, 1968
Bellis perennis L.—English daisy	—	—	—	Pasture	±68	Chippendale and Milton, 1934
Bidens frondosa L.—black beggar's tick	—	—	—	Buried	16	Toole and Brown, 1946
Calendula officinalis L.—calendule	Lab	10	28	—	—	Goss, 1937
Carduus crispus L.	—	—	—	Under church	±600?	Ødum, 1965
Centaurea cyanus L.—cornflower	Lab	10	28	—	—	Goss, 1937
C. gymnocarpa Mous & de Not.—dusty miller	Lab	10	11	—	—	Goss, 1937
C. imperialis Hort.—sweet sultans	Lab	10	24	—	—	Goss, 1937
C. nigra L.—knapweed	—	—	—	Meadow	±31	Brenchley, 1918
Chrysanthemum carinatum L.—chrysanthemum	Lab	10	28	—	—	Goss, 1937

Species	−4°C, 7% H₂O	15	58			Reference
C. cinerariaefolium Vis.—pyrethrum	—	—	—	—	—	Barton, 1966b
C. leucanthemum L.—oxeye daisy	—	10	—	Buried	39	Toole and Brown, 1946
C. segetum L.—corn marigold	Lab	10	10	Pasture	±24?	Goss, 1937; Chippendale and Milton, 1934
Cirsium arvense (L.) Scop.—Canada thistle	—	—	—	Buried	25	Madsen, 1962
C. lanceolatum (L.) Hill—bull thistle	—	—	—	Forest	±36	Peter, 1893
Cnicus arvensis Hoffm.	—	—	—	Forest	±100	Peter, 1893
Erechtites hieracifolia (L.) Raf.—fireweed	—	—	—	Forest	±80?	Livingston and Allesio, 1968
Erigeron spp.	—	—	—	Forest	±85	Oosting and Humphreys, 1940
E. annuus (L.) Pers.—daisy fleabane	—	—	—	Forest	±35	Livingston and Allesio, 1968
E. canadensis L.—horseweed	—	—	—	Forest	±112?	Oosting and Humphreys, 1940
E. strigosus Muhl.—rough daisy fleabane	—	—	—	Field	±10	Muenscher, 1955
Filago minima Fries	—	—	—	Forest	±45	Peter, 1893
Gnaphalium obtusifolium L.	—	—	—	Forest	±35	Peter, 1893
G. purpureum L.—purplish cudweed	—	—	—	Forest	±85	Oosting and Humphreys, 1940
G. sylvaticum L.	—	—	—	Forest	±18	Peter, 1893
G. uliginosum L.—low cudweed	—	—	—	Pasture	±50	Prince and Hodgdon, 1946
Grindelia squarrosa (Pursh) Dunal—broad-leaved gum plant	—	—	—	Buried	10	Toole and Brown, 1946
Helipterum incanum DC.	Lab	10	5	—	—	Ewart, 1908
H. roseum Benth.	Lab	10	5	—	—	Ewart, 1908
Hieracium aurantiacum L.—orange hawkweed	—	—	—	Field	±10	Muenscher, 1955
H. auricula L.	—	—	—	Forest	±32	Peter, 1893
H. pilosella L.—mouse ear hawkweed	—	—	—	Forest	±32	Peter, 1893

TABLE II (*Continued*)

Plant species	Dry storage			Soil storage		
	Environment[a]	Age (yr)	Germ.[b]	Environment[c]	Age[d] (yr.)	References
H. pratense Tausch—yellow hawkweed	–	–	–	Forest	±80?	Livingston and Allesio, 1968
Krigia virginica (L.) Willd.	–	–	–	Forest	±58?	Oosting and Humphreys, 1940
Lactuca sativa L.—lettuce	–4°C, 8% H_2O	20	86	–	–	Barton, 1961
Lapsana communis L.—nipplewort	–	–	–	Forest	±18	Peter, 1893
Leontodon hispidus L.	–	–	–	Forest	±36	Peter, 1893
Madia angustifolia ?	Lab	40	1	–	–	Schjelderup-Ebbe, 1936
Matricaria inodora L.—chamomile	–	–	–	Buried	19	Madsen, 1962
Onopordon acanthium L.—Scotch thistle	–	–	–	Buried	>39	Toole and Brown, 1946
Pectis papposa Gray—chinchweed	–	–	–	Desert	±20	Went, 1957
Rudbeckia hirta L.—black-eyed susan	–	–	–	Buried	>39	Toole and Brown, 1946
Senecio fuchsii C.C. Gmel.	–	–	–	Forest	±100?	Peter, 1893
S. jacobaea L.—stinking Willie	–	–	–	Pasture	±22	Chippendale and Milton, 1934
S. vulgaris L.—common groundsel	–	–	–	Meadow	±58	Brenchley, 1918
Solidago spp.—goldenrods	–	–	–	Pasture	±50	Prince and Hodgdon, 1946

S. graminifolia (L.) Salisb.—narrow-leafed goldenrod	—	—	—	Field	±10	Muenscher, 1955
S. rugosa Mill.	—	—	—	Forest	±35	Livingston and Allesio, 1968
Sonchus arvensis L.—perennial sow thistle	—	—	—	Forest	±45	Peter, 1893
S. oleraceus L.—common sow thistle	—	—	—	Forest	±100	Peter, 1893
Taraxacum officinale Weber.—dandelion	—	—	—	Pasture	±68	Chippendale and Milton, 1934
T. vulgare?	—	—	—	Under church	±600?	Ødum, 1965
Xanthium pensylvanicum Wallr.—cocklebur	—	—	—	Buried	16	Toole and Brown, 1946
Zinnia angustifolia HBK.—Haagaene zinnia	Lab	10	40	—	—	Goss, 1937
Z. elegans Jacq.—zinnia	Storage, sealed	11	80	—	—	Kempf, 1970

[a] Moisture of seed is on a fresh weight basis, sealed implies the seed was dried first.

[b] Germination percentage is given if cited; "+" means author gave no percentage; n.s.l. means the author reported no significant loss in germination at the year reported.

[c] Buried means the seed was buried at a known time, otherwise the seed was recovered from the soil ecology mentioned, below the soil surface.

[d] The age is unqualified if known exactly, preceded by ± if the age is based on past history of the particular soil, and followed by ? if the age cited may be seriously doubted because age was determined by associate objects or by only one or two germinating seeds of the species.

seeds in soil. The seeds may be dormant because of lack of light needed for germination, too high CO_2 or ethylene or too low oxygen in the soil atmosphere, or for other reasons still undetermined.

Several experiments have shown that even within a species there are differences among cultivars in seed longevity under identical storage conditions. One difficulty with most of these experiments has been that conditions during seed development were not identical, for a collection of cultivars was used. Since these cultivars were grown at different times and places, effects between cultivar and conditions during seed development may be confounded in the data. To avoid these complications, James et al. (1967) obtained seeds of fifty-two cultivars of six species of vegetables from three successive crop years, and found significant difference in seed longevity among cultivars of all six species under the same storage conditions. Thus, they confirmed work of others that there are differences in seed longevity not only among species but also among cultivars.

B. Effect of Storage on Genetic Change

Plant breeders have often expressed the fear that during long-term storage there will be a genetic drift in stored seeds, resulting after many years in a population genetically different from the original one. Their fears are grounded in fact, but if proper storage practices are followed the dangers can be mostly avoided.

The genetic drift can be ascribed to at least two causes. First, as a stored lot ages, many seeds die, and it is assumed that the few that survive will not have the same average genetic composition as the original lot had. The life of an individual seed in a seed lot is a function of the environment during its development and preparation for storage, as well as genetic propensities for longevity. It is, therefore, as likely that the early death of some seeds in a lot is due as much to their physiology and injuries as to their genetic composition. Even so, it is possible that a seed lot which loses a major percentage of its germination may be subject to genetic drift. The answer to this form of genetic drift is to store the seeds so well that little or no germination is lost during seed storage. Much of the discussion in this chapter will be to show how, by proper techniques of storage in keeping seed dry and cool, it is possible to keep seed lots without loss of germination for many years.

Genetic drift in seeds is also caused by an increase in the proportion of mutations as storage time increases. This is well documented for several species. Nawaschin (1933), working with aging Crepis seeds, was one of the first to report that chromosome aberrations increased as the seeds aged. It appears that mutation rates are minimized by the best storage conditions of low seed moisture and cool temperature.

With perennials it is possible to promote genetic purity by maintaining individual plants or clonal lines, but this is more expensive than storing seeds. With annuals and biennials it is necessary to grow a seeded crop each year or every other year if long-term storage is not practiced. The amount of genetic drift in annual or biennial seed production of breeding material is very great, because of climatic differences, recombination, outcrossing, and careless help.

Thus, proper long-term seed storage, where high germination is maintained, is far superior to alternative methods in terms of cost and, for most crops, in terms of maintaining genetic purity of the breeding material and gene plasm. Outstanding examples of properly built storages for long-term storage of germ-plasm seeds are the National Seed Storage Laboratory, U.S. Department of Agriculture, at Ft. Collins, Colorado (James, 1967), and the Japanese National Seed Storage Laboratory at Hiratsuka (Ito and Kumagai, 1969).

C. Seed Moisture Content

One of the two main factors influencing seed longevity is seed moisture content. Over most of the range, the higher the seed moisture the more rapid is the decrease in germination capacity. Yet, at extremely low moisture contents of seeds, a slight increase in rate of loss in germination also occurs. This section examines environmental influences on seed moisture contents. It also discusses the physiology and biochemistry of seeds in relation to the effects of seed moisture content on seed longevity.

1. RELATION TO AMBIENT RELATIVE HUMIDITY

Since seeds are hygroscopic, their moisture content comes to equilibrium with the relative humidity of the atmosphere around them.

Relative humidity is a measure of water vapor in air relative to the amount that air can hold at saturation. For example, at 20°C, 1 kg of dry air can hold 14.8 gm of water vapor. If it contains only 7.4 gm of water vapor, then the RH is 50% (7.4/14.8 = 50%). As air warms, however, the amount of water vapor it can hold increases rapidly (see following tabulation):

Temperature (°C)	Water vapor/kg dry air at saturation (gm)
0	3.8
10	7.6
20	14.8
30	26.4

It can be seen that in this range the amount of water that a kilogram of air can hold at saturation approximately doubles for each 10°C rise in temperature. Thus, if the air in the example above is cooled to 10°C the RH will rise to 97% (7.4/7.6–97%), or if it is heated to 30°C the RH will drop to 28% (7.4/26.4–28%). If this air is cooled much below 10°C the air will exceed the saturation point and water will condense as liquid water on a surface (such as a seed surface). The temperature at which condensation of water occurs is called the dew point. As long as the moisture content of the seeds is below equilibrium with the relative humidity, a gradient will exist, with water vapor moving from the atmosphere to the seeds, lowering the relative humidity of the atmosphere and raising the seed moisture content. Likewise, if the moisture content of the seeds is above equilibrium with the relative humidity, a reverse gradient will exist, with water vapor moving from the seed to the atmosphere, lowering the seed moisture content and raising the relative humidity.

If the volume of air is very large in relation to the volume of seeds, as in a well-ventilated seed storage, then seed moisture will come to equilibrium with the relative humidity of the air, and there will not be a measurable change in air relative humidity. However, if the seeds are in a sealed moistureproof container of small volume, then the seed moisture content will remain essentially constant, with the enclosed air reaching a relative humidity in equilibrium with the seed moisture. The reason for this apparent contradiction is that, in an open storage, there is a relatively small amount of water to change in the seed, in proportion to an almost infinite amount of water in the atmosphere. In a sealed container containing 1 kg of seeds, the weight of air might only be 1 gm. At 20°C and 25% RH, this air would then contain only 0.004 gm of water, compared with approximately 60 gm of water in the seeds. Thus, a major shift in the water content of the air in the container would result in an unmeasurable shift in the water content of the seeds. Hence, depending on the relative quantities of water available, either the seed moisture content or the air relative humidity will determine the equilibrium values reached.

Even at a specific temperature, different seeds do not have the same equilibrium moisture contents at the same relative humidity. This is illustrated in Table III. Different seed chemicals have different hygroscopic equilibria, as discussed by Hlynka and Robinson (1954). Relevant to the present discussion is the fact that proteins are the most hygroscopic compounds in seeds, with cellulose and starch less so, and lipids essentially hydrophobic. Thus, at the same relative humidity, a seed high in protein or starch content and low in oils will have a much higher moisture content than will a seed high in oil content. In Table III it is possible to compare at 45% RH, for example, sorghum seed (with an equilibrium moisture content of 10.5%) with flaxseed (having one of only 6.5%). Thus, if the approximate oil content of a seed is known, its moisture content at a given relative humidity can be approximated from Table III even though the particular species is not listed.

There have been many exact determinations of the moisture content of seeds of a number of species in equilibrium with a wide range of relative humidities. The values are frequently given to 0.01%. Such precise moisture determinations are not actually possible with present methods of measuring seed moisture. Further, and even more important, several important variables render unrealistic such a precise value of equlibrium moisture for seeds of a given species. During the development of a seed the environment will influence its chemical composition, and any change in the percentage of oils or proteins or thickness of the

TABLE III

APPROXIMATE MOISTURE CONTENT OF SEEDS IN
EQUILIBRIUM WITH AIR AT VARIOUS RELATIVE HUMIDITIES[a]

Seeds	Relative humidity (%)				
	15	30	45	60	75
Cereals (starchy)					
Rye	7.0	8.5	10.5	12.0	15.0
Rice (milled)	6.5	9.0	10.5	12.5	14.5
Sorghum	6.5	8.5	10.5	12.0	15.0
Corn (maize)	6.5	8.5	10.5	12.5	14.5
Wheat	6.5	8.5	10.0	11.5	14.5
Barley	6.0	8.5	10.0	12.0	14.5
Oats	5.5	8.0	9.5	12.0	14.0
Vegetables (starchy)					
Spinach	7.0	8.0	9.5	11.0	13.0
Pea	5.0	7.0	8.5	11.0	14.0
Bean, snap	5.0	6.5	8.5	11.0	14.0
Oil seeds					
Soybean	–	6.5	7.5	9.5	13.0
Flaxseed	4.5	5.5	6.5	8.0	10.0
Vegetables (oily)					
Tomato	6.0	7.0	8.0	9.0	11.0
Carrot	5.0	6.0	7.0	9.0	11.5
Cucumber	6.0	7.0	7.5	8.0	9.5
Lettuce	4.0	5.0	6.0	7.0	9.0
Cabbage	3.5	4.5	6.0	7.0	9.0

[a]25°C, moisture content wet basis, in percent.

seed coat will result in a different moisture equilibrium value. Variations in seed size will change the proportion of the compounds and, therefore, give different moisture equilibrium values. Hysteresis effects are even more important in influencing moisture equilibrium values. Seeds and many other hygroscopic materials have higher equilibrium moisture values when they are dried from a high moisture content than when they regain moisture from a low moisture content. Such hysteresis effects are supposed to be caused, during desorption or drying, by a molecular shrinkage which reduces the availability of sorptive sites on resorption. The hysteresis effect can account for as much as a 2% moisture difference between extremely dry corn seeds and freshly harvested high-moisture seeds brought to moisture equilibrium with the same RH in the range of 30 to 70%.

The data of Table III also imply temperature effects in seed moisture content at equilibrium with a given relative humidity. The table gives moisture equilibrium values for 25°C. At cooler temperatures the equilibrium moisture content will be slightly higher at the same relative humidity. This is so because the water molecules have less energy at the cooler temperatures and more water molecules are adsorbed on the macromolecules and capillary surfaces of the seed. The difference in equilibrium moisture can be as great as 1% with extremes in storage temperature.

For all of these reasons the values in Table III can be only approximate. With extreme examples, the moisture equilibrium value at a given relative humidity may vary as much as 1.5% from the value stated in the table.

The moisture equilibrium curve as a function of relative humidity for a given substance is sigmoid or, if data of Rockland (1969) are substantiated, really a combination of three straight lines. There is a rapid rise in equilibrium moisture between 0 and 25% RH, a gentle rise between 25 and 70% RH, and a final rapid rise from 70 to 100% RH. The shape of the moisture equilibrium curve is related to the way the water is held by the seed. Rockland (1969) summarizes the three types of water related to the three parts of the equilibrium moisture curve:

1. Monolayer, frozen or iceberg, hydrate localized, polar site, bound, or oriented water. In this type the water molecules are regarded as being bound to ionic groups, such as carboxyl and amino groups, and exist as a monolayer around macromolecules.

2. Multilayer, chemisorbed, intermediate water. In this type, water molecules are regarded as hydrogen-bonded to hydroxyl and amide groups and loosely bonded for several layers above the monolayer.

3. Mobile, free, capillary, or solution water. In this type the unbound,

free water is held by capillary forces, and solute constituents cause a lowering of vapor pressure.

2. PROBLEMS OF STORING SEEDS OF HIGH-MOISTURE CONTENT

The problems of maintaining seed germination increase with seed moisture content. This can be generalized as follows:

Seed moisture above 40–60% — germination occurs
Seed moisture above 18–20% — heating may occur
Seed moisture above 12–14% — fungi grown on and in seed
Seed moisture above 8–9% — insects become active and reproduce

Thus, the problems of high seed moisture include germination during storage, fungal attack, and insect attack besides the serious longevity-halving effect with each 1% increase in seed moisture.

Normally seeds are stored with a moisture content low enough so that germination, with resulting death of the seed, will not occur in storage. If germination has progressed very far, the seed cannot be redried without killing both seed and seedling. Unfortunately, even though the seed is too dry to germinate when placed in storage, conditions of storage can be such that the seed moisture level rises to the level where germination occurs. Obviously, in areas of high rainfall, if the storage is not protected against leaks, water may enter, wetting the seed to a germination level. Even when the storage is built with protection against direct entrance of water, the walls or floor of stone or concrete may be so porous as to allow water penetration. A common problem in large seed storages is movement of water vapor from a warm area to a cool area. For example, along outside walls in cold climates, where cooling is appreciable, the dew point is reached and water condenses on the walls and adjacent seeds. Such accumulations of moisture are prevented by proper construction and proper ventilation of storage facilities.

In the seed moisture range of 18 to 14%, heating can occur — a most serious problem. Heating is caused by the respiration of seeds, of fungi and bacteria in and on the seed, and of insect populations, which may build up rapidly in such a moist environment. High temperature and high moisture will kill seeds rapidly, as will invasion of microorganisms and insects. Further, the heat generated may be so great as to cause charring and even combustion of seeds. Sometimes fires result, which may destroy the seeds and the seed storage as well. To avoid overheating, it is necessary to avoid storing of seeds with a high-moisture content and to use ventilation when necessary.

At seed moisture contents that are lower but still too high for safe seed storage (e.g., 12–20%), the activity of microorganisms, particularly fungi, can be great. The higher the moisture content in this range the more rapid is the growth of organisms and the greater is the danger that they will destroy the capacity of seeds for germination. Christensen and Kaufmann (1969) have published a thorough and comprehensive review of storage fungi. They state that storage fungi can grow on seeds in equilibrium with an RH of 70 to 90%. These fungi are almost entirely species of *Aspergillus* and *Penicillium,* which are almost ubiquitous on decaying organic matter. Although these organisms generally do not invade seeds in the field, they rapidly attack stored seeds of high moisture contents. As seen in Table III, in the 70–90% RH range, starchy seeds have moisture contents of 13 to 20%, and oily seeds have contents of 10 to 19%. Since it is almost impossible to keep seeds free of storage fungi, the easiest and best alternative is to keep seeds dry—in equilibrium with an RH of 65% or drier. Under such conditions, these organisms cannot damage seeds.

At moisture levels below those conducive to fungus activity, insects continue to be active and to breed. At seed moisture contents below 8 to 9%, little or no insect activity occurs and no insect reproduction will take place. An exception is the kapra beetle (*Trogederma granarium*). Although this insect is active and can breed at seed moisture contents below 8%, Mookherjee (1962) found that if the air around the seed dropped below 14% in oxygen content the kapra beetle could not reproduce. If seed moisture content is maintained below 8% and, in areas of kapra beetle infestation, the storage is partially or completely sealed so that respiration reduces the oxygen content below 14%, insects cannot survive in the seed. Insects affecting stored seeds are discussed in more detail in Chapter 4 of this volume.

3. PROBLEMS OF STORING SEEDS OF LOW-MOISTURE CONTENT

When Boswell *et al.* (1940) showed that dry seeds survived longer than high-moisture seeds in storage, there was some speculation that further drying to 0% moisture content would be even better. However, even though extremely dry seeds can be stored for a long time without loss in germination capacity, an intermediate-moisture content gives maximum seed longevity. Kosar and Thompson (1957) first illustrated this with lettuce. Seeds stored at 10°C at RH below 25% showed a more rapid loss in capacity for germination (a loss of less than 10% in 4 years) than did seeds stored at 34 to 58%. At 67% RH or higher, all seeds were dead in 4 years.

The explanation of this anomaly had been found by research on food dehydration sometime before the effect was noted in seeds. Lea (1962) discussed the oxidative deterioration of food lipids. At low-moisture contents in the monomolecular moisture range, as further drying occurs, a part of the monomolecular water layer is removed from the macro-molecules, removing a layer that is protective against oxidative processes. This oxidation can be the result of oxygen penetration, ultraviolet light, and metallic ions present in the seed. The most serious oxidation is the autoxidation of lipids, which, once started, can continue as a chain re-action. The seed contains natural antioxidants, the tocopherols, but these are used up during storage as oxidation occurs, and no more are pro-duced in very dry seeds. The free radicals, produced by autoxidation of the lipids, are very reactive even in seeds of low-moisture content. They can combine with proteins, destroying enzymes and lipoprotein mem-branes and, with nucleotides, destroying DNA and cell reproductive ability. Thus, cells are destroyed, and if enough cells in the meristematic regions are dead the seed can no longer germinate.

It has been found that the breaking of the monomolecular water layer begins at about 25% RH and becomes increasingly severe as the RH is decreased below 25%. Thus, the longevity of seeds is maximum in the RH range of 20 to 25%.

D. Seed Temperature

The other main environmental factor affecting seed longevity is storage temperature. The cooler the temperature the more slowly seed vitality declines. This rule apparently continues to apply even at temperatures below freezing. Since tissues containing free water develop ice crystals in long-term storage below freezing, seed in such storage should have a moisture content in equilibrium with 70% RH or lower, below the seed moisture level containing free water.

Ways of ventilating, insulating, and refrigerating storage facilities to maintain temperatures as low as possible are discussed in Section IV,C. Just as with high relative humidities, high temperatures are conducive to the activity of microorganisms, especially insects. At 5°C and below, insects become inactive.

The role of high temperature in speeding seed deterioration is not known. It is generally assumed that the high seed respiration at high temperatures is related in some way to rapid loss in germination, but it is quite obvious that the cause of death is not depletion of stored foods. Further, the speed of deterioration has a Q_{10} of around 4, as Harrington's rule-of-thumb indicates, compared with a Q_{10} of 2, which is considered normal for respiration. Q_{10} value is the quotient of two biological rates at temperatures differing by 10°C. Some enzymic reactions have a Q_{10} of 4, so possibly enzymic reactions are involved. The effect of temperature on seed longevity remains a fertile field for future research.

IV. Storage of Seed

A. Storage from Physiological Maturity until Planting

It is customary for seedsmen, plant breeders, botanists, and others interested in storage of seeds to give primary attention to rooms or buildings labeled as seed storages. However, it should be restated with emphasis that seeds are in storage from the moment they reach physiological maturity until they germinate, or, if the storage is poor, until the moment they are thrown out because they are dead. This storage period includes the time on the plant between maturity and harvest, where weathering is often a serious factor. In many crops the seed has lost much of its vigor and is already low in germination before it is harvested. The operations of harvesting and cleaning may injure the seed. But, beyond this injury, the period of harvesting and cleaning is frequently one of high temperatures. During these times seeds may still have a high-moisture content. Seed deterioration can be rapid during these periods. Transport from field to storage, between storages, and from storage to the planting site involves periods of storage also, during which deterioration can be serious. For example, seeds in a railroad boxcar on a siding for a few days in a hot humid climate may lose germination capacity. A grower in the Imperial Valley of California may have the seeds in the back of a pickup truck for a day, with air temperature as high as 44°C and the seeds reaching a temperature of 60°C. Thus, in considering effects on germinative capacity, the complete storage life of the seed is involved rather than only the time spent in a storage facility. The principles of good seed storage apply equally well to all stages in the life of the seed, although the discussion below centers on storage buildings.

B. Need for Long- or Short-Term Storage

Prior to storing seeds a decision must be made on how long it will be necessary to maintain the germination capacity of seed lots. because longer storage requires more exacting storage conditions. The two kinds of problems that must be dealt with in seed storage concern pests (including humans, birds, rodents, insects, and microorganisms) and the environment (e.g., temperature, relative humidity, light, radiation, oxygen, carbon dioxide). The control required will depend largely on the length of storage desired. As the requirements become more exacting, the cost of the storage facility per unit of seed stored increases rapidly. The types of storage needed can be related to the time of storage expected and can be placed in four types: (1) storages for commercial seeds; (2)

storages for carryover seeds; (*3*) storages for foundation, stock, and enforcement sample seeds; and (*4*) storages for germ-plasm seeds. These are discussed in the order of length of storage involved.

1. COMMERCIAL SEEDS

About 80% of the world's need for seed storage is for keeping commercial seeds from harvest to the next planting season, a period ranging from a few days to 8 or 9 months. For most species the requirements for seed storage are relatively simple. If the region of storage is one without rainfall and of low relative humidity from harvest to planting, as in the Central Valley of California or the Nile Valley of the United Arab Republic, a smooth bare place of ground may be reasonably satisfactory. In fact, bare ground is used as a seed storage in both areas. However, loss from stealing (by man, birds, and rodents) can be considerable, and germination capacity can decline, at least in the surface seed, from heat damage and ultraviolet rays. Since most of the world is not endowed with such a favorable storage climate, stored seeds normally need shelter. Storage facilities vary greatly, ranging from the Pusa bin of India (Pradhan and Mookherjee, 1969) and woven bags of seeds hanging from the rafters in many rural houses of the world, to elevators containing thousands of tons of seeds and grains in many rural towns in the Midwest of the United States.

To store seed successfully from harvest to planting, these various storages must have certain features in common. First, seeds placed in storage must be cleaned to be free of trash, which may harbor insects or fungi and prevent free circulation of air. It must also be undamaged, to minimize decline in vigor and germination. It must be dried to a moisture content of less than 14% for starchy seeds and less than 11% for oily seeds. Thus, proper seed storage begins with sound seeds, on the smallest farm or with the largest seed handler.

Second, the storage must be constructed so that rain cannot enter and no serious gain in moisture will occur during this short storage period. The low humidity of warm air passing from the cooking fire through the seeds hung from the rafters in a rural home keeps the seeds dry. The Pusa bin is lined with plastic which is sealed to keep seeds dry, and large elevators have only small open areas in proportion to the volume of seeds they hold, to that seeds remain dry.

Third, the storage must not be a heat trap, allowing seed to reach excessively high temperatures. An example of a heat trap unfit for seed storage is a steel bin exposed to the sun in a hot climate.

Fourth, efforts must be made to prevent attacks of rodents and insects by closing all openings after the storage is filled. Unfortunately, many structures used for storage harbor rodents and insects in earth floors or cracks in walls (especially wooden walls) or allow entry via loose-fitting doors and windows.

2. CARRYOVER SEEDS

Most of the remaining 20% of stored seeds is carried over through one growing season to the second planting time following harvest of the seed. This storage period is usually between 1 year and 18 months. In many dry regions of the world, such as the arid western United States, and in many cool regions, such as England, seeds of most cultivated species will maintain a high germination capacity for as long as 18 months with only the bare shell of a storage. However, in warm humid areas, such as the southeastern United States and the monsoon area of Southeast Asia, seeds deteriorate rapidly and will lose capacity for germination if kept as carryover seed in a simple seed warehouse. Also, seeds of soybeans, cotton, onions, and several flower and tree species lose germination capacity rapidly under warm humid storage and may deteriorate in periods as short as 2 or 3 months in common storage under these climatic conditions.

Therefore, in warm and humid areas, and for seeds of these species, more substantial storages are needed. The first requirement is some insulation to keep the storage as cool as possible. One possibility is a false ceiling with ventilation between ceiling and roof. Heat inflow will be reduced by thick stone or brick walls or by a layer of insulation in walls or ceiling. Ventilating fans to bring in cool night air can help, if it is not so humid as to raise the seed moisture contents to a level that allows attack by storage fungi.

The second requirement is to keep the seed dry. This can be done by proper bulk storage, with only the seeds close to the walls exposed to the possibility of excessive hydration. If bags are used, storage on pallets will keep the seed from direct contact with a damp floor. Sealing the floor against moisture penetration is also useful. Storage of seeds in steel bins with tight-fitting lids or in moistureproof bags will solve the problem of moisture penetration providing the seeds are already dry enough for sealed storage.

3. FOUNDATION, STOCK, AND ENFORCEMENT SAMPLE SEEDS

This group of seeds may need to be kept for several years since genetic drift is minimized by reproducing foundation or stock seeds as seldom as is practical. Enforcement seed samples must be kept for a year or more with, as near as possible, a germination percentage the same as when the sample was drawn. These stringent specifications call for much better seed storage facilities than are needed for commercial or carryover seeds.

Since the quantities of seeds involved are not large, the storage room is only a small part of the total storage area and, in fact, is often a small room within a large seed warehouse. Required for this storage is a combination of relative humidity and temperature that will maintain germination without loss for 3 to 5 years. This is usually accomplished with an RH of about 25% at 30°C or less or an RH of about 45% at 20°C or less.

The required relative humidity can be achieved by making the room moistureproof and using a dehumidifier. An alternative with many small seed lots is to use moistureproof metal boxes or polyethylene bags, at least 7 mils thick, maintaining the proper relative humidity with a desiccant in the box or bag. Since these containers would be opened from time to time the desiccant would have to be redried at intervals.

The proper temperature for seed storage occurs naturally in some areas, but in hot areas, refrigeration is necessary. If refrigeration is to be efficient the room must be insulated.

4. GERM-PLASM SEEDS

Germ-plasm seeds are meant to be kept for many years, perhaps centuries. Basic requirements are the coldest temperatures economically possible and seed moistures in equilibrium with 20 to 25% RH. It should be remembered that seeds deteriorate slightly faster in equilibrium with an RH below 20% than at 20 to 25% RH. Germ-plasm storages built so far have rooms which can be maintained at 5° and −10°C and 30% RH. In addition, the stored samples are dried to the proper moisture level and kept in moistureproof containers.

C. Construction of Seed Storages

It should be clear that the longevity of a seed lot is intimately related to its storage facility. There must be protection against theft and against rodents, birds, insects, and fungi that might enter the storage and destroy the seeds from outside. There must also be control of temperature and relative humidity to minimize biochemical destruction of the seeds. This section discusses methods of seed storage construction that will give the necessary protection and control. Figure 2 illustrates a desirable common seed storage, and Fig. 3 illustrates the construction needed for long-term storage of seeds.

1. PROTECTION AGAINST THEFT, RODENTS, BIRDS, AND INSECTS

Fortunately, a good seed storage should have no windows and only one door, thus minimizing the chances of theft. If the door is sealed properly against rodents and insects and locked when the storage is not in use, theft is not likely.

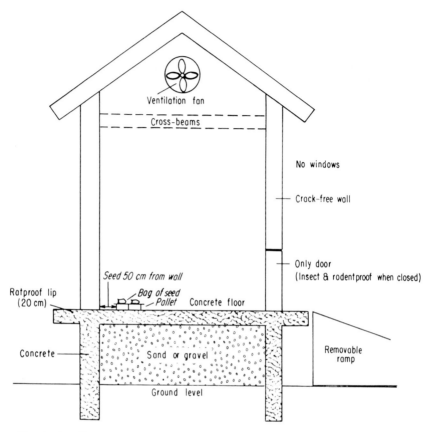

FIG. 2. A well-constructed common storage for seeds. Note the fan for ventilation, the above-ground floor and pallets for better moisture control, the concrete lip and removable ramp for rodent control, and the concrete floor and crack-free walls for better insect control.

Rodents, usually rats, destroy a tremendous amount of grains and seeds. A study in India arrived at an estimate of a 9% loss of grains and seeds to rodents between harvest and use. Rats are common in storages in every country in the world. Seed storages should, therefore, be constructed to be as rodentproof as possible. Wooden construction is less desirable than brick, stone, concrete, or metal, but even wooden storages are acceptable if the stone or concrete foundation extends 3 ft above the ground. The entrances (preferably only one) should also be 3 ft above the ground, with a removable driveway. There should be a lip around the building at the 3 ft height extending out 6 in. Such construction makes entrance by rodents through the walls virtually impossible as long as the foundation remains uncracked. Rodents sometimes enter up through dirt

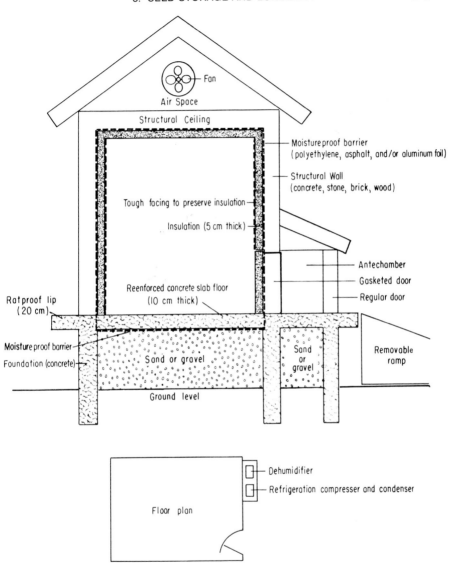

FIG. 3. A plan of a storage for long-term storage of seeds. It contains all the features of the common storage, shown in Fig. 2, plus insulation and moistureproofing with a refrigeration unit and dehumidifier for temperature and moisture control.

floors or between cracks in concrete or stone floors. This means the floor must be paved, and any cracks that may develop must be repaired. Of course, the door must be rodentproof. If these construction precautions are taken, rodents can enter the storage only through the open door.

Unfortunately, this occurs too often when loads of seeds, empty bags, or other material are hauled into the storage. Care can be taken to prevent rodents entering in this way.

If the seed storage is constructed to be rodentproof, if care is taken to prevent rodents from entering in materials brought into the storage, and if proper sanitation is practiced, then rodents seldom if ever invade the storage. Rodents in seed storages are a much bigger problem than they have any reason to be. If, in spite of all precautions, rodents invade the storage, control measures become necessary. These call for absolute cleanliness in the storage, no piles of loose seed, no piles of rubbish, all cracks filled, and no torn sacks, so there are no nesting places. Finally, rodenticides may be required.

Storage insects can be even more damaging than rodents and are more difficult to control. In addition certain species infest seeds in the field and enter storage in the seed. Storage insects are less of a problem in temperate latitudes than in the tropics. Even in the tropics, however, a storage can be built that minimizes the chance for infestations, and measures can be used that will eliminate them.

Construction of the floor, walls, and ceiling of the storage should be such that no cracks exist which can harbor insects, Particularly difficult to prevent are cracks between floor and walls, and between ceiling and walls. Wooden construction is especially prone to insect-harboring cracks, but brick and stone construction may also have them, and cracks can even develop in concrete walls and floors if the foundation is not properly firm and reinforcing is insufficient. Plaster, insulation, and plywood, properly applied, can be used to minimize cracks. Ventilation openings should be screened against insects. All cracks around openings, such as electric conduits, ventilation openings, and doors, should be thoroughly sealed.

The first requirement for insectfree seed storage is a building that itself cannot harbor insects. In a regular maintenance program, any insect-harboring cracks which develop should be repaired immediately.

As with rodents, constant sanitation is necessary. There should be no place for insect breeding, such as piles of floor sweepings or other litter. A yearly painting of the interior with a residual insecticide will further minimize the possibility of the building itself being a source of infestation of the seed. Sanitation should extend outside the building. Discarded seed and cleanings should be hauled away, not dumped just outside the door and left to harbor storage insects.

Once the seed storage is completely free of insects, the most serious source of reinfestation is infested seed which is brought in. Seed may be brought from the field already infested or it may be transferred from an infested storage. Such infestation is controlled by fumigation. Rather than fumigate the whole storage it is better to have a fumigation room or to fumigate the seed on a concrete slab under a tarpaulin before it enters the main storage room. The fumigation room should have its own door to the outside and only after fumigation should the seed be brought into the storage area.

Numerous effective fumigants are available, including carbon disulfide, hydrocyanic acid gas, methyl bromide, aluminum phosphide, and naphthalene. Cotton (1956) discusses methods of fumigation of grains and seeds for insect control. It must be mentioned here that fumigation (particularly repeated fumigation) may seriously reduce vigor and even the germination capacity of seeds. This is particularly true of seeds with high moisture content. Seeds with moisture contents greater than 14% should be dried to below this value before fumigation. High temperature also increases damage to seeds by fumigants. Hence, fumigation is used only with entering seed, and all other measures are used to maintain insect control in the stored seed. Prophylactic biweekly or monthly fumigation of seed storages could lead to serious germination problems. It is far better to build an insectproof storage and make certain that seeds, bags, or anything else placed in the storage are insectfree. Of course, if a storage containing much seed does become infested, then fumigation is required.

2. TEMPERATURE CONTROL

One of the two most important environmental factors influencing seed longevity is temperature. The lower the temperature the longer the seeds maintain germination capacity. Thus, temperature control is an important consideration in building a seed storage. The longer germination must be maintained, the more important is a cool or cold storage. In most parts of the world, however, storage becomes increasingly expensive as storage temperature is reduced. Thus, commercial seed storages can afford only minimal temperature control—which is normally all that is needed. For storage of germ-plasm seeds, in contrast, maximum longevity is essential, requiring lowest temperatures and maximum control despite the expense. Temperature control may be achieved by ventilation, insulation, and refrigeration. These methods are not mutually exclusive and are normally used to supplement each other.

a. VENTILATION. It would seem that whenever the outside temperature is below the temperature of the seeds in storage, ventilation should be in order. The situation is somewhat more complicated, however, and the relative humidity of the outside air in relation to the seed moisture content must also be considered. It does little good, and may be harmful, to lower the temperature if at the same time the seed moisture content is being raised. Thus, proper ventilation of a seed storage requires consideration of the relative humidity of the ventilating air as well as its temperature.

Following is an illustration of the problems involved in proper ventilation of a seed storage. Because of hot weather, wheat seed in storage has reached 30°C. Its moisture content is 12.5%, in equilibrium with about 65% RH. At night the outside air cools to 20°C with a RH of 80%. Is it then desirable to ventilate the seed? Or will the high humidity increase the seed moisture even while cooling the seed? The cool air will warm as it absorbs heat from the seed, approaching the seed temperature, and its RH will drop to 45%. Thus, both cooling and drying will occur, and ventilation under these circumstances will be highly desirable. If, in contrast, the night air is at 25°C and 90% RH, its RH upon heating to 30°C would have dropped to only 68%, so any reduction in seed temperature would be accompanied by an increase in seed moisture.

Frequently ventilation can be used to reduce both seed temperature and seed moisture content. Ventilation can be used to reduce seed temperature if the ventilating air is dry enough so that seed moisture content is not seriously affected. Ventilating air can also be used to dry seed further, even though its temperature may be the same as the seed temperature, or even somewhat higher. Two precautions are important: (1) moisture content of the seed should not be allowed to increase to a value in equilibrium with air RH above 65%; and (2) the seed temperature should not increase to above 33°C for more than a few hours.

Seed temperature and temperature of the outside air can be measured with thermometers. Relative humidity of the seed storage and the outside air can be measured with a psychrometer. The ventilating fans can be turned on whenever the outside air temperature and relative humidity are low enough to benefit the seeds by either reducing seed temperature or seed moisture content. Either manual or automatic operation is possible.

Ventilation also serves two other purposes. If the moisture content of seed is high or if there are pockets of high-moisture seed in a bulk storage, heating can result. A slow flow of air through the seeds (1–2 m³ air/m³ seeds) will dissipate the heat, preventing hot spots from developing. A dangerous condition can develop in unventilated storages, particularly in poorly insulated ones. If it is cold outside but the sun heats one side of the storage, or if seeds are warm in the center of the storage, convection currents of air will be created as the air expands in the warm areas and contracts in the cool ones along the cold sides or floor. The warm air will have low relative humidity, and moisture will move from the seeds to the warm air. This warm air will then move to a cool area and become cooled. This increases its relative humidity, and in this area the moisture then moves from the now high relative humidity air into the seeds, increasing their moisture content. Under these conditions, moisture content may be too low for storage fungi on the average but high enough in the cooler areas for rapid growth of fungi. Again, a slow flow of air by ventilation fans will prevent the formation of convection air currents and tend to maintain a uniform seed moisture content and temperature.

b. INSULATION. Heat flows from a warmer area to a cooler one, lowering the temperature of the warmer area and raising the temperature of the cooler area until the temperatures are equal. A seed storage is insulated to reduce the flow of heat from the hot outdoors into the storage. Insulation sometimes is undesirable in retarding flow of heat from the storage to the out of doors. By judicious ventilation, however, as discussed above, heat flow to a cooler external environment can be increased even in a well-insulated storage. Under refrigeration, the storage temperature is almost always lower than that outside, so insulation is imperative or the cost of refrigeration becomes prohibitive.

Heat moves from one material to another by three different physical means: (1) conduction, in which the energy of warmer molecules transfers to cooler molecules in materials, conducting heat from a warmer to a cooler place; (2) convection, in which warm or cool molecules are physically moved from one place to another, as in the movement of air, an extremely important factor in cooling by ventilation or refrigeration; and (3) radiation, in which thermal waves travel through space from a warm material to a cool one. Smooth materials reflect these waves and reduce radiation heating. Aluminum foil is an excellent reflector of heat radiation.

A seed storage has four kinds of heat that must be controlled. Three of these—field heat, heat of respiration, and incidental heat—can be controlled by ventilation and refrigeration. They are discussed under refrigeration control of temperature. The fourth kind of heat of concern is heat gained by leakage. In a seed storage this is the flow of heat from the warmer outside through the walls, roof, and floor of the storage to the cooler air and seeds in the storage. Heat flow through a material is measured as thermal conductivity, which is a function of the temperature difference between two places in the material. Heat flow is twice as fast with a 20° temperature difference as with a 10° difference. Thermal conductivity is also a function of the distance the heat must flow. Heat flows twice as fast through 1 in. of insulation as through 2 in. of the same material. These are the most important factors influencing heat flow through materials. Table IV gives the relative thermal conductivity of several common materials.

TABLE IV
RELATIVE THERMAL CONDUCTIVITY OF SEVERAL MATERIALS

Material	Relative thermal conductivity
Good insulators	
Air	0.016
Mineral wool	0.023
Glass wool	0.024
Corkboard	0.025
Foam polystyrene	0.028
Fiberboard	0.030
Sawdust	0.030
Fair insulators	
Straw, chopped, dry	0.06
Wheat	0.08
Wood	0.09
Maize 13% H_2O	0.12
Sand, dry	0.19
Poor insulators	
Brick	0.4
Glass	0.5
Stone (marble)	1.2
Concrete	1.2
Conductors	
Steel	26.
Aluminum	117.
Copper	224.

Although air is the best insulator listed in Table IV, it has a serious defect. If the air space is more than a fraction of an inch convection currents occur, moving heat from the warmer to the cooler surface. Therefore, if air is to be a useful insulator, it must be trapped so it cannot move. The main reason why some of the materials listed in Table IV are good insulators is that the solid parts of these insulators trap a multitude of small air spaces. All the good insulators listed are essentially equal in insulating value.

It should be noted also that seeds, as represented by wheat and maize, are also fairly good insulators. This can be advantageous or disadvantageous, depending on seed temperature, type of storage, length of storage desired, and outside temperature.

As can be seen in Table IV, the good insulators are about 15 times as good as brick and over 40 times as good as stone or concrete. This means that a 2.5-cm-thick slab of glass wool has essentially the same insulating value as over 37 cm of brick or 1 m stone or concrete. In economic terms, insulating the walls and roof of a storage with 5 cm of one of the good insulators, reduces to one-third the amount of refrigeration needed to remove heat leaking into the storage, This, of course, is an approximation. A refrigeration engineer can calculate more exactly the heat leakage to be expected for a specific storage under prevailing external and desired internal temperatures. He can calculate the reduction in heat leakage to be expected from insulating the storage. Heat leakage into a storage is dramatically reduced by a combination of good conductive insulation and radiation-reflecting insulation.

c. REFRIGERATION. For long-term storage of seed and even for carryover seed in hot climates, refrigeration may be necessary to keep the storage temperature below the usual ambient temperatures. A storage kept at 5°C requires refrigeration for part of the year in all crop-growing areas of the world, and in tropical areas needs year-around refrigeration. Except in the extremely long-term storage of germ plasm, an alternative to refrigeration is storing the seed dry, either by using dehumidification or by drying and storing in sealed containers. These methods are discussed below.

Heat from four sources must be removed by refrigeration. The major source is heat leakage from outside. The greater the temperature gradient, the more serious is this source. Insulation, as discussed above, reduces heat leakage but does not eliminate it. The better and thicker the insulation, of course, the less refrigeration is required because leakage of heat is reduced.

The second greatest source is "field heat," or heat that must be removed from the seed or anything else placed in storage, reducing its temperature to that of the storage. Refrigeration capacity must be sufficient to remove field heat in a reasonably short time, probably within a month if the storage is filled at one time.

The third source is heat of respiration. With fresh vegetables and fruits, that is a major source of heat in storage, whereas in seed storage it poses only a minor problem because dry seeds have an extremely low respiration rate.

The fourth source is incidental heat, which includes heat from electric lights and external heat that enters a storage when the door is opened. Since this can be a major source if the door is left open for long periods, employees must be trained to open and close the door each time they enter, rather than blocking the door open. The heat of respiration of people in the storage, which is usually included in incidental heat, may be appreciable. In determining the amount of refrigeration needed, refrigeration engineers usually calculate the amount of heat from leakage through insulation, from field heat, and from respiration, and then arbitrarily add 10% more as incidental heat.

Refrigerating a seed storage depends on the same principle used in the home freezer, refrigerator, or air conditioner. A gas is used which can easily be compressed and condensed to a liquid, such as Freon or ammonia. The compression and condensation must be done outside the storage area, because heat is released in the process. The heat is dissipated by air or by water cooling. The liquid is pumped into the storage area, where it is allowed to expand and vaporize in the cooling coils. Both of these physical processes absorb heat from air blown over the coils. The air is thus cooled and circulates through the storage, thereby cooling the seed. If the storage RH is 65%, the air temperature is 25°C, and the cooling coils are − 5°C, the air will be cooled below its dew point of 18°C and moisture will be condensed on the cooling coils as frost, which must be periodically melted and drained out of the storage. Thus, some dehumidification is achieved at the higher temperatures. If the refrigerating unit is operated at − 5°C to maintain a 5°C storage temperature, in theory the equilibrium relative humidity could be a minimum of 40% if the refrigerator has the capacity to run nearly 24 hours per day and there is no moisture leakage from the outside. The practical minima of relative humidity at different temperatures are given in the following tabulation.

Storage temperature (°C)	Minimum RH by refrigeration (%)
32	30
27	35
23	40
21	45
19	50
16	60
14	70

Below about 15°C the relative humidity possible with refrigeration alone is too high for proper seed storage. Over a period of time, at 15°C and lower, seeds will gain moisture and reach an equilibrium moisture content which may not be harmful at these low temperatures but will lead to rapid deterioration when the seeds are removed from storage and returned to ambient temperatures. Hence, refrigeration alone is not considered sufficient for storage of seed. Refrigeration storage must be used in combination with dehumidification or with sealing the dried seeds in moistureproof containers before they are placed in refrigerated storage.

3. CONTROL OF SEED MOISTURE

It has been pointed out that the lower the seed moisture content (down to equilibrium with 20 to 25% RH) the longer the seed will survive. Even commercial seeds, before they are stored, must be dried to the moisture level below which storage fungi can grow (below equilibrium with 65 to 70% RH). It is equally important to keep seeds at this low moisture content or lower for all of their storage life. Therefore, control of seed moisture content is extremely important. Such control can be achieved by several methods, including judicious ventilation, moistureproofing the storage, dehumidification, use of sealed moistureproof containers, or the use of desiccants.

a. VENTILATION. This is an effective technique for reducing the temperature of the storage and of the seeds in it. However, as discussed under Temperature Control, if the outside air reaches an equilibrium relative humidity in the storage that is higher than the equilibrium moisture content of the seed, ventilation can be detrimental to seeds even when cooling them. This is particularly true if the seed moisture content is raised enough to foster the development of fungi.

However, throughout much of the world, seeds are harvested in late summer or autumn. The outside temperature is declining at that time of year, and ventilation can, therefore, usually effectively cool and further dry the seed. For example, seed harvested in late summer might enter storage with a temperature of 30°C, and in temperate climates autumn night temperatures frequently fall to 5°C or lower. Even if the night air is at the dew point at 5°C, this air will have an RH of only 20% at 30°C and can cool the seed to 13°C before the RH of the air exceeds the critical 65%. In many areas of the world, for commercial or carryover seeds, a storage with some insulation can keep seeds cool and dry by proper ventilation alone. In other areas, such as Tampa, Florida, or Calcutta, India, ventilation alone will not be adequate.

b. MOISTUREPROOFING. It is useful to moistureproof a storage and relatively inexpensive — much cheaper, in fact, than insulating a storage. The three most common moistureproofing materials are polyethylene, asphalt, and aluminum foil. To be effective polyethylene should be 10 mils thick, asphalt should be 3 mm thick, and aluminum foil should be bonded by a moisture-resistant plastic to some surface (such as paper) that will keep the foil from cracking. Whichever material is selected, the entire storage must be moistureproofed. This includes placing a layer of the moistureproofing underneath the concrete floor. This should be overlapped with the wall moistureproofing, which, in turn, should overlap the ceiling moistureproofing. All seams should be overlapped, of course. The door or doors should be moistureproofed and gasketed like cold-storage doors. If ventilation openings are included there should be moistureproof coverings over them when they are not in use. All nails and openings for pipe and wire should be sealed against moisture penetration. A storage that is properly moistureproofed will allow essentially no moisture leakage from outside.

To prevent moisture from entering the room as seeds are brought in or removed, an antechamber is desirable at the entrance. The seeds are brought into the moistureproof antechamber, and the outside gasketed door is closed. Only then is the inside gasketed door opened into the main storage. Keeping moisture from entering through the door is even more important than stopping heat, because it costs more to remove moisture.

In constructing a seed storage, moistureproofing and insulating are done together. It is essential that the insulation be on the dry side of the moistureproofing. Water is a good conductor of heat, and if the insulation becomes damp it loses much of its insulating value. Normally, the floor is insulated only if the storage is a room above ground level. The order of materials in the wall, from outside to inside, are the structural materials (concrete, stone, wood, brick, steel sheeting), moistureproofing, insulation, and an interior covering to protect the moistureproofing from physical damage. It is possible to bond the insulation to the structural wall with a moistureproof mastic, such as asphalt, but care must be taken to prevent skips or thin spots. If the storage has only a roof with no ventilation attic, then it is possible to apply the moistureproofing there as the outside layer—as, for example, an asphalt felt or a hot tar roof. The most important consideration is to be sure there are no cracks or pinholes where moisture can penetrate.

c. DEHUMIDIFICATION. As mentioned, refrigeration can also be a means of dehumidification, and if the storage temperature is maintained at 20°C or higher, refrigeration provides an efficient and economical means for maintaining low relative humidity in the storage. Refrigeration may be all that is needed for dehumidification of carryover seed or stock seed in hot regions. Where cool temperatures and low humidities are required, however, other means of dehumidification are needed in addition to refrigeration.

For dehumidification, either solid or liquid spray desiccants can be used. The common solid desiccants are silica gel and activated alumina. Both are nontoxic, reasonably indestructible, cheap, and efficient. Silica gel is a glass (SiO_2) gel made porous by driving off water with heat. This leaves a large amount of microcapillary surface that adsorbs a monomolecular layer of water, as do the macromolecules in seeds. Silica gel can absorb water up to 40% of its dry weight. Table V gives the amount of water that silica gel can absorb at various relative humidities. A bed of silica gel is dried with heat at about 175°C, which removes all of the absorbed water. It is then cooled, and air from the storage is blown through the bed. The bed absorbs moisture from the air until it reaches equilibrium with the relative humidity of the air. The silica gel is then reheated and moisture is driven off to the outside with hot air. There may be two beds in a unit—one absorbing moisture from the storage while the other is being reactivated or a single bed may be rotated so part of it is dehumidifying the storage while another part is being reactivated. The absorption of water releases heat, and the cycle is usually so fast that the silica gel has not completely cooled by the time it is again dehumidifying the storage. Therefore, the discharge temperature is about 10°C above the storage air temperature. This discharge air can be recooled by running it through a radiator containing circulating cold water or running it over the coils of a refrigerating unit. The dehumidifier should be located outside of the seed storage.

TABLE V

MOISTURE CONTENT OF SILICA GEL IN EQUILIBRIUM
WITH SEVERAL RELATIVE HUMIDITIES

% RH	% H_2O adsorbed
0	0.0
5	2.5
10	5.0
15	7.5
20	10.0
25	12.5
30	15.0
35	18.0
40	22.0
45	26.0
50	29.0
55	31.5
60	33.0
65	34.0
70	35.0
75	36.0
80	37.0
85	38.0
90	39.0
95	39.5
100	40.0

It is obvious that the desiccant becomes less efficient as the relative humidity is lowered, because it takes less water out of the air during each cycle. Hence, with all other conditions constant, maintaining 25% RH requires a unit 2–3 times the size needed for 45% RH. For the dehumidifier to work efficiently, in removing moisture from seed or maintaining seed moisture at a low level, the storage must be completely moistureproofed, as described above. Otherwise, the dehumidifier will be working to dehumidify the whole outdoors as moisture seeps into the storage. As many seedsmen have found to their dismay, this is an impossibility.

The other type of dehumidifier depends on use of a saturated salt solution with a low relative humidity equilibrium (e.g., lithium chloride, about 14% RH at 10°C). The air from the storage is blown through a refrigerated spray of the salt solution, which absorbs the moisture in the air down to the relative humidity equilibrium of the solution. The moisture absorbed by the solution is driven off by heating the solution. The advantage of the liquid spray desiccant is that the air returning to storage is of uniformly low relative humidity compared with that of the solid desiccant system, in which the relative humidity of the air is reduced less and less as the desiccant becomes more nearly saturated. In further contrast, air from the liquid spray desiccant returns to the room cold, and its relative humidity drops still further as it warms, whereas the air from the solid desiccant must be cooled, increasing its relative humidity. Therefore, the liquid spray desiccant system is more efficient. Two serious drawbacks must be considered, however. Corrosion is a serious problem and care must be taken to replace corroded parts before a breakdown occurs. Also, lithium chloride is toxic.

d. SEALED CONTAINERS. Another very effective way of controlling seed moisture is to store the seed in sealed moistureproof containers. However, empirical observations have indicated that moisture contents above 12% for starchy seeds and 9% for oily seeds lead to deterioration that is faster in sealed containers than in containers through which exchange of gases and moisture occurs. The moisture range of 9 to 12% for starchy seeds and of 7 to 9% for oily seeds, over which sealed storage is probably better than open storage, is still a controversial matter. Further research is needed to resolve the question of the maximum moisture content for seeds in sealed containers and to study the basic reasons for the change from desirability of sealed storage at low moisture contents of seeds to undesirability at high ones.

If seeds are just dried to safe moisture levels and then stored sealed in moisture-vaporproof containers, the low moisture content of the seed will be maintained even under storage conditions of high relative humidity. Harrington (1963) demonstrated the value of moistureproof containers in increasing seed longevity under high-humidity conditions, and showed the relative resistance of various containers to moisture-vapor penetration. Moistureproof containers include sealed tin or aluminum cans, glass Mason jars with gasketed lids, and pouches of aluminum foil laminated to Mylar or polyethylene.

Figure 4 illustrates a tin can sealed with an aluminum foil–polyethylene laminate and protected with a lid. The laminate can be easily punctured for removal of the seed. Figure 5 illustrates a gasketed glass jar and a gasketed plastic bottle. Moisture-resistant containers include various heat-sealed plastic pouches or bags of a thickness equivalent to 3-mil high-density polyethylene. Cloth and paper bags have no moisture resistance unless laminated to aluminum foil or some moisture-resistant plastic. Larger containers, such as fiberboard drums properly laminated with aluminum foil, and steel boxes or bins, all with gasketed lids, are moistureproof.

Fɪɢ. 4. A tin can for seed packaging. The can is sealed with an aluminum foil–polyethylene laminate which is protected by the lid. The laminate can be easily punctured to remove the seed. Only seeds dried to a moisture level safe for placing in a sealed container should be packaged in this can. (Courtesy of the Rockefeller Foundation, New Delhi, India.)

FIG. 5. A glass jar and a thick plastic bottle both with gasketed lids. These are satisfactory for a long-term storage of seeds dried to a moisture level safe for storage in sealed containers. (Courtesy of the Rockefeller Foundation, New Delhi, India.)

It should be clear not only that properly dried seed stored in moisture-proof containers can be stored in areas of high humidity, but also that storage temperatures are less critical for such seeds. Because the germination capacity of such packaged seeds remains high under adverse temperature and relative humidity, the need for frequent retesting for germination capacity is eliminated. Hence, when moisture-resistant containers are used, the California Seed Law (now followed by the Federal Seed Act and legislation of other American states) permits longer periods up to 3 years between germination tests of seeds packaged below certain maximum moisture contents. Such storage is also being used in maintaining foundation seeds by certifying agencies and in maintaining stock seeds by several seed companies. Seeds so packaged can be shipped through and into tropical areas.

e. USE OF DESICCANTS. Since moistureproof containers are difficult to open and reseal, they are not useful for plant breeders and seed-control officials, who must store many small samples that must be readily accessible. Such samples could be stored in a dehumidified refrigerated room.

Such rooms are in use, though they are expensive. Also, in many areas of the world, electricity is often unreliable, so refrigeration and dehumidification may not always work. A third possibility which can be equally successful in maintaining seed longevity and yet allow ready access to many stored samples is the use of metal boxes with gasketed snap-on lids, with desiccant (such as silica gel) enclosed with the seed samples (see Fig. 6). Silica gel is available with all or some of the granules treated with cobalt chloride. The usual cobalt chloride-treated silica gel turns from blue to pink at about 45% RH. Thus, a quantity of silica gel (1 kg/10 kg of seeds and packets) is dried and enclosed with seeds in the metal box. When the indicator granules turn pink, the silica gel is removed, reactivated by drying in an oven at 175°C, cooled in a sealed container, and returned to the metal box. The seeds are thus kept below equilibrium with 45% RH, a moisture content desirable for several years of storage in a temperature range of 20 to 25°C. The metal box has other advantages. It is rodent- and insectproof as well as moistureproof. The boxes, which are not very expensive, are easily stacked on shelves in a small area (see Fig. 7). Also, seeds in equilibrium with RH of 45% will not be damaged by stored fungi. The only care required is periodic inspection to make certain that the indicator silica gel remains blue.

FIG. 6. A gasketed metal box for storage of seeds in small bags or packets with a desiccant such as silica gel to keep them dry. (Courtesy of the Rockefeller Foundation, New Delhi, India.)

Fɪɢ. 7. Shelving of gasketed metal boxes containing seed samples and silica gel. Thousands of packets of breeding lines, official control samples, or germ-plasm seeds may be safely stored in a small area in this manner.

D. Control Devices

A good seed storage extends the longevity of seeds. Various control devices can be used to tell whether the storage is functioning as it should. These include the primary tests of germination and vigor on samples of the seeds themselves, as well as controlling and checking the seed temperature and seed moisture contents—the two most important factors influencing seed longevity.

1. Gᴇʀᴍɪɴᴀᴛɪᴏɴ ᴀɴᴅ Vɪɢᴏʀ Tᴇꜱᴛꜱ

Seed testing, which is discussed fully in Chapter 5 of this volume, is here mentioned only briefly as a control tool for seed storage. Seeds are relatively worthless when they lose capacity for germination. When the germination capacity of a seed lot falls below legal minima, customers will not buy it. Some vegetable growers are demanding that seeds not only germinate well but have high vigor. Therefore, any storage program, whether for commercial seed or germ-plasm seed, needs to have a seed testing program as well. The storer of commercial seeds needs to know which lots will begin to deteriorate most rapidly, so he can sell them first. The director of a germ-plasm storage must be able to ascertain when to "grow out" a seed lot before it is seriously reduced in germination capacity or is completely lost.

Certainly, every lot of seed should be germination-tested when it is received for storage, so that seed quality is recorded. Subsequent periodic checks can be made to detect deterioration. If deterioration occurs, storage conditions should be checked to discover the reason and corrective measures taken.

Vigor tests are more difficult to evaluate with present knowledge of what constitutes vigor. At the same time, a seed analyst who uses uniform germination procedures knows when a lot is less vigorous than other lots of the same species. Since a decrease in vigor usually precedes loss of germinative capacity, measurement of vigor can be a very useful check that indicates whether something is wrong with the storage controls or whether seeds are about to end their useful life under the conditions of storage. Although there is need for much more basic research on seed vigor, vigor tests of stored seeds should be more widely used.

2. MEASUREMENT AND CONTROL OF TEMPERATURE

The thermometer is a simple, inexpensive, and (if properly calibrated) accurate instrument for measurement of temperature. Thermometers should be used for periodic checks of all other temperature-sensing devices, such as thermographs and thermocouples. The simple thermometer has the disadvantage that it has to be in a place where visual observations can be made. Thermographs and recording thermometers and thermocouples give a continuous temperature record. In addition thermocouples can be placed in otherwise inaccessible areas, such as various spots in bulk storages, outlet ducts of dehumidifiers, refrigeration compressors, or even high elevations in storage warehouses. Continuous records indicate problem temperature spots, although they do not control the temperature in those spots. Thermostats, in contrast, can be set to start ventilating fans if hot spots develop in stored seed, or to shut off dehumidification or refrigeration machines if high temperatures develop in their working parts, or to turn off refrigeration when the desired minimum storage temperature is reached. Temperature controls vary with the sophistication of a storage facility, from a single thermometer hanging in the middle of a room to programmed recorded temperature controls costing thousands of dollars.

Regardless of how sophisticated the temperature-sensing equipment may be, it is of little value if the sensing points are not properly located.

Temperature next to an uninsulated wall is likely to be completely different from that 2 m inside a bulk storage bin of seeds (seeds are a good insulator). Also, the temperature 2 m from refrigeration coils is likely to be completely different from that in a part of the storage which may be cut off from the circulation of air from the refrigerating unit. Thus, design of the storage, the placing of seeds, and the sensing equipment must be correct if the temperature-control system has actually to control seed temperature.

3. MEASUREMENT AND REGULATION OF SEED MOISTURE

Since seed moisture content is a major factor influencing seed longevity, moisture content should receive periodic checks during the time a seed lot remains in storage. This may be done directly (by moisture tests) or indirectly (by measuring the relative humidity around the seed).

Seed moisture content may be determined variously — the methods used differ in accuracy, simplicity, and time involved. There is a problem of just what is to be measured in determining moisture content. As discussed above, water in a seed is in various states, from free water to chemically bound water. More energy is required in removing the more tightly held water. In addition, attempts to remove tightly held water may remove other volatile compounds or lead to oxidation, giving values of moisture content that are too high or too low. It is generally agreed, however, that the most accurate measure of seed moisture content is the Karl Fischer reagent method (Hart and Neustadt, 1957), in which the reagent combines chemically with the water. Hence, that is the standard with which other methods are compared. The Karl Fischer method requires a chemist as well as time and expensive reagents, so it is used only as a reference in research on moisture testing.

Two methods of determining seed moisture content which compare favorably with the Karl Fischer method are the air-oven method and oil-distillation method. The air-oven method at 130°C for 1 hour is acceptable for most seeds under the International Rules for Seed Testing, 1966 (International Seed Testing Association, 1966). Prescribed for seeds containing volatile compounds, including all tree seeds, is the air-oven method at 105°C for 16 hours or the toluene distillation method. In the air-oven method the loss in weight is considered to be the water that was in the seed. In the oil-distillation method the volume of water distilled from the seed is measured directly. The percentage moisture is obtained by dividing the water content by either the wet or dry weight of the sample. Used most frequently in seed and grain commerce in expressing moisture content of seeds is the wet weight or original weight basis. This differs from the method of calculation used by agricultural engineers and many biologists, who use the dry weight basis. Table VI can be used for wet weight–dry weight conversions. The two values are fairly close at low seed moisture contents, and far apart at high moisture contents. For example, a seed with a 20% moisture content on a wet weight basis (20/100) would have a 25% moisture content on a dry weight basis [20/(100 − 20)]. Thus, the method of calculating moisture content should always be specified. Seed moisture contents can be determined by the air-oven method or the oil-distillation method by trained technicians. These determinations can be made rather rapidly and inexpensively.

To obtain almost instantaneous moisture readings, various electric moisture-measuring machines have been invented. If properly calibrated for seeds being tested, such devices are fairly accurate when seed moisture contents are above 10% and less accurate when moisture contents are less than 10%. In addition, the machines must be checked against standard methods to correct for drift.

TABLE VI
WET WEIGHT–DRY WEIGHT CONVERSION TABLE

Wet weight	Dry weight	Wet weight	Dry weight	Wet weight	Dry weight	Wet weight	Dry weight
1.0%	1.0%	9.4%	10.4%	14.0%	16.0%	19.5%	24.2%
2.0	2.0	9.6	10.6	14.2	16.6	20.0	25.0
3.0	3.1	9.8	10.9	14.4	16.8	20.5	25.8
4.0	4.2	10.0	11.1	14.6	17.1	21.0	26.6
4.5	4.7	10.2	11.4	14.8	17.4	21.5	27.4
5.5	5.8	10.4	11.6	15.0	17.6	22.0	28.2
6.0	6.4	10.6	11.9	15.2	17.9	22.5	29.0
6.2	6.6	10.8	12.1	15.4	18.2	23.0	29.9
6.4	6.8	11.0	12.4	15.6	18.5	23.5	30.7
6.6	7.1	11.2	12.6	15.8	18.8	24.0	31.6
6.8	7.3	11.4	12.9	16.0	19.0	25.0	33.3
7.0	7.5	11.6	13.1	16.2	19.3	26.0	35.1
7.2	7.8	11.8	13.4	16.4	19.6	27.0	37.0
7.4	8.0	12.0	13.6	16.6	19.9	28.0	38.9
7.6	8.2	12.2	13.9	16.8	20.2	30.0	42.9
7.8	8.5	12.4	14.2	17.0	20.5	35.0	53.8
8.0	8.7	12.6	14.4	17.2	20.8	40.0	66.7
8.2	8.9	12.8	14.7	17.4	21.1	45.0	81.8
8.4	9.2	13.0	14.9	17.6	21.4	50.0	100.0
8.6	9.4	13.2	15.2	17.8	21.7	60.0	150.0
8.8	9.6	13.4	15.5	18.0	22.0	70.0	233.3
9.0	9.9	13.6	15.7	18.5	22.7	80.0	400.0
9.2	10.1	13.8	16.0	19.0	23.5	90.0	900.0

The infrared moisture meter provides a fast air-oven method with an infrared lamp as a heat source, with the seed dried on the pan of a balance so that loss in weight can be read directly. This method will give results in 15 to 20 minutes, but, because of the high temperature, it compounds the problems of volatile loss and oxidation which occur in the air-oven method. Other variables are loss of accuracy of the balance and aging of the infrared lamp. However, for rapid determinations of moisture content for seeds of low-moisture content, the infrared method appears to be more accurate than electrical methods.

Recent research indicates that electron spin resonance (ESR) signals (Randolph *et al.,* 1968) and nuclear magnetic resonance (NMR) (Toledo *et al.,* 1968) may be useful in determining moisture content in biological materials, including seeds. Because of the high cost of the machines, it is doubtful that these methods will have practical use in moisture determination of seeds in storage.

Measurements of relative humidity of the air around the seeds gives a fairly reliable indication of the seed moisture content. Relative humidity depends on flow of air into the storage as well as on the moisture content of seeds. Also, the moisture content of seeds in bulk bins or in large stacks of bags can vary from place to place in the bin or stack, and the relative humidity of the air might indicate the moisture content of only the surface seeds. However, in a sealed storage where the seed moisture has reached equilibrium with the relative humidity of the air, determination of relative humidity provides a good estimate of seed moisture content (Table III).

Since the relative humidity of the air influences the moisture content of seeds, it is important to know the relative humidity of the air in the storage. A psychrometer is an accurate method of determining relative humidity. It depends on the principle of wet- and dry-bulb temperatures. The dry bulb gives the ambient temperature, and the wet bulb is cooled in proportion to the relative humidity of the air. A sling psychometer is inexpensive and is easy to use. Recording psychrometers are available. A hygrothermograph records both relative humidity and temperature, but the relative humidity recorder must be adjusted each time the chart is changed otherwise it will quickly be giving greatly erroneous readings. A simple and inexpensive estimation of relative humidity is made by means of a cobalt chloride card with several spots that change from blue to pink at progressively higher relative humidities. These cards should be used only for approximations of relative humidity, but they can be easily placed in sealed metal boxes, in bins, or in various places in a storage to indicate major changes in relative humidity. Lithium chloride cells accurately measure relative humidity by electrical conductivity as long as they do not become contaminated or saturated with water. These cells are useful for programmed control of relative humidity. They are more reliable than the usual humidistat.

In humidity and seed-moisture control in a seed storage, checking and regulating devices range from the simple to the complex, as with temperature control. As the control system for seed temperature and moisture becomes more sophisticated its cost rises, the training needed by the operator increases, and the maintenance requirements and chance of breakdown increase. In return, the ability to extend seed longevity also increases.

V. Overview of Seed Storage in Relation to Needs of Storer

For those seeds that can be dried to equilibrium with low relative humidities, including most seeds, it is within the power of the person wishing to store seeds to decree the longevity of such seeds. The quality of the storage then becomes a question of how long germination is to be maintained and what is the ambient climate in the storage area. Yet, in spite of the fact that information is available on requirements for storage for given seed longevity, a considerable amount of seed is lost each year. Therefore, an education program is necessary. Even when the storer knows he needs a better storage than he has, funds to build one are often lacking. Many plant-breeding lines have, therefore, been lost even though the breeder knew what was needed. It must also be realized that seed storage facilities in Calcutta, India, or Rio de Janeiro, Brazil, must be much better than those in Sacramento, California, or Oslo, Norway. The low humidity of Sacramento and the low temperatures of Oslo favor seed longevity.

To be useful, a seed storage must control birds, rodents, and insects, and the seeds stored must be dry enough that storage fungi cannot grow. Beyond that, the degree of control of seed moisture and seed temperature depends on the seed longevity desired. With our present state of knowledge, an ideal storage should be insulated and moistureproof, without cracks or crevices. The seeds should be in equilibrium with an RH of 20 to 25%, either by maintaining the storage at this relative humidity by dehumidification or by drying seeds and packaging them in moistureproof containers. The temperature should be cooled by refrigeration to at least 5°C, and probably to some temperature below 0°C. Seeds so stored should maintain capacity for germination for hundreds of years. The cost of building and maintaining such an ideal storage is too high except for valuable germ plasm. Therefore, the storer must scale down his specifications from such an ideal storage to one in a cost range compatible with the value of his seed and the seed longevity he requires.

REFERENCES

Anderson, J. A., and Alcock, A. W. (1954). "Storage of Cereal Grains and their Products." Amer. Ass. Cereal Chem., St. Paul, Minnesota.

Bajpai, P. N., and Trivedi, R. K. (1961). Storage of mango seedstone. *Hort. Advan.* **5,** 228–229.

Barton, L. V. (1939). Storage of elm seed. *Contrib. Boyce Thompson Inst.* **10,** 221–233.

Barton, L. V. (1943). The storage of citrus seeds. *Contrib. Boyce Thompson Inst.* **13,** 47–55.

Barton, L. V. (1953). Seed storage and viability. *Contrib. Boyce Thompson Inst.* **17,** 87–103.

Barton, L. V. (1960). Storage of seeds of *Lobelia cardinalis* L. *Contrib. Boyce Thompson Inst.* **20,** 395–401.

Barton, L. V. (1961). "Seed Preservation and Longevity." Leonard Hill, London.

Barton, L. V. (1965). Viability of seeds of *Theobroma cacao* L. *Contrib. Boyce Thompson Inst.* **23,** 109–122.

Barton, L. V. (1966a). Effects of temperature and moisture on viability of stored lettuce, onion, and tomato seeds. *Contrib. Boyce Thompson Inst.* **23,** 285–290.

Barton, L. V. (1966b). Viability of pyrethrum Seeds. *Contrib. Boyce Thompson Inst.* **23,** 267–268.

Barton, L. V. (1967). "Bibliography of Seeds." Columbia Univ. Press, New York

Becquerel, P. (1906). Sur la longévité des graines. *C. R. Acad. Sci.* **142,** 1549–1551.

Becquerel, P. (1934). La longévité des graines macrobiotiques. *C. R. Acad. Sci.* **199,** 1662–1664.

Bitters, W. P. (1970). University of California, Riverside, California (personal communication).

Boswell, J. G., Toole, E. H., Toole, V. K., and Fisher, D. F. (1940). A study of rapid deterioration of vegetable seeds and methods for its prevention. *U.S., Dept. Agr., Tech. Bull.* **708,** 1–47.

Brenchley, W. E. (1918). Buried weed seeds. *J. Agr. Sci.* **9,** 1–31.

Brown, E. O., and Porter, R. H. (1942). The viability and germination of seeds of *Convolvulus arvensis* L. and other perennial weeds. *Iowa, Agr. Exp. Sta., Res. Bull.* **294,** 473–504.

Bruch, E. (1961). 1932 – California Department of Agriculture Buried Seed Project – 1960. *Calif., Dep. Agr., Bull.* **50,** 29–30.

Campbell, C. (1970). University of Florida, Homestead, Florida (personal communication).

Cardosa, M., Zink, E., and Bacchi, O. (1966). Estudo sôbre conservação de sementes de seringueira. *Bragantia* **25,** Suppl., XXXV-XL.

Carruthers, W. (1911). On the vitality of farm seeds. *J. Roy Agr. Soc Engl.* **72,** 168–183.

Chacko, E. K., and Singh, R. N. (1958). Studies on the germination and longevity of fruit-tree seeds – *Citrus* spp. *Indian J. Hort.* **25,** 94–103.

Child, R. (1964). "Coconuts." Longmans, Green, New York.

Childs, J. F. L., and Hrnciar, G. (1949). A method of maintaining viability of citrus seed in storage. *Proc. Fla. State Hort. Soc.* **61,** 64–69.

Chippendale, H. G., and Milton, W. E. J. (1934). On the viable seeds present in the soil beneath pastures. *J. Ecol.* **22,** 508–531.

Christensen, C. M., and Kaufmann, H. H. (1969). "Grain Storage, the Role of Fungi in Quality Loss." Univ. of Minnesota Press, Minneapolis.

Clausen, K. E. (1965). Yellow (*Betula alleghaniensis*) and paper birch (*Betula papyrifera*) seeds germinate well after 4 years storage. *U.S., Forest. Serv., Res. Note* **LG-69**, 1–2.

Clay, D. W. T. (1964). Germination of the kola nut (*Cola nitida*). *Trop. Agr. (London)* **41**, 55–60.

Coker, W. C. (1909). Vitality of pine seeds and the delayed opening of cones. *Amer. Natur.* **43**, 677–681.

Cotton, R. T. (1956). "Pests of Stored Grain and Grain Products." Burgess, Minneapolis, Minnesota.

Crocker, W. (1909). Longevity of seeds. *Bot. Gaz.* **47**, 69–72.

Crocker, W. (1938). Life-span of seeds. *Bot. Rev.* **4**, 235–274.

Crocker, W. (1945). Longevity of seeds. *J. N. Y. Bot. Gard.* **46**, 26–35 and 48.

Crocker, W. (1948). Life span of seeds. *In* "Growth of Plants," pp. 28–60. Von Nostrand-Reinhold, Princeton, New Jersey.

Crocker, W., and Barton, L. V. (1953). Storage and life span of seeds. *In* "Physiology of Seeds," pp. 140–151. Chronica Botanica, Waltham, Massachusetts.

Darlington, H. T., and Steinbauer, G. P. (1961). The eighty-year period for Dr. Beal's seed viability experiment. *Amer. J. Bot.* **48**, 321–325.

de Fluiter, H. J. (1939). Waarnemingen betreffende het bewaren van zaadkoffie *Bergcultures* **13**, 1506–1512.

Dent, T. V. (1942). Some records of extreme longevity of seeds of Indian forest plants. *Indian Forest.* **68**, 617–631.

Dillman, A. C., and Toole, E. H. (1937). Effect of age, condition, and temperature on germination of flax seed. *J. Amer. Soc. Agron.* **29**, 23–29.

Duvel, J. W. T. (1906). The storage and germination of wild rice seed. *U.S., Dep. Agr., Bur. Plant Ind., Bull.* **90**, 5–14.

Ewart, A. J. (1908). On the longevity of seeds. *Proc. Roy. Soc. Victoria* [N.S.] **21**, 1–120.

Fancourt, E. (1856). Vitality of seeds. *Gard. Chron.* **16**, 39.

Filter, P. (1932). Untersuchungen über die Lebensdauer von Handels- und Andere Saaten. *Landwirt. Vers.-Sta.* **114**, 149–170.

Fivaz, A. E. (1931). Longevity and germination of seeds of *Ribes*, particularly *R. rotundifolium* under laboratory and natural conditions. *U.S., Dep. Agr., Tech. Bull.* **261**, 1–40.

Forest Service. (1948). Woody-plant seed manual. *U.S., Dep. Agr., Misc. Publ.* **654**.

Glasscock, H. H., and Wain, R. L. (1940). Distribution of manganese in the pea seed in relation to marsh spot. *J. Agr. Sci.* **30**, 132–140.

Glendening, G. E., and Paulsen, H. A. (1955). Reproduction and establishment of velvet mesquite as related to semi-desert grasslands. *U.S., Dep. Agr., Tech. Bull.* **1127**, 1–50.

Godwin, H. (1968). Evidence for longevity of seeds. *Nature (London)* **220**, 708.

Goss, W. L. (1937). Germination of flower seeds stored for ten years in the California State Seed Laboratory. *Calif., Dep. Agr., Bull.* **26**, 326–333.

Guha, S., and Maheshwari, S. C. (1966). Cell division and differentiation of embryos in the pollen grains of *Datura in vitro*. *Nature (London)* **212**, 97–98.

Guppy, H. B. (1912). "Studies in Seeds and Fruits." Williams & Norgate, London.

Haarer, A. E. (1962). "Modern Coffee Production." Leonard Hill, London.

Haferkamp, M. E., Smith, L., and Nilan, R. A. (1953). Studies on aged seeds. I. Relation of age of seed to germination and longevity. *Agron J.* **45**, 434–437.

Halma, F. F., and Frolich, E. (1949). Storing avocado seeds and hastening germination. *Calif. Avocado Soc. Yearb.* pp. 136–138.

Hamilton, R. A. (1957). A study of germination and storage life of *Macadamia* seed. *Proc. Amer. Soc. Hort. Sci.* **70**, 209–217.

Harlan, H. V., and Pope, M. N. (1922). The germination of barley seeds harvested at different stages of growth. *J. Hered.* **13**, 72–75.

Harmond, J. E., Brandenburg, N. R., and Klein, L. M. (1968). Mechanical seed cleaning and handling. *U.S., Dep. Agr., Agr. Handb.* **354**, 1–55.

Harrington, J. F. (1959). Drying, storing, and packaging seeds to maintain germination and vigor. *Proc. Short Course Seedsmen, State Coll. Miss.* pp. 89–108.

Harrington, J. F. (1960a). Germination of seeds from carrot, lettuce, and pepper plants grown under severe nutrient deficiencies. *Hilgardia* **30**, 219–235.

Harrington, J. F. (1960b). Thumb rules of seed drying. *Crops Soils* **13**, 16–17.

Harrington, J. F. (1963). The value of moisture-resistant containers in vegetable seed packaging. *Calif., Agr. Exp. Sta., Bull.* **792**, 1–23.

Harrington, J. F. (1970). Seed and pollen storage for conservation of plant gene resources. *In* "Genetic Resources in Plants — Their Exploration and Conservation" (O. H. Frankel and E. Bennett, eds.), pp. 501–521. Blackwell, Oxford.

Hart, J. R., and Neustadt, M. H. (1957). Application of the Karl Fischer method to grain moisture determination. *Cereal Chem.* **34**, 26–36.

Heinis, F. (1943). Zur Kenntnis der Lebensdauer resp. Keimfähigkeit von Samen. *Ber. Geobot. Forschungsinst. Rubel* p. 75.

Heit, C. E. (1967a). Propagation from seed. VI. Hardseededness — a critical factor. *Amer. Nurseryman* **125**, 10–12 and 88–96.

Heit, C. E. (1967b). Propagation from seed. X. Storage methods for conifer seeds. *Amer. Nurseryman* **126**, 14–15, 38, 40, 42–44, 46, 48, 50, 52, and 54.

Heit, C. E. (1967c). Propogation from seed. XI. Storage of deciduous tree and shrub seeds. *Amer. Nurseryman* **126**, 12–13 and 86–94.

Hemsley, W. B. (1895). Vitality of seeds. *Nature (London)* 5–6.

Hlynka, I., and Robinson, A. D. (1954). *In* "Storage of Cereal Grains and their Products" (J. A. Anderson and A. W. Alcock, eds.), pp. 1–45. Amer. Ass. Cereal Chem., St. Paul, Minnesota.

Holmes, G. D., and Buszewicz, G. (1958). The storage of seed of temperate forest tree species. *Forest. Abstr.* **19**, 313–322 and 455–476.

Huxley, P. A. (1964). Investigations on the maintenance of viability of robusta coffee seed in storage. *Proc. Int. Seed Test. Ass.* **29**, 423–444.

International Seed Testing Association. (1966). International rules for seed testing, 1966. *Proc. Int. Seed Test. Ass.* **31**, 1–152.

Ito, H., and Kumagai, K. (1969). The national seed storage laboratory for genetic resources in Japan. *Jap. Agr. Res. Quart.* **4**, 32–38.

James, E. (1961). An annotated bibliography on seed storage and deterioration. A review of 20th century literature reported in the English language. *U.S., Dep. Agr., ARS* **ARS 34-15-1**, 1–81

James, E. (1963). An annotated bibliography on seed storage and deterioration. A review of 20th century literature reported in foreign languages. *U.S., Dep. Agr., ARS* **ARS 34-15-2**, 1–30.

James, E. (1967). Preservation of seed stocks. *Advan. Agron.* **19**, 87–106.

James, E., Bass, L. N., and Clark, D. C. (1964). Longevity of vegetable seeds stored 15 to 30 years at Cheyenne, Wyoming. *Proc. Amer. Soc. Hort. Sci.* **84**, 527–534.

James, E., Bass, L. N., and Clark, D. C. (1967). Varietal differences in longevity of vegetable seeds and their response to various storage conditions. *Proc. Amer. Soc. Hort. Sci.* **91**, 521–528.

Jensen, L. A. (1971). Observations on the viability of Borneo camphor (*Dryobalanops aromatica* Gaertn). *Proc. Int. Seed Test. Ass.* **36**, 141–146.

Johannsen, W. (1921). Orienterande Forsøg med Opbevaring af Agern og Bøgeolden. *Forestl. Forsoksv. Dan.* **5**, 372–390.

Jones, H. A. (1920). Physiological study of maple seeds. *Bot. Gaz.* **69**, 127–152.

Joseph, H. C. (1929). Germination and vitality of birch seeds. *Bot. Gaz.* **87**, 127–151.

Kannan, K., and Balakrishnan, S. (1967). A note on the viability of cinnamon seeds (*Cinnamomum zeylanicum* Nees). *Madras Agr. J.* **54**, 78–79.

Kempf, A. N. (1970). Bodger Seeds, Ltd., El Monte, California (personal communication).

Kester, D. (1970). University of California, Davis, California (personal communication).

Khan, R. H., and Laude, H. M. (1969). Influence of heat stress during seed maturation on germinability of barley seed at harvest. *Crop Sci.* **9**, 55–58.

Kjaer, A. (1940). Germination of buried and dry stored seeds. *Proc. Int. Seed Test. Ass.* **12**, 167–188.

Kondo, M., and Kasahara, Y. (1940). (Some examples of the year's storage of seed.) *Crop Sci. Soc. Jap. Proc.* **12**, 21–24. (In Japanese.)

Kondo, M., Kasahara, Y., and Akita, S. (1950). Germination of hemp seeds stored for 19 years and their growth. *Nogaku Kenkyu*, **39**, 37–39.

Koopman, M. J. F. (1963). Results of a number of storage experiments conducted under controlled conditions (other than agricultural seeds). *Proc. Int. Seed Test. Ass.* **28**, 853–860.

Kosar, W. F., and Thompson, R. C. (1957). Influence of storage humidity on dormancy and longevity of lettuce seed. *Proc. Amer. Soc. Hort. Sci.* **70**, 273–276.

Large, J. R., Fernholz, D. F., Merrill, S., Jr., and Potter, G. F. (1947). Longevity of tung seed as affected by storage temperatures. *Proc. Amer. Soc. Hort. Sci.* **49**, 147–150.

La Rue C. D., and Muzik, T. J. (1951). Does the mangrove really plant its seedlings? *Science* **114**, 661–662.

Lea, C. H. (1962). *In* "Lipids and their Oxidation" (H. W. Schultz, ed.), pp. 3–28. Avi Publ. Co., Westport, Conn.

Leggatt, C. W. (1948). Germination of boron deficient peas. *Sci. Agr.* **28**, 131–139.

Libby, W. F. (1951). Radiocarbon dates. II. *Science* **114**, 291–296.

Livingston, R. B., and Allesio, M. L. (1968). Buried viable seed in successional field and forest stands, Harvard Forest, Massachusetts. *Bull. Torrey Bot. Club* **95**, 58–69.

Lopez, M. (1938). Storage and germination of large-leaf mahogany seeds. *Philipp. J. Forest.* **1**, 397–410.

McAlister, D. F. (1943). The effect of maturity on the viability and longevity of the seeds of western range and pasture grasses. *J. Amer. Soc. Agron.* **35**, 442–453.

McClelland, T. B. (1944). Brief viability of tropical seeds. *Proc. Fla. State Hort. Soc.* **57**, 161–163.

MacGillivray, J. H. (1953). "Vegetable Production," pp. 364–366. McGraw-Hill (Blakiston), New York.

Mackay, D. B., and Tonkin, J. H. B. (1967). Investigations in crop seed longevity. I. Analysis of long-term experiments with special reference to the influence of species, cultivar, provenance, and season. *J. Nat. Inst. Agr. Bot.* **11**, 209–225.

Madsen, S. B. (1957). Investigation of the influence of some storage conditions on the ability of seed to retain its germinating capacity. *Proc. Int. Seed Test. Ass.* **22**, 423–446.

Madsen, S. B. (1962). Germination of buried and dry stored seeds. III. 1934–1960. *Proc. Int. Seed Test. Ass.* **27**, 920–928.

Mills, E. A. (1915). "The Rocky Mountain Wonderland." Houghton, Boston, Massachusetts.

Mirov, N. T. (1943). Storage and germination of California cork-oak acorns. Forest Research Notes. U.S. Dep. Agr., Calif. Forest Range Expt. Sta. 36, 1–18.

Mirov, N. T. (1946). Viability of pine seed after prolonged cold storage. *J. Forest.* **44,** 193–195.

Mookherjee, P. B. (1962). Susceptibility of the khapra beetle: *Trogoderma granarium* Everts to oxygen deficiency and increased CO_2 concentration in the air. *Curr. Sci.* **31,** 474–475.

Moss, E. H. (1938). Longevity of seed and establishment of seedlings in species of *Populus. Bot. Gaz.* **99,** 529–542.

Muenscher, W. C. (1936). Storage and germination of seeds of aquatic plants. *N.Y., Agr. Exp. Sta., Ithaca, Bull.* **652,** 1–17.

Muenscher, W. C. (1955). "Weeds," 2nd ed. Macmillan, New York.

Myers, A. (1940). Longevity of seed of native grasses. *Agr. Gaz. N. S. W.* **51,** 405.

Nawaschin, M. (1933). Altern der Samen als Ursache von Chromosomenmutationen. *Planta* **20,** 233–243.

Ødum, S. (1965). Germination of ancient seeds. Floristical observations and experiments with archaeologically dated soil samples. *Dan. Bot. Ark.* **24,** 1–70.

Oosting, H. F., and Humphreys, M. E. (1940). Buried viable seeds in successional series of old fields and forest soils. *Bull. Torrey Bot. Club.* **67,** 253–273.

Owen, E. B. (1956). The storage of seeds for maintenance of viability. Commonw. Agri. Bureaux, *Commonw. Bur. Pastures Field Crops Bull.* **43,** 1–81.

Paton, R. R. (1945). Storage of tuliptree seed. *J. Forest.* **43,** 764–765.

Peter, A. (1893). Culturversuche mit "ruhenden" Samen. *Nachr. Kgl. Ges. Wiss. Goettingen. Math-Phys. Kl.,* **17,** 673–691.

Porsild, A. E., Harington, C. R., and Mulligan, G. A. (1967). *Lupinus arcticus* Wats. grown from seeds of Pleistocene Age. *Science* **158,** 113–114.

Pradhan, S., and Mookherjee, P. B. (1969). Pusa bin for storage of grain. *Indian Counc. Agr. Res., Tech. Bull.* **21,** 1–11.

Prince, F. S., and Hodgdon, A. R. (1946). Viable seeds in old pasture soils. *N. H. Agr. Exp. Sta., Tech. Bull.* **89,** 1–16.

Ramsbottom, J. (1942). Duration of viability in seeds. *Gard. Chron.* **111,** 234.

Randolph, M. L., Heddle, J. A., and Hosszu, J. L. (1968). Dependence of ESR signals in seeds on moisture content. *Radiat. Bot.* **8,** 339–343.

Rees, B. (1911). Longevity of seeds and structure and nature of seed coats.*Proc. Roy. Soc. Victoria* **23,** 393–414.

Robertson, D. W., Lute, A. M., and Kroeger, H. (1943). Germination of 20-year-old wheat, oats, barley, corn, rye, sorghum, and soybeans. *Amer. Soc. Agron. J.* **35,** 786–795.

Rocha, F. F. (1959). Interaction of moisture content and temperature on onion seed viability. *Proc. Amer. Soc. Hort. Sci.* **73,** 385–389.

Rockland, L. B. (1969). Water activity and storage stability. *Food Technol.* **23,** 11–21.

Roe, E. I. (1948). Viability of white pine seed after 10 years storage.*J. Forest.* **46,** 900–902.

Rossman, E. C. (1949). Freezing injury of maize seed. *Plant Physiol.* **24,** 629–656.

Rowntree, L. (1930). Longevity of seeds in the desert. *Horticulture* **8,** 270.

Schjelderup-Ebbe, T. (1936). Uber die Lebensfähigkeit alter Samen. *Skr. Nor. Vidensk.-Akad. Oslo, Ma.-Naturvidensk. Kl.* pp. 1–178.

Schloesing, A. T., and Leroux, D. (1943). Essai de conservation de graines en l'absence d'humidité d'air et de lumiére. *C. R. Acad. Agr. Fr.* **29,** 204–206.

Schubert, G. H. (1952). Germination of various conifer seeds after cold storage. *Forest. Res. Notes, U.S. Dep. Agr., Calif. Forest Range Exp. Sta.* **83,** 1–7.

Scott, D. H., and Draper. A. D. (1970). A further note on longevity of strawberry seed in cold storage. *HortScience* **5**, 439.

Sifton, H. B. (1920). Lonvevity of the seeds of cereals, clovers, and timothy. *Amer. J. Bot.* **7**, 243–251.

Simpson, D. M. (1946). The longevity of cottonseed as affected by climate and seed treatments. *Amer. Soc. Agron. J.* **38**, 32–45.

Sivori, E., Nakayama, F., and Cigliano, E. (1968). Germination of achira seed (*Canna* sp.) approximately 550 years old. *Nature (London)* **219**, 1269–1270.

Steward, F. C., Kent, A. E., and Mapes, M. O. (1966). The culture of free plant cells and its significance for embryology and morphogenesis. *Curr. Top. Develop. Biol.* **1**, 113–154.

Storey, W. B. (1969). Macadamia. *In* "Handbook of North American Nut Trees" (R. A. Jaynes, ed.), pp. 321–335. N. Amer. Nut Growers' Ass., Knoxville, Tennessee.

Takagi, I. (1939). (On the storage of mulberry seeds.) *Tokyo Imp. Seric. Col. Res. Bull.* **2**, (In Japanese.)

Toledo, R., Steinberg, M. P., and Nelson, A. I. (1968). Quantitative determination of sound water by NMR. *J. Food Sci.* **33**, 315–317.

Toole, E. H., and Brown, E. (1946). Final results of the Duvel buried seed experiment. *J. Agr. Res.* **72**, 201–210.

Tourney, J. W. (1930). Some notes on seed storage. *J. Forest.* **28**, 394–395.

Tschirley, F. H., and Martin, S. C. (1960). Germination and longevity of velvet mesquite seed in soil. *J. Range Manage.* **13**, 94–97.

Turner, J. H. (1933). The viability of seeds. *Bull. Misc. Inform., Roy. Bot. Bard.* pp. 257–269.

Uebersezig, M. (1947). Successful storage of slash pine seed for fifteen years. *J. Forest.* **45**, 825–826.

United States Department of Agriculture. (1961). "Seeds, the Yearbook of Agriculture." US Govt. Printing Office, Washington, D.C.

Verret, J. A. (1928). Sugar cane seedlings. *Rep. Ass. Hawaii Sugar Technol.* **7**, 15–23.

Visser, T., and Tillerkeratne, L. M. (1958). Observations on the germination and storage of tea pollen and seed. *Tea Quart.* **29**, 30–35.

Went, F. W. (1957). "The Experimental Control of Plant Growth." Chronica Botanica, Waltham, Massachusetts.

Went, F. W., and Munz, P. A. (1949). A long term test of seed longevity. *El Aliso* **2**, 63–75.

Widmoyer, F. B., and Moore, A. (1968). On the effect of storage period, temperature and moisture on the germination of *Aesculus hippocastanum* seeds. *Plant Propogator* **14**, 14–15.

Winters, H. F., and Rodriguez-Colon, F. (1953). Storage of mangosteen seeds. *Proc. Amer. Soc. Hort. Sci.* **61**, 304–306.

Wollenweber, H. W. (1942). Über die Lebensdauer von Kartoffelsamen. *Angew. Bot.* **24**, 259–260.

Woodroof, J. G. (1967). "Tree Nuts: Production, Processing, Products," Vol. II. Avi Publ., Westport, Connecticut.

Youngman, B. J. (1952). Germination of old seeds. *Kew Bull.* **6**, 423–426.

Zink, E., and Ojirna, M. (1965). Influência das condições de armazenagem no poder germinativo das sementes de nêspera. *Bragantia* **24**, Suppl., IX-XII.

4

INSECTS ATTACKING SEEDS DURING STORAGE

R. W. Howe

I. Introduction

A seed is essentially a plant embryo plus a supply of food that enables the young plant to start to grow. Some species of animals inevitably use seeds as food, insects being among the earliest to do so and man one of the most recent.

In the ordinary way seeds need moisture to germinate and most seeds sooner or later are shed into an area that is moist or becomes so. A small proportion may accidentally reach a niche protected from moisture, and more may be collected by animals and placed in a sheltered store. Most of these seeds will remain inert and be available as food for one or more years. This chapter deals with the insects that attack dry seeds of this kind. These insects are adapted to live on foods of low-moisture content and most are unable to subsist at high-moisture contents. They have assumed economic importance because man has learned to grow, harvest, and store large quantities of seeds in which enormous numbers of these insects can develop.

The insect species of economic importance are well known, and have been associated with man for centuries (Fig. 1). No doubt, there are little known species associated with seeds that man does not value or store, and it is possible that some of these may be able to utilize seeds stored by man if they chance to be introduced into international trade.

FIG. 1. *Sitophilus granarius* collected in 1950 from the tomb of Queen Ichetis (ca. 2500 B.C.) in the Step Pyramid at Saqqarah, Egypt.

There have been many speculations about the origin of insects in stores, and it is usually assumed that these insects have always fed on the seeds with which they are now associated. This may not be justified, for man has greatly increased the size of the seeds of some of the plants he grows. I have suggested, for instance (Howe, 1965b), that the granary weevil *Sitophilus granarius* might have been breeding in acorns at the time that primitive man began to cultivate his earliest cereal crops. The size of these small grass seeds places a physical limit on the size of the insect developing inside them and the number of eggs that these insects can lay is usually dependent on the size of the insect. However, Joubert (1966) claims that the gnatlike adults of the moth *Sitotroga cerealella* that develop in the tiny seeds of species of the grass *Setaria* in South Africa are of normal fecundity. This could well be the original home and one of the original foods of this moth.

Other species that attack seeds externally were, probably, and often still are in the wild state – scavengers of vegetable, fungal, and animal detritus.

There is no sharp demarcation between field and storage species of insects. The limiting moisture levels, both maximum and minimum, differ from species to species. Some species, such as the maize weevil *Sitophilus zeamais,* can attack young developing seeds on the growing plant as well as the stored seeds, but the great majority of storage species are not known from growing seeds. A few species, such as *Bruchus pisorum,* are commonly found in stores but are not storage species. These have developed in the growing and ripening pea seed in the field but have emerged in storage. They are unable to grow in stored peas. A few species, such as *Bruchidius atrolineatus,* attack the developing seed on a growing plant and can pass another generation in store but then fail to continue possibly because the seeds become too dry.

Most of the seeds that are stored are kept as food for man or farm animals. They may be processed before consumption. Food problems strictly fall outside the scope of this book which is concerned with seed development, germination, and protection. The problems of storage of food grains and of seeds, however, are very similar. The differences concern germination since planted seeds are without value unless they germinate, whereas dead seeds are reasonably adequate as a food unless the grain is to be malted. A further difference concerns the length of storage.

Food grains may be eaten immediately after harvest or they may be stored for any period up to several years; seed grains must be kept in condition until the time for sowing and if not used then, until the next sowing time. A third difference is the quantities stored. Food grains may be stored in bulks of several thousand tons, whereas the amount of seed stored is fixed by the demand. Losses expressed as a percentage of the original weight are usually more extensive in small lots, so seeds are in greater hazard, although large bulks incur the special risk of heating to a greater extent. The need for protection is greater for seed grain than for feed, and the risks from insects and from control measures are correspondingly greater. Germination may be lost either from insect damage or from chemical treatment applied to kill the insects.

Most of the studies on seed insects have been made on those attacking cereals, pulses, or spices but some attention has been given to seeds of forest trees. Small seeds, especially if hard, suffer relatively little damage and, although some fruits are attacked by storage pests, the damage is unimportant since these plants are rarely propagated by seeds. Infestation of stored material of any kind is a matter for concern if the insects are able to spread to and damage seeds.

II.   Damage Caused by Insects

Reliable information concerning the losses caused by insects in stored seeds and food is scarce. I have attempted to review this topic previously (Howe, 1965a) and this section is a precis of that paper. With seed grains, any damage that prevents germination is paramount whether the embryo is eaten or killed in any other way. Farmers, foresters, horticulturists, and maltsters all demand a germinative capacity close to 100% and samples of even germination. A few lapses from these standards will lose the seed merchant the good will of his customers. The seeds must also be of good appearance, free of live insects, holed grains, and other obvious contamination such as webbing or animal droppings. Live insects are especially unwelcome because they spread readily in the warehouse to other produce.

With food grains, weight losses are especially important but contamination even by dead insects or pieces of them may be serious. The destruction of the embryo diminishes the vitamin content of the food. The insects may cause other chemical changes in the food and may transmit diseases or parasites.

Figures 2 to 7 illustrate damage to stored seeds caused by insects.

A.   Germination

It is the insect species that are not adapted to living in seeds that present the greatest hazard to germination because so many selectively eat the embryo (Fig. 2). Thus Waloff and Richards (1946) estimate that a caterpillar of *Ephestia elutella* eats forty-eight embryos of wheat before pupating. Adults and larvae of beetles, such as *Cryptolestes,* also move from one embryo to another as do many mites, and a simple count of damaged embryos in a sample gives a minimal estimate of the germination loss. Species adapted to live in seeds feed mostly in the cotyledons or endosperm (Fig. 3) and the damaged seeds may germinate. These species may, of course, damage the embryo when laying eggs or by adult feeding but larval feeding seldom damages it directly. Howe (1952) recorded a

FIG. 2. Damage to the embryo. (a) eggs of *Cryptolestes ferrugineus* deposited on pericarp over the embryo of wheat kernel; (b) adult of this beetle emerging from embryo; (c) emergence hole; (e) kernels of wheat with embryo eaten away by *Trogoderma granarium*; (d) kernels of barley with embryo eaten away by *Oryzaephilus surinamensis*. (f) blowup of (a).

FIG. 3. Radiograph of wheat samples. The lower lot contain the late stages of the life cycle of a weevil.

loss of some 2 to 3% from 97% in controls when between ten and fifteen *Sitophilus oryzae* larvae were present per 100 wheat kernels at 21°C, the loss rising to 8% after the adults had emerged. In a sample of a 100 holed kernels chosen after adult emergence, 27 still germinated. The rate of germination was retarded by weevil larvae—compared with 90% that germinated in 2 days in controls, only 78% had done so in samples containing ten to 15 weevil larvae.

Pingale (1953) performed a similar experiment in mud-plastered bamboo bins in India starting with a much lower density of insects, ten pairs of weevils on 22.5 kg of wheat, but at a higher temperature, 26.5° ± 1°C. There was a marked fall of germination of some 13% during the third month of storage and during the next 3 months germination fell to 10%. In spite of the vast increase in numbers of insects, this loss of germination is unlikely to be a direct result of insects feeding within kernels, for under these conditions even lots of wheat as small as 20 kg develop heating. The loss of germination in jowar sorghum caused by *S. oryzae* reported by Venkat Rao *et al.* (1958) can also be attributed to heating because only 61% of kernels were holed, but 95% failed to germinate.

As a rule the smaller cereal seeds, wheat, rye, barley, oats, rice and sorghum support only a single developing weevil at a time. Maize and most of the pulses, however, can sustain several larvae of seed-adapted species simultaneously. These larger seeds are widely stored in the tropics in small lots but when they suffer multiple attack they are very likely to heat (see Chapter 3 of this volume). Thus, although the number of insects developing exceeds the numbers of damaged seeds, the number failing to germinate could well be higher still. Pingale (1953) stored *Phaseolus aureus* for 6 months in bamboo bins, some free of insects and some infested with bruchids. He took samples at monthly intervals to estimate weight loss, germination, numbers of holed beans and numbers of adult beetles and found that the loss of germination was always higher than the percentage of holed seeds. The difference between these values reached a peak after 3 months when the numbers of live beetles was also maximal, and then remained steady until the germination fell to zero.

Summing up, it is evident that a very low level of infestation by the true seed pests causes little loss of germination although it may render the rate of germination less consistent, but an increase in insect attack quickly spoils the germinative capacity of seed probably because of heating. It is likely that seedlings developing from holed seeds will lack vigor partly because of the loss of endosperm eaten by the insect and partly because storage insects contaminate their food with mold spores (Agrawal *et al.,* 1957).

B.   *Weight Loss*

Compared with the loss of germination, any other damage to seed grains is trivial and has importance only if it can be linked to loss of germination or to the goodwill and reputation of the seed merchant. Naturally merchants wish to avoid damage that causes a loss of weight before they sell, or, if they sell by volume, lowers the bushel weight of their seeds. This, along with visible insect emergence holes (Fig. 4) or pupal "windows" is an indicator of poor quality.

Weight loss in food grains after harvest is of major importance because by then it is too late for the plant to compensate the loss by further growth. Furthermore, the insect and mite pests that cause the loss are

FIG. 4. Emergence holes in pulses; (a) in haricot beans caused by *Acanthoscelides obtectus;* (b) in lentils caused by *Callosobruchus chinensis* — these seeds are covered with the egg shells from which the feeding larvae hatched.

inconspicuous until the damage is considerable by which time they may also have caused secondary damage by heating. Finally because the insects and their excrement may remain within the grains, the losses may be undetected or underestimated especially if the produce gains in moisture content during storage.

Unfortunately, it is difficult if not impossible to determine weight losses very accurately because the problem of collecting representative samples is so intractible. Most large-scale estimates are hopeful guesses expressed as round figures. Several experimental efforts have been made but these have two kinds of defect. First, artificially introduced insects often become established less readily and so do less damage. Second, experimental stacks must be small if the work involved is not to be excessive, and, usually, damage is relatively greater in small stacks.

In general there are so many environmental variables influencing weight losses that no sample can be typical. Losses increase with length of storage because population increases tend to be exponential; they are greater the higher the temperature and humidity because these factors encourage population growth; they are greater the larger the initial insect population, and they vary according to the seeds and to the insects concerned. Some seeds, such as coffee, soybeans, and rice paddy are resistant to insect attack; others, such as wheat and groundnuts, are susceptible. Some species such as *Rhizopertha dominica* and *Trogoderma granarium* destroy a great deal per individual; others, such as *Tribolium castaneum* may be exceedingly abundant without obviously causing a corresponding amount of damage. Moths, such as *Ephestia cautella,* may eat all the food available to them in a few months and increase no more, whereas *Sitophilus* spp. continues to increase in numbers. Heating may ensue from insect attack and, although causing other kinds of damage, may restrict insects to the periphery of a stack and so limit the extent of weight loss.

Perhaps, this can be best summarized, as in my review (Howe, 1965a), that losses are greatest in the underdeveloped tropical countries where they can be least afforded and that in the developed temperate countries there are no excuses for significant losses in storage.

C. Holed Seeds

The proportion of holed seeds in a sample is a statistic that is very easily determined or estimated. It is valuable because it can be assumed that even when holed seeds germinate the plant that grows will be a weakling; hence it is a better measure of seed quality than weight loss. Also because it must always be greater than the percentage loss of weight, it is a more disquieting figure for food grains. However, as has been pointed out above, it may underestimate loss of germinative capacity if there has been the slightest risk of heating.

Holed seeds provide a refuge for insects especially during mechanical cleaning. They also provide a source of food for these species that cannot attack sound seed.

D. *Heating*

The topic of heating is discussed extensively in Chapter 3 of this volume, but since it has already been mentioned several times in this chapter, I will deal with it briefly. Seeds are poor conductors of heat and hence the temperature within a mass of seeds is slow to change. If there is a source of heat within the mass of seeds, however, this heat does not escape readily and the temperature therefore rises in the vicinity of this source and this phenomenon is called heating (Fig. 5). The sources of

Fig. 5. Typical appearance of a bulk of wheat that has heated. The bulk remains standing as a cliff because dampness and fungi cause the grains to stick together as a cake.

heat may be chemical or biological, and the most important biological sources are fungi and insects. Since most fungi require a moist environment, fungal heating is often called damp-grain heating, whereas insects are associated with dry-grain heating. Insects may take advantage of fungal heating but do not always do so, but fungi always appear during insect heating because the shell around the heating zone always gets damper. This is principally caused by moisture migration down a vapor pressure gradient but is assisted by the liberation of metabolic water by a large population of insects. Fungal hot spots move upward through a bulk of grains because moisture is carried upward by convection, but insect hot spots are spread out laterally and occasionally even downward by the migration of insects away from the zone they have made uncomfortably hot.

E. Contamination and Indirect Damage

There are two aspects of contamination of importance with seeds. Fragments of insects, frass, and similar contaminants matter little except in their influence on good will. The exception is the silk spun by moth caterpillars and a few beetle larvae which may bind seeds together in clumps. This interferes with the smooth running of seeds and thus with its sowing. It is probable in any event that many of the seeds spun together will be damaged by feeding but some of the webbing is spun only to make a cocoon in which to pupate.

The other form of contamination is that arising from treatments applied to remove, control, or exclude insects. The chief dangers arise from mechanical treatments and fumigation. Mechanical injury may prevent germination or it may facilitate the feeding of insects. At least its extent should be easy to estimate and measures should be taken to avoid it as

Fig. 6. Damage to nutmeg caused by *Araecerus fasciculatus.*

FIG. 7. Damage to groundnuts. By *Ephestia cautella*, (a) Larval cocoon on seed. By *Caryedon serratus*, (b) Larva feeding within seed; and (c) emergence holes in and cocoons on pods.

soon as it is discovered. Fortunately, mechanical methods for exclusion of insects do not cause damage to seed. Even a simple jute sack offers considerable protection to cereals from *Sitotroga cerealella* and to peanuts from *Caryedon serratus,* both of which are discouraged by a simple covering of the produce that restricts the free air space. In general, cotton, paper and laminate bags, plastic liners, bags made of layers of several materials, drums of various kinds, and sheets for covering stacks all offer some protection but none guarantee freedom from insects unless the oxygen level in the intergranular air falls to 3% or less. At this level of oxygen the seed remains alive provided the moisture content is not high enough for anaerobic fungi, yeasts, or bacteria to be active. Pixton (see Pixton and Hill, 1967) has stored both English and Manitoba wheat, insect- and mold-free, in metal bins at normal and low (5%) oxygen tensions for 12 years, and in the worst instance germination was as high as 88%.

The risks of chemical treatment are difficult to evaluate for a balance must be struck between the dosages needed to kill or exclude insects and those harming the seeds. A rise in the tolerance of insects or a fall in that of the seeds could rule out the use of an insecticide. The tolerance of seeds depends considerably upon sorption especially for gaseous insecticides or fumigants. Sorption is a complex matter. It increases with increasing moisture content and temperature, especially with oily seeds and maize. Generally it is safe to fumigate cool, dry seeds especially with methyl bromide (Page and Lubatti, 1963) but because very high dosage rates of this fumigant were used to eradicate the Khapra beetle, *Trogoderma granarium,* from the United States, a considerable amount of work has been done on the influence of this and other fumigants on germination of many kinds of seeds (Strong and Lindgren, 1959, 1961).

Contamination is of great importance for food grains. If it is obvious, then customers are reluctant to buy, but even insects hidden inside seeds to be ground into flour are a hazard to the miller subject to pure food regulations. In many parts of the world, meals used for human and animal food contain extremely high numbers of chitinous fragments of insect. There is, at present, little evidence as to the level of contamination by insect and mite fragments that can be tolerated in food, and there is no agreement among research workers as to whether or not the fragments cause alimentary disorders. There is similar disagreement about the influence of uric acid and other waste products in excreta, but it is widely accepted that insect attack accelerates the breakdown of oils into free fatty acids.

There is also a small but real risk of storage insect harboring parasitic worms and disease organisms and also of their causing allergies.

III. Factors Influencing Insect Infestation

Limitation of damage to seed depends considerably upon excluding insects. Since the growth of insect populations tends to be exponential, should insects gain entry, the buildup of population is determined by the period of storage as well as the suitability of the environment for the insects.

A. Preharvest Infestation

A major hazard is the infestation of the growing crop in the field by species able to multiply in stores. Fortunately both the species of plants affected and the insects concerned are limited in number. Of the cereals, only maize (corn) is attacked in the field (Fig. 8) and, although bruchid beetles often infest ripening seeds in the field, cowpeas are the only pulse crop of widespread importance to be vulnerable.

Field infestation is of dual importance because it limits the benefits of storage practices that otherwise exclude insects from stores and it prolongs the period available for insect multiplication. It is unfortunate that the improved varieties of maize developed by plant breeders to give a greater yield are also more susceptible to field infestation by storage insects. This is because increased yields result from longer cobs bearing more or larger seeds, but the sheaths in these varieties are not long enough to protect the ripening seeds from weevil attack. The principal enemy of this crop in the field is the maize weevil, *Sitophilus zeamais*. The literature concerning this problem, along with observations of their own in Kenya, is assessed by Giles and Ashman (1971). They establish that the sole sources of weevils invading the growing crop are nearby stores or associated dumps of rubbish. The weevil does not breed in growing wheat or any wild plant. It is unable to breed on seeds or cores of maize cob that are buried in the field after harvest because these rot, but it can breed on any that remain on the surface. They confirmed the observations of Powell and Floyd (1960) in Louisiana that weevils can lay eggs in developing seeds that have a moisture content (on a wet weight basis) of about 70% and that these seeds dry to about 25% moisture content by the time these eggs grow to the adult stage.

The developing maize seeds are soft and are more vulnerable to some species in the field than they are in storage. Only fairly strong fliers can reach the plant in the field and these include the beetles *Cathartus quadricollis*, *Tribolium castaneum*, and *Cryptolestes ferrugineus* and the moth *Sitotroga cerealella*. All the other stored products moths fly weakly. They may be carried from one building to another by wind, but attack of

FIG. 8. Damage to maize cobs in the field caused by (a) *Mussidia nigrivenella*, (b) *Sitotroga cerealella*, and (c) *Sitophilus zeamais*.

produce in the field is rare. A few moths, such as *Spectrobates cera-toniae* and *Mussidia nigrivenella* that need wetter produce, as well as *Ephestia calidella* and *Ephestia figulilella*, attack ripe seeds or fallen fruits and seeds in the field.

Many species of stored products beetles are wingless or do not fly. Other species fly rarely, only in warm humid weather and often then only in bright sunlight or at dusk. A few species including those mentioned above, *Lasioderma serricorne*, *Carpophilus* spp., the bruchids, and *Ptinus* spp., though only males in many species of this last genus, fly quite strongly and regularly. Other important pests such as *Oryzaephilus surinamensis* may make very short flights (Giles, 1969).

Among the bruchids, *Acanthoscelides obtectus* is able to breed actively in growing beans in Germany. It cannot pass the winter in the open but, having been introduced into the field by sowing, it attacks the new crop and infested seeds are harvested (Zachariae, 1959, 1960). No doubt, it also flies from stores. It is also found in the field in other parts of Europe and presumably is able to survive in the field in the warm temperate and subtropical parts of Africa and America.

In the tropics, *Callosobruchus maculatus* attacks ripening cowpeas in the field, and it is possible that the life history of this species is especially adapted to facilitate transfer between the store and the growing crop. There is a form of this species, with an easily recognized variant of the pattern of colored scales on the elytra (Utida, 1953), which is considerably more active than the normal form. Caswell (1960) considers this to be a distributional phase because it flies much more readily and appears unable to lay eggs until it has flown. It lays considerably fewer eggs, apparently developing the flight muscles instead, and the period needed for larval development is some 25% longer than for the normal form. Sano (1967) found that if the constant temperature at which larvae were developing was raised by 5°, 10°, or 15° to 35°C, some 30–50% of the insects emerged as the active form. This sudden temperature rise is typical of the phenomenon known as "heating" which results from a high density of larvae in a quantity of seeds. A high environmental temperature induces about 5% of this form.

Booker (1967) recorded a 2% infestation in the field of seeds of cowpea by bruchids in northern Nigeria, mostly by *Bruchidius atrolineatus* which died out in store in 3 or 4 months. Booker states that this species persists through the dry season on wild species of *Vigna* that may retain some pods for more than a year. In addition to this species of bruchid and *Callosobruchus maculatus,* Booker also recorded *Callosobruchus rhodesianus,* and Phelps and Oosthuizen (1958) recorded *Callosobruchus chinensis* in freshly harvested cowpeas in South Africa. Prevett

(1961) recorded a weevil, *Piezotrachelus varium*, but this is a true field pest.

Prevett (1966a) divides species of bruchids found in northern Nigeria into three possible categories — the true field pests able to develop only on green pods, those that attack green pods and subsequently attack dry pods, and those that lay only when the pods are nearly dry and continue in store. In the second category, he mentions *Bruchidius baudoni* that lives in *Acacia* spp. and a *Spermophagus* from *Ipomoea*. In the final category of storage pests that start to infest seeds in the field, he mentions *Callosobruchus subinnotatus* which attacks the Bambarra groundnut *Voandzeia subterranea*. Another important bruchid genus, *Caryedon*, includes several species fitting these two categories (Prevett, 1966b). The majority are in the final category, although few of the host seeds or fruits are of economic importance. The groundnut seed beetle *Caryedon serratus* lays on dry pods (Fig. 7). Although it has not been recorded on peanuts before harvest, these being subterranean, it does attack the seed pods of *Tamarindus* and *Piliostigma* which remain on the tree throughout the year. The palm kernel borer, *Pachymerus cardo*, causes loss of viability of oil palm seeds in natural conditions (Rees, 1963).

B. Invasion of Seeds in Storage by Insects

The species of insects that can fly into fields to attack the ripening seed can also, of course, fly into the buildings in which the harvested seed is stored. But generally seeds are more subject to man-made risks. Indeed, the crop is seldom infested in the field unless the storage buildings in the neighborhood harbor a large population of pests. These insects may be in cracks in the building fabric, in produce that has been stored for food until the new harvest, in surplus stocks of seeds, in rubbish left in stores, or in and around machinery where cleaning is difficult or impossible. They are often left behind in the bags, sacks, or bins in which the seeds are placed — even in new sacks or bags if these are left in infested premises. Indoors the adults of insects that cannot fly disperse widely, especially at night, and the larger larvae can also move some distance though they are less likely to move from one room to another by their own efforts.

Storage insects, nevertheless, are spread most frequently and over wider distances by man. Each parcel of seeds placed in an infested store or container may acquire insects, which under favorable conditions, may breed. The disturbance caused by handling induces some of the insects to leave and, perhaps, infest the transporting vehicle or other produce nearby, but some will eventually reach the destination, be it a few yards or thousands of miles away. Fortunately seeds are usually handled in small quantities and have a recognized value so that they are usually protected from these hazards. Farmers often keep a store especially for seed grain which is cleaned more thoroughly than other stores and which houses no machinery. The debris removed by a seed-cleaning machine should certainly not be stored alongside the seeds, especially if it contains insects. Unless it has some value, this debris should be destroyed.

C. Multiplication of Insects on Stored Seeds

A simple formula for the rate of multiplication of a population with a stable age distribution is $r = \log_e (Es\female)/d$, where r is the logarithm of the rate of increase in a given unit period, E an expression for rate of egg laying, s the proportion of females surviving to egg laying adult, $\female$ the proportion of females, and d the developmental time from birth to adult egg laying measured in the given unit periods (Howe, 1953). Though a stable age distribution is never achieved in practice, this formula is a useful guide to the pest potential of storage insects (Howe, 1965c). It is clear that the rate of development $(1/d)$ has the greatest influence on the rate of multiplication when this is rapid, so species with a short life cycle have a great advantage.

Many of the species found indoors are tropical species in which generations follow on in rapid succession and eventually overlap; only a few species have only one generation a year and are tied to seasons by a resting stage. Under the most favorable conditions, some species can complete a generation in 3 weeks, and population increases of the order of 50 times a month are feasible. Keeping seeds in good condition in the tropics once insects have gained entry is clearly a difficult problem.

The rates of development, oviposition, and mortality are influenced considerably by temperature, humidity, and moisture content and by the nutritive value of the seed. There is no evidence that sex ratio is influenced consistently by the environment. Other environmental variables influence behavior and, hence, may alter the multiplication rate of populations.

1. TEMPERATURE

Most storage insects will develop within a range of about 20°C from a poorly defined minimum threshold to a fairly sharp maximum. There is an optimum for the rate of increase about 5°C below the maximum. Over a range of some 10°C from 2° to 3°C above the minimum to 2° to 3°C below the optimum the curve relating the rate of development $(1/d)$ to temperature is approximately straight; if this line is extended to lower temperatures, it chances to give a good estimate of the minimum threshold. In fact, from minimum to optimum, the curve relating temperature and rate of insect development is sigmoid. The minimum is the lowest *constant* temperature at which growth is completed. If the temperature is raised and then lowered, growth continues at temperatures below the threshold but gradually decelerates. In an environment in which the temperature varies across the threshold, a considerable fraction of the growth occurs at temperatures below the minimum.

Temperatures in buildings, especially within lots of produce, are buffered, and the daily and seasonal variations are much smaller than in the open. Insects, however, tend to live near the surfaces of produce or in the fabric so are seldom at constant temperatures. Although constant temperatures are convenient for laboratory work, they are not "natural." The optimum determined in the laboratory under constant conditions will not be the optimum in the warehouse, especially if this is determined as a mean either of maximum and minimum daily reading or of more frequent ones. The range of variation is of great importance. Provided the temperature does not rise to lethal levels, for a given mean temperature, an insect will develop quicker with a larger range of variation (Messenger and Flitters, 1959).

The optimum temperature for one process differs from that for another. The optimum for the rate of oviposition is lower than that for development — that for the total number of eggs laid is correlated with the weight of the freshly emerged females and is, perhaps, only 5°C above the developmental threshold. The statistical weight of developmental rate, however, is such that the optimum for multiplication may be safely taken as that for development.

In the laboratory a great number of species have optima for multiplication in the region of 32° to 33°C, and even higher for *Latheticus oryzae* and *Trogoderma granarium* (Howe, 1965c). These are the tropical species that are abundant in areas where the mean temperature is about 28°C. A number of important species have laboratory optima at about 27° to 28°C, and these, *Sitophilus* spp., *Ephestia elutella, Anagasta kuehniella, Stegobium paniceum, Cryptolestes turcicus* are probably native to areas of Mediterranean climate. Only a small number of species have lower optima and are clearly cool temperate species, and these, *Ptinus tectus, Endrosis lactella,* and *Hofmannophila pseudospretella,* all develop relatively slowly at their optima.

2. HUMIDITY AND MOISTURE CONTENT

Humidity and moisture content are closely interrelated so that seeds and the air surrounding them tend to come into equilibrium (Pixton, 1967). Whether air or seeds are the controlling influence depends upon the relative mass. The relative humidity of the air in a sack of seeds is controlled by the moisture content of the seeds, but if this sack is stored in a large room, the relative humidity of the air will eventually determine the moisture content of the seeds (see Chapter 3 of this volume). Generally the humidity and moisture content must be kept low to prevent germination and growth of fungi. A value of 70% relative humidity (RH) is usually accepted as the maximum level permissible if fungi are to be discouraged.

It is generally convenient to consider equilibrium relative humidity (ERH) rather than moisture content, because the latter differs so widely between oily and more starchy seeds. However, the minimum ERH at which insects can develop tends to be lower for seeds with a high moisture content (Howe, 1956). A number of storage species are limited by their moisture needs, especially the temperate species (Howe, 1965c) which seldom encounter low humidity in warehouses. Stored products insects are remarkable, however, in that so many are able to grow at the lowest ERH obtainable in the laboratory and that many more can grow at an RH of 40% or less. Development is slightly retarded as the RH decreases from about 90% toward the lower limits, but it is not until the humidity is about 10% above the minimum limit that the developmental period and mortality are much affected. Mold growth makes laboratory experiments at humidities above 80% RH difficult to perform. Oviposition is more drastically affected by low humidity than is development. Most adult beetles and moths will drink free water if they can find it, and the availability of drinking water greatly increases the numbers of eggs laid by some beetles. However, *Tribolium* appears to lay fewer eggs if given water, though it may, in fact, lay more but acquire the habit of eating eggs after drinking. Larvae also will drink water when it is available.

Tolerance of low humidity is greatest at the optimum temperature and least at limiting temperatures, especially the maximum. Temperature gradients in stored produce must be avoided in all but the driest seeds, because a temperature gradient automatically causes a vapor pressure gradient. Water vapor will move down this gradient raising the moisture content at the cool end. The ERH of intergranular air in seeds of a fixed moisture content rises very slightly with temperature, a rise of $10°C$ being accompanied by an increase of about 3% RH.

3. FOOD

Plant breeding to date has mainly been concerned with characteristics of the growing plant. However, in the past decade, Doggett (1958) in Uganda and Wilbur (VanDerSchaaf *et al.*, 1969) in Kansas have been comparing resistance to insect attack of a number of varieties of cereal seeds. The differences among cereals do not seem to be very large, but wide variation of susceptibility among pulses is much more probable. Soybeans are seldom infested by insects and are thought to contain a toxin, yet some laboratory workers breed their stocks of bruchids on soybeans.

Differences of susceptibility may be attributed to physical causes such as hardness, convexity, smoothness, and thickness of pericarp or to chemical reasons such as the presence of toxins or attractants. It is probable that the preference of some insect species for particular hosts arises from chemical attractants.

However, given a crop of a variety of seed, the extent of insect damage is determined by such factors as the amount of mechanical damage and broken seeds caused by harvesting and cleaning, the amount of dust and dockage, and the moisture content which influences softness of the grains. The infestation of feed grain also depends upon processing it may undergo. Rice paddy is almost immune from insect attack unless the rough husk is broken during harvesting (Breese, 1960, 1964). Oats, and to a lesser extent barley, are also quite resistant but wheat, rye and, to some extent, maize are very susceptible. Nevertheless, species not adapted to attack seeds will cause more damage if there is a pentiful supply of dust, dockage, and broken kernels on which the young larvae can feed.

Pulses are mostly very susceptible to bruchid attack especially in the tropics. Peanuts present a difficult storage problem. If kept in the shell they are vulnerable to the seed beetle, *Caryedon serratus,* and, if decorticated, they are soft and vulnerable to a wide variety of insects.

The size of parcel has some influence on the extent of risk. Except for the very small ones, stored products insects tend to remain near the periphery of a stack or a bag. The amount of damage, therefore, tends to be proportional to surface area rather than volume. It is much easier, though, to protect small quantities and should a population gain entry the damage caused to seeds stored in large quantities could be very serious. This is because seeds generally are poor conductors of heat. Metabolic heat from insect activity raises the temperature of the produce because it dissipates slowly. This heat both kills seeds and generates moisture gradients that encourage fungi, mites, and other insects.

4. LIGHT

Of the factors that influence behavior of storage insects, light is the most important. The daily cycle of light change induces a rhythm of activity in the pests, and light itself discourages oviposition and activity in most though not all storage insects. In the dark and in dimly lit warehouses during the day, insects are easily seen on surfaces and flying. When the light intensity increases the insects retreat into the darker places, within stacks and machinery and into cracks in the floors and walls. If the sequence of light and dark is maintained, with many species pupation, adult emergence and mating, all tend to take place at the same time each day often at dusk or shortly thereafter, with a surge of activity at that time, often of larvae as well as of adults. Activity is induced by changes of light intensity, temperature, and humidity, especially by a decrease of each of the first two and an increase in humidity. These changes occur together at a rapid rate at dusk, but the reverse changes at dawn also stimulate some activity. The maize weevil, *Sitophilus zeamais,* becomes active earlier in the day in some places—in mid-afternoon in Kenya (Giles, 1969) and just before noon in Japan. We do not know how local strains of insects differ nor what changes take place in the stores to induce weevils to walk out of doors, but once they are in the sun they soon fly and disperse either to a growing crop of maize or to another store.

5. LENGTH OF STORAGE

Insects lay more eggs in a dark and damp store than in a light dry one. These eggs develop fastest in a warm and fairly damp store, especially if the seeds in which they are growing are not disturbed. Not only do the insect populations tend to increase exponentially, but the accumulation of metabolic heat raises the temperature of the seeds and accelerates the increase in numbers of insects.

The consequence of these factors is that each succeeding day, week, and month of storage is increasingly hazardous if living insect, mite, or fungal pests are present. Since little can be done to shorten storage by changing the date of sowing, the period of hazard must be shortened by delaying as long as possible any exposure to pests.

IV.   Prevention and Control of Insect Infestation in Stores

A.   Storage Facilities

Prevention of insect attack during storage, rather than control of insects, is even more important for seeds than for other stored produce. For this reason seeds should be separated from feed and kept in a special building or room used only for seeds. Furthermore, on farms, home-grown seeds should be segregated from purchased seeds, as far as possible in different buildings or rooms, especially in those areas and for those crops that do not suffer field attack by storage pests. Segregation is essential to minimize the spread of insects from one lot of seeds to another, a risk that inevitably increases as seeds are moved from one place to another and from one owner to another. A corollary to this is that seeds should be given the best storage facilities and buildings available and that these should not house seed-cleaning machinery, drying plants, piles of bags, or any other incidentals that insects could use as a hiding place. The building should be cool, dry, light and airy.

Generally seeds should be stored in small units, possibly in amounts of the order of 50 kg to avoid the risks of physical gradients, particularly vapor pressure, which arise in large bulks. The risks of direct insect damage to seeds is at its greatest in these small quantities, which should not be kept in loose piles. Seeds must be kept in bins or bags made of some fabric that insects do not easily penetrate. If metal bins are used these must not be subject to sharp temperature changes that might cause the deposition of dew; if wooden bins are used, these should provide a minimum of crevices and should be carefully inspected to ensure that insects are absent. The surface should be covered by a lid since the free surface is especially vulnerable to invasion.

When seeds are kept in bags, jute and cotton are best avoided since many insects can easily penetrate them or lay eggs through them and may invade new bags kept in readiness for seeds. Second-hand bags are a notorious source of infestation. The choice between the various kinds of paper and plastic, of bags with multiple layers of various fabrics or liners and of fiber and metal drums, must depend on cost, their convenience for handling, and so on. In particular the bags should be permeable to water vapor if the seed is initially stored at a moisture content near the upper permissible limit, or impervious if the area is a very damp one during the seed storage season. Insects are much more adept at boring out of a bag than into it but many species can lay eggs through small holes. It is extremely important not to have gaps in seams or stitch holes, nor to use foldover or tied closures in which insects can accumulate and lay eggs, and from which freshly hatched larvae may find their way into the bags. The most successful bags for excluding insects usually contain a layer that has been treated with an insecticide (Highland, 1967) usually a pyrethrin which is also a repellent.

B. Prevention of Infestation

Prevention of infestation is achieved either by mechanical exclusion of insects or by rendering the environment unsuitable for insects. Exclusion by management and mechanical barriers has already been mentioned above, but it is possible to reinforce the barriers by covering bins or stacks of bags with insectproof plastic sheets or tarpaulins.

1. COOLING

The need for stored seeds to be dry has already been stressed, but it must be emphasized again that damp seed is unlikely to escape damage by fungi and is very susceptible to attack by a large range of insect pests.

In the temperate parts of the world, storage difficulties can also be minimized by keeping the seeds cool. Temperature has more influence on the rate of multiplication of pests than any other single factor of the normal environment and after moisture is the key factor in storage. Insects multiply only slowly up to about 5°C above the minimum threshold of the species so temperatures up to 22°C can be tolerated for periods of up to a year. But at 30°C the numbers of some species increase fiftyfold during each month of storage (Howe, 1965c). Consequently, exceedingly low initial densities of insects, densities far below those that can be found by normal methods of inspection, have the potential to reach damaging levels in 3 to 6 months. The artificial drying of produce without adequate subsequent cooling introduces an unnecessary risk and may even create a greater hazard than there was before drying. Temperature and moisture are inextricably related, and Burges and Burrell (1964) provide a useful chart showing safe and unsafe combinations of the two.

If seeds are stored in small lots, they will quickly cool to the ambient temperature and thereafter need to be insulated against temperature rises. Bigger bulks can be cooled by circulation of untreated air in climates as warm as the Mediterranean area of Israel (Navarro *et al.,* 1969). In the tropics the circulated air would need to be cooled by refrigeration; the expense and the risks involved are such that other methods of environmental manipulation are more promising.

The main risk associated with cooling is that of raising the moisture content of the seeds in some places. If vapor pressure remains unchanged then humidity rises as temperature falls and, hence, the moisture content may rise if the water vapor is not carried away. On no account should dew be allowed to form. Vapor pressure gradients are undesirable, but when there are temperature gradients such as are deliberately introduced by cooling, they are inevitable. The danger points are always the zones of highest temperature or highest moisture content, and within these the dampest seeds, because a moisture content estimation of a sample is merely the average for a number of kernels which may vary widely (Oxley, 1948).

2. AIRTIGHT STORAGE

Variations of the composition of the atmosphere within produce are never large enough to incommode storage pests unless the pests are so numerous as to cause exceptional damage. In this event the suppression of the pest may come too late to be of value. The volume of air space within produce is only some 40–50% of the volume of the container (Jones, 1943) so if the interchange of this air with the air outside can be prevented or restricted, the oxygen on which the pests depend will be used up. Bailey (1965) has shown that, if oxygen levels fall below

3%, the insects of cereals die rapidly, but an increase of carbon dioxide above levels normally caused by aerobic respiration is not lethal. It is evident, therefore, that if the level of oxygen can be kept below 3%, pest insects will not live long especially in warm climates. This can now be done by adding an inert gas or by circulating carbon dioxide or nitrogen through storage bins (Jay and Pearman, 1969).

In tropical areas where the risk of infestation is high the insects themselves will, in closed bins, remove the oxygen by respiration and this is the basis of airtight storage. The greater the number of insects the more rapidly the oxygen is depleted and, hence, the sooner the insects are killed or at least immobilized. The major problem with airtight storage is the risk of a leak that allows in sufficient air to maintain some insects or aerobic fungi. Such a leak is difficult to detect if the oxygen concentration remains low or the carbon dioxide concentration remains high. The rate of oxygen leakage into the storage container would under these circumstances determine the size of the pest population and the amount of damage caused by it.

To date the problem has largely been the engineering one of building airtight structures at reasonable cost. This is often most easily achieved by using natural pits or digging holes but the advantage this gives of low temperature is replaced by the need for waterproofing. Fortunately there are now strong gasproof fabrics like butyl rubber (Hyde, 1968) that can be used, and these may be very suitable for seeds stored in fairly small quantities.

C. Chemical Methods of Protection and Control

Chemical insecticides may be used as repellents, as a means of minimizing the insects in a building or container, or may be applied directly to seeds, either to protect them from infestation or to kill insects already present. A repellent material is unlikely to be widely used unless it is also toxic to insects but some widely used insecticides do have some repellency.

In general, insecticides can be divided into those that kill insects more or less immediately and those that persist and kill for some time after their application. The former group consists of mostly gaseous fumigants and the latter of materials applied as sprays of some kind. (See Parkin, 1963.)

1. FUMIGANTS

Toxic gases are of value in killing insects already present in storage facilities or in seeds when harvested or purchased. Fumigants have a residual action only if the gas is highly adsorbed and is liberated slowly after fumigation as may happen with ethylene dibromide or dichlorvos. Extensive adsorption carries with it a risk of damage to the germ, and ethylene dibromide should not be used for seeds. Adsorption also incurs a risk to the consumer of food, and dichlorvos is not approved for this use in the United States.

There should be no need for fumigation of crops that are not attacked before harvest and are stored after harvest in insectfree premises. The fumigants most likely to be used by commercial pest control operators for seeds are methyl bromide and phosphine. The influence of the former on germination has been widely studied (Strong and Lindgren, 1959), and a fumigation can be completed in a day. Phosphine is scarcely adsorbed at all so would not be expected to harm the germ, but it is most efficiently used at low concentration with a long period of exposure – as long as 2 weeks. Possibly this fumigant which is applied as tablets, pellets, or powder, is best applied at the start of storage with the gasproof covers left in place. On farms, fumigations may be performed by the farmer with mixtures of ethylene dichloride and carbon tetrachloride. This mixture is handled as a liquid which can be poured onto the seeds in bins which are then covered with tarpaulins. This is satisfactory for feed grains but must carry some risk for seeds. Carbon disulfide has also been used in this way but is best avoided.

2. CONTACT INSECTICIDES

The poisonous effect of fumigants disappears as the gas disperses. A solid insecticidal dust or a film of liquid on surfaces retains its influence for a long time and is likely to kill any insect remaining on it. A film applied to the surfaces of buildings, bins, and sacks will, thus, offer some protection but since the insects can rapidly move onto the seeds the best protection is offered by covering as much as possible of the surface area of the seeds. Chemical seed protectants are unpopular nowadays especially those that kill or harm birds or other wildlife that eat some of the sown seed. There is no need to list the many possible insecticides; the use of several of them is prohibited over much of the world, and the details of dosage rates, formulations, and methods of application of the available insecticides are always supplied by manufacturers. The two insecticidal materials that may be used safely for some time are the pyrethrins and malathion.

Pyrethrins rapidly paralyze insects but death is slow. This material is also a repellent. It is usually applied as a solution of 0.5% in a technical white oil or a refined kerosene and may be sprayed or dispersed as a mist. Because it rapidly interferes with the coordination of the flying and walking of insects, it is appropriately sprayed at dusk or at any other time when insects are active. Then instead of being able to spread and invade seeds they are "knocked down" and trapped. Pyrethrins are broken down by light and by oxidation so have a relatively short life. This group of insecticides is effective against beetles and adult moths but not against caterpillars. It is synergized by piperonyl butoxide. In seeds its best use would be as a quickly breaking emulsion of pyrethrins and piperonyl butoxide sprayed directly on the seeds, for there is no deleterious effect on germination. It can also be applied as a dust admixed with seeds.

Malathion is an organophosphorus compound with low toxicity for birds and mammals. It is an effective insecticide against seed beetles and may be admixed with seeds as a dust or sprayed directly onto the seeds. Dust formulations may break down rapidly if incorrectly formulated. This rules out formulation of the dust in some areas where local diluents cause breakdown. Malathion can also be applied to seeds as a quickly breaking emulsion.

No other insecticide is in wide use in food stores. In experimental work in warehouses (Green *et al.*, 1968), dichlorvos has shown great promise but because of doubts about its safety it has not been placed on the approval list of the United States Department of Agriculture. It is toxic to many seed pests at low concentrations, disperses well through buildings, and acts both as a contact poison and a fumigant. Because it seemed not to be persistent, one proposed use for it was automatic nightly fogging of warehouses that were freely ventilated during working hours. It is widely marketed as a slow emission resin strip, and it is possible with these for rather higher concentrations of dichlorvos to accumulate in closed rooms.

D. Other Methods of Control

The use of chemicals on foodstuffs destined for consumption by humans or by domestic or farm animals has always been unpopular, and in recent years their use on seeds, some of which might be eaten by wild animals and birds, has also become unpopular. Nonchemical means of control are, therefore, constantly under review, but, in addition, the basic ecological, physiological, and epidemiological work on the life and behavior of pests inevitably yields information about the conditions that are unfavorable or lethal to the pests. Some aspects of this approach have already been mentioned above when storage at low temperatures and under low-oxygen tensions were recommended as means of preventing damage by pests. Exclusion of oxygen is as often a control as it is a preventative measure, but cold, to be a control measure needs to be much more severe than it is as a preventative. As a general rule, the use of intense cold is too expensive a way of killing pests in either food or seed grains which seldom have a high monetary value. This points to the real problem of physical methods of control — they need expensive equipment or power. Natural cold can be and is used to control insects in Canada

(Watters, 1965) and other regions of similar climate, but these regions have storage pest problems only in artificially heated buildings.

Heat treatment of seed is superficially attractive especially as seed and food grain often need to be dried with heat before they can be safely stored. However, the safety margin between temperatures that will kill insects and those that harm germination is narrow. Because the thermal conductivity of seeds is low, the seeds must be heated in shallow layers to avoid the risk that the side near the heat source will be hot enough to kill the seeds whereas that away from it is not hot enough to kill the insects. In any event, to kill the majority of species of insect storage pests in their most resistant stages in a few minutes, a temperature of 60°C (140°F) is needed, whereas germination may be deleteriously affected by 50°C (122°F). Although this risk rules out heat as a safe method for killing insects in seed, it might be used for food grains if a cheap source of heat can be found along with a method for handling the seeds at a reasonable rate. Some radiations have the property of heating produce uniformly through a layer. These generate high temperatures by the dielectric heating of seeds placed in an alternating high-frequency electric field. In theory, insects in and among the seeds should reach a higher temperature than the grain in such a field, but in practice an efficient apparatus has not been devised yet.

Watters (1965) reviews the whole field of control of storage insects by physical methods with some optimism. He refers to a report of sound depressing the oviposition of the moth *Piodia interpunctella,* although no mention is made on the effect of these waves on humans. Light also depresses the oviposition of many pests but it does not seem feasible to eliminate dark areas in produce. Light also attracts insect pests in some circumstances but not in sufficient numbers to act as a control measure. Mechanical forces can be used to kill insects in meals but not in seeds which would be damaged and so become more susceptible to later insect attack. Moving air has been used as a curtain to prevent the influx of flying insects.

One method that can certainly kill insects in seeds is the use of ionizing radiations from a gamma or accelerated electron beam source but this is largely confined by cost to large central depots. A sensible appraisal of this usage is given by Bailey (1966). Low pressure also kills storage insects, possibly through oxygen starvation, but once again we await the provision of a cheap practical method of lowering pressure and keeping it low.

Biological methods of control may offer more hope of providing control at a cost similar to that of chemical insecticides. The use of parasites and predators in warehouses is not practical, because these and parasitized pests are just as objectionable in food as ordinary pest insects and the level of infestation at which control is effective is too high. There are similar objections to the introduction of pathogenic bacterial diseases of insects even when these bacteria do not affect humans. The idea of introducing disease organisms into food is abhorrent and again the method is unlikely to be highly efficient when the insects are sparse.

The most promising lines of biological attack nowadays aim at disrupting the growth and reproductive cycle of the insect. One method that has had success in other fields is the liberation of sterile males into populations to compete with the resident fertile males in mating with the females. This method requires a number of characteristics that are unusual in storage insects. Many of the seed insects are long-lived and mate several times and even the shorter-lived moths and bruchids mate more than once. The chances of a female laying fertile eggs when sterile males are present is therefore high. Furthermore the mobility of many species in warehouses is limited. Some cannot fly *(Sitophilus granarius, Trogoderma granarium, Ptinus tectus)*, others seldom fly, and few are powerful fliers. For there to be any chance of sterile males of warehouse insects to mix with the natural population, these sterile individuals must either be dispersed widely rather than at a limited number of points and at the time of day and in the conditions conducive to active movement. Rearing on a large scale involves problems of disease and of avoiding overheating and then an irradiation plant for sterilization. This again adds up to an expensive technique and a trial made with *Ephestia kuehniella* was unsuccessful. Chemosterilants have been used experimentally on bruchids for theoretical rather than practical aims. The conclusive objection to the sterile male technique in seed storage is that large numbers of insects must be introduced, insects that will damage the seed even though they have no progeny, and will contaminate the product with frass and eventually with dead bodies.

There is some possibility that chemicals with an influence on development resembling that of a juvenile hormone can be introduced to control populations of pest insects. To be of value this chemical must not only prevent the reproductive stage appearing but must ensure that affected larvae eat no more than untreated ones. It is perhaps early to decide whether this form of chemical poisoning is less dangerous to other life than the insecticides now in use, but it is unlikely to be introduced if current research produces any evidence of risk.

Interference with behavior especially in relation to reproduction may provide a new means of pest control. The production of pheromones, chemicals liberated by one individual and influencing the behavior of another at some distance away, has been demonstrated for a wide variety of insects found in warehouses. These have at least two kinds of function related to reproduction, to assemble mates from a distance and to act as an aphrodisiac on nearby mates. The first property might be used to bait traps if the substances responsible could be manufactured, and, if the bait attracted virgin females, it would quickly be extremely effective. An attractant for males would act more slowly but should succeed.

The value of identifying and making an aphrodisiac is less clear, but possession of such a substance would stimulate an interesting program of research. The pervasion of the environment by an aphrodisiac perhaps in unusual concentrations could well confuse and disrupt behavior patterns and drastically limit reproduction. How readily an insect might adjust and whether "resistant" strains might appear from the mating of individuals less dependent on this chemical signal are questions we cannot answer yet. Wright (1970) explains how the time needed for and the cost of a search for olfactory stimulants for pest insects can be cut.

V. Insect Pests of Seeds

Insects attacking seeds are of two kinds — those adapted for life in seeds that are more or less restricted to feeding in seeds and those that are general scavengers and are able to deal with seeds along with organic debris. The latter group, though often wrongly called secondary pests, include species that can cause considerable damage to seeds.

A. Species Developing within Seeds or Pods

A number of storage species are typically associated with seeds. The larvae feed entirely within the seed and usually the pupae are also formed there (Fig. 7). The adult is the only stage of the insect likely to be found outside the seed. The egg may be laid in or on the seed or close to it. This group includes two important families of beetles, the Curculionidae or weevils, many of which feed in seeds of a large variety of plant families and the Bruchidae which are more restricted, breeding principally in the pulses, peanuts, and palm seeds. The weevils of economic importance attack cereal seeds. These belong to the genus *Sitophilus* and in addition to seeds will breed in fleshy starchy tubers, such as cassava, in manufactured products, such as macaroni or pasta, and in compressed flour. The bruchids appear to breed only in seeds or pods, possibly because of chemical attractants in the seeds but they will grow in laboratory made pellets.

One species of the wood-boring family of beetles Bostrychidae, *Rhizopertha dominica,* is able to attack cereal grains although its original habitat may have been fleshy stems or roots such as bamboo. It is of special

importance as it is the only species capable of gaining entry into the undamaged grains of paddy rice, but it is also capable of feeding in flour.
One species of moth, *Sitotroga cerealella,* is restricted to developing
inside cereal grains. No insects of other orders are of any great economic
importance but seed chalcids which are really field pests occasionally
emerge in stores.

B. General Feeders

Individuals of species that are not especially adapted to feeding on
seeds may depend upon mechanical damage at harvest or later, or upon
the previous attack of the species mentioned above, but they are often
able to damage sound seed. The misconception that these insects are
secondary pests that cannot attack sound seeds arises from laboratory
experiments with newly hatched larvae. First-instar larvae are small
enough to find minute holes through which they can enter a seed but
seldom have mandibles powerful enough to break the seed coat. Older
larvae of species, such as *Lasioderma serricorne* and *Ephestia* spp. can
bite through the seed coat and the larger adults of beetles such as *Tene-
brioides mauretanicus* also can do so.

Seeds are most liable to attack by these insects when seed moisture
content is high and the seeds relatively soft. They may also be vulnerable
when dry enough for the seed coat to crack and expose the embryo.
Under these circumstances the embryos of many seeds may be completely and cleanly removed, while the endosperm remains untouched.
The embryo is attacked by many mites and by moth larvae of the family
Phycitidae. Except in very dry grain when they are restricted to the
embryo, beetle larvae especially those of *Cryptolestes* spp. and *Oryzaephilus* spp., *Trogoderma granarium,* and occasionally of *Tribolium* spp.,
invade the endosperm from the embryo. However, spider beetle larvae,
especially of *Ptinus tectus,* eat the bran from grains which they spin
together while feeding.

C. Methods of Study

The insect pests of seeds in storage have been studied extensively both
in warehouses and in the laboratory. The value of warehouse studies is
limited more or less to the areas in which the studies were made and to
areas of similar climate. Laboratory studies have progressed from observations made in working laboratories to those made under controlled
conditions from which predictions of fair reliability can be made for all
parts of the world. Ideally warehouse and laboratory work should be
done side by side, for the former includes measurement of environmental
variables and summing up of the problems. These must be turned into

questions that can be investigated in the laboratory and then the conclusions tested against field observation. A great deal of laboratory work must be directed toward solving questions of technique to determine how these influence the results of laboratory experiments.

On the whole, simple laboratory experiments have proved satisfactory. It is usually easy to get storage insects to lay eggs that hatch well and the larvae to feed, grow, and pupate with little mortality. The experimenter has the choice of following a limited number of isolated insects or accepting the influences of crowding by using more insects in batches. The defect of most experimental work is the paucity of controlled conditions used, either through lack of facilities or lack of time; this often leads to false statements about the optimum temperature for some characteristic of the species, but no harm is done if "optimum" is read to mean "the best of the conditions used in this work." Rather more work has been done on the rates of insect development than on rates of oviposition. Fortunately the former is the more important in ranking insects as pests.

Warehouse studies are of special importance in places where some species have seasonal interruptions of the life cycle, whether caused by lack of food or by climate, or there is some adaptation of the life cycle to cope with these interruptions. Many storage species are still partially adapted to the environment in which we presume they evolved even though the triggers for these life cycle patterns are no longer very strong. It is important to work out how far these patterns still concentrate the emergence of adults and, where food supplies are seasonal, how the generations of pests are related to the climate. To date not much attention has been paid in the laboratory to factors other than the obvious ones of temperature, humidity, and food, and neither the importance of photoperiod nor of gradients in the environment has been assessed.

D. Some Important Species of Beetles

The principal problems arising in dealing with individual pest species in a work of this kind relate to identification and nomenclature. A taxonomic description of each species is not appropriate and a short general one is useless. Since it is difficult for experienced entomologists to identify many of the seed pests with certainty, descriptions will not be given here (see, however, Figs. 9 to 13). Specimens collected that seem to present a hazard and cannot be readily identified should be sent to a specialist through an advisory entomologist. Nomenclatorial problems arise through strict application of rules that may lead to a change in the accepted specific name of a common pest, or through the widespread disregard of these rules, or simply because knowledge is gained through study. Every decade one or two well-known species prove to be a com-

FIG. 9. A. Adult bruchid beetles found in stored peas: (a) Typical female of *Calloso-bruchus maculatus,* (b) the flight form of the female of this species, (c) female of *Callosobruchus analis.*

B. Adult bruchids that attack beans: (a) *Acanthoscelides obtectus* and (b) *Zabrotes subfasciatus.* (c) Adult bruchid that attacks groundnuts: *Caryedon serratus.*

plex of species usually with differences in ecology and one or two names well-known to the layman fall out of use. Here I deal with a number of species in an arbitrary order.

1. BRUCHID BEETLES

This family is the best example of the range of adaptation from life in growing seeds to life in stored seeds. The majority of species in this family live in seeds especially of legumes, but even with wild plants many live in seeds that hang on the plant for a considerable time when ripe. The species of major economic importance live in cultivated crops of peas, beans, and groundnuts. There is some specificity of host for bruchid species, though this is less fixed than was once believed. There seem to be quite large differences in susceptibility to insect attack between varieties of a crop. Susceptibility depends both on chemical and physical factors. The chemical ones include both nutritional adequacy and toxins or repellents – a seed may be relatively free from insect damage because it lacks certain essential nutrients or because it contains something unpleasant. Many varieties of soybean may contain a toxin – in any event, many are not very vulnerable to insects in store. Physical factors include smoothness and convexity which determine the attractiveness of seeds as an oviposition site, and thickness of cuticle and hardness of endosperm which may prevent the young larva from establishing itself. The division into pea weevils and bean weevils used below is convenient but is no longer valid.

a. BRUCHIDS OF PEAS. There is no very serious storage pest of the garden or field pea, *Pisum sativum*, of temperate areas. The pea weevil, *Bruchus pisorum*, is a field pest though it is often in the seed at harvest and may remain within it through the winter. It may hibernate indoors or in sheltered places out of doors and emerges when the earliest peas begin to bloom. A good account of this species in the United States is given by Brindley *et al.* (1958).

However, tropical peas are liable to severe damage by bruchids, especially cowpeas by *Callosobruchus* spp. These can breed in *Pisum* but are much less successful, mortality being much greater and development slower in general and much more variable. The important species of *Callosobruchus* are *C. chinensis, C. rhodesianus, C. maculatus,* and *C. analis* (Fig. 9). The first two species are less tropical in their requirements having constant temperature minima around 17.5°C as against 20°C for the other two. None are cold hardy. *Callosobruchus* spp. attach their eggs either to dried pods or dried seeds and the larvae on hatching bore directly into seeds. In crowded conditions many eggs may be laid on a seed and several larvae can successfully develop and emerge from a single seed. Generally, however, the adults avoid laying eggs on seeds that already bear eggs or contain larvae. All four species can pass a generation in 30 days, and all but *C. analis* in 25 days at temperatures of 30°C or a little higher. At 25°C the cycle requires 6 weeks or, for *C. analis*, about 7 weeks. These species are not especially prolific in egg laying, each female depositing up to about 100 eggs. They are tolerant of low humidity both for oviposition and development and so are not uncommon in crops in the dry areas of the world.

These species are not easily identified so there is some doubt about their exact distribution and pest status. *Callosobruchus rhodesianus* is probably a pest only in southern Africa although it is found elsewhere in Africa, and *C. analis* is a pest only in the Indian subcontinent and Southeast Asia. *Callosobruchus chinensis* is possibly diminishing in importance and is most likely to be a problem in the Mediterranean, southern United States, China and Japan, and in places of like climate. *Callosobruchus maculatus* is undoubtedly the most important, possibly because it invades the growing crop more readily than do the other species. From time to time this species develops an active flying form. Although the females of this form lay only about ten eggs they disperse to the growing crop. The reason for the appearance of this form is not clear as yet but is probably associated with high developmental temperature (Utida, 1965; Sano, 1967). The temperature may be high because of the climate or because of heating caused by crowding. Utida (1968) suggests that eggs laid by older parents are more likely to yield the active form of adult and that later emergences are more often active, but these observations may be consequences of rises in temperature in cultures and of the longer developmental period of the active form. Later, Utida (1969) claimed that both continuous darkness and continuous light favored induction of this form, but neither light-regime is likely to have any practical implication.

b. BRUCHIDS OF BEANS. *Callosobruchus* spp. are serious pests of the several *Phaseolus* spp. with small seeds, known as gram, but are not pests of *P. vulgaris, Vicia faba,* and others with larger seeds known as beans.

The common bean weevil of temperate climates *Bruchus rufimanus* is a field pest that may remain in harvested seeds in store without breeding. There are two serious bean bruchids — one tropical, *Zabrotes subfasciatus* and one warm temperate, *Acanthoscelides obtectus* (Fig. 9B). The former, like the *Callosobruchus* spp. attaches its eggs to the seed or pod, but the latter lays its eggs loose or more commonly in slits in the pod or seed, and the first stage of its larva is capable of wandering among seeds to find a means of entry into a seed.

Zabrotes subfasciatus is an American species which has spread into Africa and possibly into Southeast Asia. The range of suitable temperatures is 20°–35°C, very similar to that of *Callosobruchus analis* and *Callosobruchus maculatus,* so it can be assumed to have a similar potential range to those species. It is not cold hardy. It passes a generation in 25 days at 30°C on haricot beans, but, although spreading in distribution, it seldom does any great damage. *Acanthoscelides obtectus* is of greater importance because it can breed in the growing crop even in temperate areas. This is also a South American species but it has spread throughout the world and can be damaging in Europe and North America. The optimum temperature for development of this species is also close to 30°C, although it needs about 27 days, but both developmental thresholds are lower than any of the species previously mentioned, being slightly below 15° and 35°C. It is not very cold hardy but survives moderate cold better than the other species and well enough to overwinter in sheltered situations in Germany.

Zabrotes subfasciatus is a small beetle with a marked sexual dimorphism both of color pattern and size. The female weighs about 4 mg and the male about 2 mg as against about 6 mg for *Callosobruchus chinensis* and *Callosobruchus rhodesianus,* 7 mg for *Callosobruchus analis* and *Callosobruchus maculatus,* and 8 mg for *Acanthoscelides obtectus.* It also lays fewer eggs, about forty per female, whereas *A. obtectus* females on average lay about seventy eggs but exceptional individuals may lay 150. All these bruchids are short-lived in the adult stage—less than a week at 35°C, about a month at 17.5°C, and 2 months at 15°C. *Acanthoscelides obtectus* lays eggs in batches or singly among seeds. *Zabrotes subfasciatus* is more likely to lay several eggs on a seed than are the *Callosobruchus* spp.; indeed, the larger seeds of haricots or butter beans can successfully support a large number of growing larvae of both species of bean weevils—as many as twenty being recorded from a single seed. As a rule the multiple simultaneous use of a seed is a sign of an exceptionally heavy infestation.

Larvae of bruchids usually complete growth close to the external wall of the seed so that the adult can emerge by pushing a hole in the seed wall. The position of fully fed larvae or pupae is usually visible as a dark mark known as a window. The habits of species differ, some eating away an area of uniform thinness against the epidermis, others eating so as to leave a ring of weakness, but if the larva does not do this properly, the adult fails to escape. *Zabrotes subfasciatus* leaves a weak ring, whereas *Acanthoscelides obtectus* weakens the whole window surface so infestations of these species can be identified by the appearance of the window alone McFarlane and Weaving, 1967).

c. Bruchids Attacking Other Seeds. There are a number of *Bruchus* spp. that attack growing crops and are brought into store with the ripe seeds, but these are not known to breed there. Two of these, *B. ervi* and *B. lentis,* attack lentils in the Mediterranean. Some species of bruchids also attack growing crops and these may achieve some limited breeding in store, possibly before the seeds dry out completely. *Bruchidius quinqueguttatus* is found on lentils in the Mediterranean, and *B. atrolineatus* on cowpeas in West Africa. *Bruchidius baudoni* attacks both the green and dry seeds of *Acacia* spp. (Peake, 1953; Prevett, 1966a), but *B. natalensis* and *B. strangulatus* attack only the green pods of *Acacia* and *Crotolaria* spp., respectively (Prevett, 1966a). *Bruchidius schoutedeni* has been found in dried seeds of *Sesbania pachycarpa* but not yet in green pods.

The ripe seeds of the Bambarra groundnut *Voandzeia subterranea* in West Africa are attacked by *Callosobruchus subinnotatus*, and seeds of oil palm in tropical South America by a number of *Pachymerus* spp. (Prevett, 1966c) — *P. bactris, P. cardo, P. nucleorum, P. lacerdae,* and *P. thoracicus. Pachymerus cardo* has spread to West Africa and one species, *P. abruptestriatus,* attacks the ripe seeds of ebony, *Diospyros* spp.

The only genus of serious economic importance to crops other than peas and beans is *Caryedon,* and this has a large number of species that attack only the pods or fruits of wild plants, especially of *Combretum, Cassia,* or *Acacia* spp. Prevett (1966b) presents his observations on these. The species that attacks groundnuts is now known as *Caryedon serratus* (Fig. 9B(c)) but is discussed in the literature under the names of *Caryedon gonagra* and *Caryedon fuscus.* It also attacks the pods of *Tamarindus indica, Piliostigma* spp., and *Cassia fistula* and many other plants (Davey, 1958). *Caryedon serratus* is the principal pest of groundnuts stored in the shell. It can also attack decorticated nuts, but the adult beetles are large (about 6 × 3 mm) and cannot penetrate far into the tighter packing lots of decorticated nuts. This species has become well established in Africa and India and, more recently, in the West Indies and South America. The optimum temperature for development is close to 30°C, at which temperature it takes about 6 weeks for a generation in groundnuts. The adult lays about 110 eggs and lives about 3 weeks at this temperature. Growth and development are quicker on tamarind than on groundnuts.

2. CURCULIONIDAE

The weevils include a large number of species that attack seeds and fruits but the only species common in warehouses belong to the genus *Sitophilus.* Most of the other weevils that are found infrequently in warehouses either attack timber or are field pests, such as *Piezotrachelus varium* in cowpeas, which are unable to persist for long in dried seeds.

There are three species of *Sitophilus, S. granarius, S. oryzae* (Fig. 11a,b), and *S. zeamais,* but the literature is complicated by problems of nomenclature. Until 1959 when the issue was settled internationally, the generic name *Calandra* was used in most European literature, whereas *Sitophilus* was used in North America. In addition there are ambiguities over the trivial name for two of the species, *S. oryzae* and *S. zeamais,* which are difficult to distinguish and are still confused. Both specific names were used in Germany up to 1939, but, since individuals were

FIG. 10. Kernels of wheat dissected to show developmental stages of *Sitophilus granarius:* (a) egg; (b) young larva; (c) old larva; (d) pupa; (e) pharate adult; (f) mature adult.

identified principally by size, there is no certainty that the names were correctly used. Elsewhere it was then thought that the difference in size was a consequence of the size of the grain in which the weevil developed, with the larger ones developing in maize. When it first became apparent that the two are distinct strains and that fertile mating between them is impossible (Birch, 1944; Richards, 1944), they were termed large and small forms, and it was not until some time later that they were accepted as distinct sibling species (Birch, 1954). Unfortunately, when Floyd and Newsom (1959) gave detailed descriptions of the two species they adopted the name *S. oryzae* for the large strain and introduced *S. sasakii* for the small form. Kuschel (1961) corrected this, calling the small form

FIG. 11. (a and b) Adults of two storage beetles; (a) *Sitophilus oryzae;* (b) *S. granarius.* (c and d) Larvae of two beetles that attack stored seeds; (c) *Trogoderma granarium;* (d) *Lophocateres pusillus* which resembles a tiny *Tenebroides mauretanicus.*

S. oryzae and the large one *S. zeamais.* However, the overall result is that with few exceptions only the trivial names *zeamais* and *sasakii* and the terms small and large strains, as used over the last 30 years, can be accepted as correctly defining the species under consideration and the use of the name *S. oryzae* is often very dubious. Recently, Boudreaux (1969) claimed that it is easy to distinguish these two species but this is probably true only for people who deal with them frequently. The differentiating characters lie in the genitalia and in puncturation and microsculpture of the pronotum.

There are, however, some quite distinct differences in ecology of *Sitophilus* spp. *Sitophilus granarius* is wingless and unable to fly, the elytra being fused together; *S. oryzae* rarely flies whereas *S. zeamais* is a strong flier able to invade the ripening maize seeds in the field. Although all three species can develop in almost all the cultivated cereal crops with large enough seeds, each has its preferences so that populations of them in different foodstuffs can exist side by side. The Latin names of the species are apt and descriptive. The optimal conditions for multiplication of the species differ very little, $0.5°C$ perhaps covering all three, yet the distribution of the species is such that *S. granarius* might be described as a temperate species and *S. zeamais* as a tropical species. This is explained in part by cold-hardiness. The adult *S. granarius* can live some 2.5 months at $0°C$ and is likely to overwinter in warehouses in temperate areas, whereas the other species rarely survive a winter. The success of *S. granarius* in the typical temperate crops, wheat and barley, may also have some significance, since rice and maize favored by the other species are subtropical or at least warm temperate crops.

The optimum constant temperature for all is about $30°C$. At this temperature the life cycle from oviposition to emergence of the offspring from the cereal grain needs about 30 days for *S. granarius* and *S. zeamais* and about 26 days for *S. oryzae*. At $15°C$ the developmental period extends to about 180 days. *Sitophilus granarius* is the most tolerant of low humidity. Although it suffers heavy mortality at 40% RH or 10.5% moisture content, this tolerance presumably explains its presence in barley in the Mediterranean coasts of North Africa. The other species are adversely affected by 50% RH, especially in their oviposition rates. *Sitophilus zeamais* lays rather more eggs per day at 70% RH than does *S. granarius* at the same temperature and the preoviposition period before egg laying starts is of the order of a day less at $25°C$. Nevertheless, all three species can lay over 300 eggs in a favorable environment.

The details of the life cycle in warehouses are similar for all three. The female bores a hole with her rostrum and lays an egg in it, then seals the hole with a gelatinous plug. The shape, hairiness, and softness of the grain as well as its position and stability, determine where eggs are laid but only one egg is laid in a grain if there is no overcrowding. In any event, it is unusual for more than one larva to develop at the same time in a wheat grain, though this is possible, but two larvae can readily develop in a maize kernel. Larval feeding and pupation are entirely inside the kernel, and the young adult after a period of quiescence actively bores out of the kernel. There is a further maturation period (4–5 days at 25°C) before eggs are laid. A number of wild species (e.g., *Sitophilus linearis*) feed in acorns, and *S. granarius* is also able to breed in them but the range of nuts and fruits susceptible to the storage species is not known. They can breed in fleshy stems and tubers, in compacted flour, and in various manufactured cereal products such as macaroni.

One other weevil that has been recorded attacking seeds in warehouses is *Caulophilus oryzae*. This species is a strong flier and can attack maize in the field, but in warehouses it is restricted to damp, unripe, or damaged seeds, or to produce such as ginger. It is said to be a pest of avocado seeds.

3. ANTHRIBIDAE

This family includes one species that infests seeds in commercial storage, *Araecerus fasciculatus,* the coffee bean weevil. The larva of this insect also feeds within seeds, the eggs being laid singly in holes bored in the seeds by the adult female, in maize in the softer parts of the endosperm but not the scutellum. The number of larval molts is not fixed; it is usually four but may be three or five.

This is an Asian species that has spread around the wet tropical belt. It is able to attack the seeds of plants in the field, probably those hanging on trees after ripening, and it is a strong flier. The usual wild hosts are *Crotolaria striata* and *Tephrosia candida,* the most important cultivated hosts are coffee, cacao, maize, and nutmeg (see Fig. 6). On cacao and perhaps coffee, it develops only on rather damp seeds. If the grading regulations for molds are maintained for cocoa, then it is too dry for *Araecerus fasciculatus*. This insect has, indeed, steadily declined in importance in West Africa and this is a sign that the grading requirements there are being met increasingly successfully. Whereas the RH limits in these foodstuffs are about 80%, for maize and nutmeg they are about 60% (El Sayed, 1935). At 27°C the life cycle can be completed in about 30 days at 100% RH on maize, but it increases steadily by some 7 days for each 10% fall of RH. On nutmeg development is some 10 days slower, and on coffee and cocoa it is still slower.

4. OTHER BEETLES

All the other beetles that attack cereal seeds in storage are able to live in other habitats. The species most typically associated with cereals is

Rhizopertha dominica, the lesser grain borer. The generic name is also spelled *Rhyzopertha.* This species belongs to the family Bostrichidae which consists principally of wood borers. It is frequently found developing in bamboo and other fleshy stems of growing plants and in the stored tubers of yams. It can also breed in flour. This is a tropical species with an optimum for speed of development near 32°C at which temperature, development in wheat is completed in less than 30 days. At the usual storage temperatures of 28°C or less, it develops more slowly than *Sitophilus* spp., needing some 40 days, rising to nearly 60 at 25°C, and nearly 90 at 22°C; consequently, it multiplies appreciably only when grain stocks are stored for a considerable time. It was a damaging pest during both World Wars when exporting countries such as Australia and Argentina were unable to deliver their wheat and maize regularly. *Rhizopertha dominica* is also a pest in rice areas since it is the most successful species in attacking paddy rice, although it is not able to penetrate the undamaged grain protected by a close-fitting husk. This beetle lays its eggs loose in cracks and crannies in batches of twenty or more, at intervals of a day or 2. The egg stage lasts about a week at tropical storage temperatures. The young larva is active and penetrates kernels through small blemishes or soft spots. The second- and later-instar larvae are fleshy and inactive, though unlike the bruchids and weevils they have thoracic legs. Also, unlike these families where there are always four instars, the number of instars varies from three to five, but usually there are four.

This beetle is also common where produce is heating, a condition that should never happen with stored seeds. Under these circumstances, this species is usually found in hotter places than are the weevils. In the advanced stages of heating, weevils congregate in the outer 10 cm of a heating bulk and *R. dominica* in the next 10-cm layer.

One or two other species of this family are occasionally found in warehouses but we know little of their ecology.

Some of the major pests of seeds stored for food are not obligatory internal feeders. They should be unimportant in seeds intended for sowing because it is the dust, dockage and debris, and the broken kernels that initially, at least, enable them to become established. However, they may do significant damage even in small numbers because they attack the embryo and some must be mentioned. Little laboratory information is available to predict the importance of these species in stored seeds. Laboratory observations on their development and oviposition are usually made with finely ground meals provided as food. These beetles grow and oviposit much more quickly on these than they do on seed, although they do well when true seed insects have bored holes and left quantities of frass and food debris behind. Nevertheless, a few species present a major hazard to stored seed.

The most feared of these is the Khapra beetle, *Trogoderma granarium* (Fig. 11c) that was eradicated at great expense from the Southwest of the United States and is excluded by quarantine from most of the world. The reputation of this insect is based upon the spectacular nature of the damage that it wreaks, for stacks of bagged produce may collapse and the cast larval skins may accumulate in layers several inches deep. The early stages of infestation are cryptic and pass unnoticed but eventually the larvae become very obvious so that infestation appears to develop overnight. This beetle is placed in the family Dermestidae. Most members of the family are associated with animal products such as dried meat, skins, and wool, and many are household insects such as the carpet beetles. Many of the other species of *Trogoderma* are found in insect nests, though one is common in dried milk products.

Trogoderma granarium is probably a native of the Indian subcontinent. Up to 1939 it was distributed from Burma to Morocco, especially in semidesert regions and also in Europe in buildings, especially maltings, where a hot dry environment was maintained. Since then it has spread to the semidesert areas of West Africa and southern and eastern Africa. In the latter region and in Nigeria it has been fairly successfully controlled, although not eradicated, and quarantine has kept it out of South America, Australia, and North America since the original eradication. It has been introduced into Japan but is unlikely to become established.

The success of this species in hot dry areas arises from its tolerance of these conditions. It has a high optimal developmental temperature and a high minimum threshold. It does not complete its life cycle satisfactorily at below temperatures of 25°C because of a tendency to enter a larval diapause. The optimum RH is about 75% but it does not compete successfully against other species at this humidity, whereas at lower humidities, especially at high temperatures, it has a great advantage over them. This species could be a problem for storage of cereal, groundnut, bean, or pea seeds on the fringes of deserts. The young larva needs some dust or flour for food but the larger ones can break the epidermis of many seeds, especially near the embryo.

The larva is the conspicuous stage of this species. The adult is short-lived, laying about fifty eggs in 3 to 4 days. The larva usually pupates in the warmest spots, and the adult is thus found in these places.

Another important pest of cereal seeds is *Oryzaephilus surinamensis* (family Silvanidae), whereas the nearly related merchant beetle, *Oryzaephilus mercator* predominantly attacks some oily seeds especially groundnuts. Both attack dried fruit and the latter species often occurs in grocery shops. Both are worldwide and infest a wide range of produce. Groundnuts are soft and easily attacked, especially as the pericarp breaks readily, but are less vulnerable if stored in the husk. *Oryzaephilus surina-*

mensis owes its importance in Europe to the spread of combine harvesting and grain drying, for it is a tropical species with a development threshold of about 18°C and an optimum close to 30°C. We do not really know what enables it to become established in warm grain; presumably there are enough broken and blemished kernels, sufficient dust, and sometimes fungal mycelia to enable populations to multiply. The saw-toothed grain beetle is very small, about 3 mm long, and can complete a generation in 3 weeks at 30°C, so it becomes established readily at this temperature if food is available. It is unlikely that sound seeds are damaged directly by this species, but heating follows rapidly on the increase of population so that damage by heat, fungi, and the insect are all eventually probable if infestation is allowed to persist. *Oryzaephilus surinamenis* is more cold hardy than *O. mercator,* but its survival in cool areas depends very much on heated premises.

Several species of the genus *Cryptolestes,* once known as *Laemophloeus,* from the family Cucujidae occupy the same niche as *Oryzaephilus surinamensis.* These are even smaller and also complete a generation in about 3 weeks. There are six species found in warehouses, each with a different distribution and ecology. The most widespread and important is *Cryptolestes ferrugineus* (see Fig. 2) which is found in stored wheat in Canada, and in groundnuts in Nigeria. This is a cold-hardy species that is also able to survive in hot semidesert areas so is very widely distributed but it cannot attack seeds unless these are soft or damaged. The remaining species of *Cryptolestes* are much more restricted in range, most requiring an RH of 50% or higher to complete development. The cold-hardy species are typically found in flour mills, although *Cryptolestes turcicus* does occur in wheat in North America, and the other species found in cereals are not cold-hardy.

There is one fairly large beetle, about 10 mm long, that damages seeds in North America. This is the cadelle, *Tenebroides mauretanicus* (Fig. 11d) of the family Ostomidae. This species is highly predatory both as adult and larva, and it can damage both hard seeds and bore into the wood of bins. However, this species also feeds preferentially on broken seeds and embryos of sound seeds, and it needs high humidities to multiply. Little is known of the biological details of this species, but Bond and Monro (1954) note that at 25°C a generation is passed on flour in 8 weeks.

The cigarette beetle, *Lasioderma serricorne,* a member of the woodboring beetle family Anobiidae is capable of damaging seeds especially if true seed pests have made some preliminary attacks. This species can successfully compete with *Callosobruchus maculatus* in cowpeas in the damp tropics. Several species of the nearly related family, Ptinidae, can live in seeds but are unlikely to cause much harm. The most abundant species is *Ptinus tectus,* known in Canada as *Ptinus ocellus.* This is a cool temperate species which succeeds only if the adult has access to free fluid, rain, dew, or urine. The larva spins kernels together and feeds superficially without selecting the germ. This beetle develops more slowly than those mentioned previously, needing some 9 weeks under optimal conditions on finely ground meals. On whole seeds the growth rate is considerably slower.

Tribolium castaneum is the most abundant beetle of stored produce, probably because it flies well and spreads rapidly in warehouses and other premises, and, in the tropics, it flies from building to building and into the field. It is not a true seed pest but it invades fields while the seeds are ripening and is able to attack them while they are soft. Maize in particular, is often infested by *T. castaneum* at harvest but as the kernels dry and harden they become less liable to damage. Groundnuts are soft enough to be susceptible but growth is slower than on flour, some 7 weeks being needed to complete development at 30°C.

E. Some Important Species of Moths

Only one of the moths attacking stored seeds, *Sitotroga cerealella*, invariably feeds inside the seeds. As with beetles there are a number of species that attack the growing seeds while they are ripening and are brought into store, but they do not then continue to breed. These include the tortricid, *Enarmonia splendana,* which develops within the nuts of chestnuts. There are also species that can continue to breed in store, such as *Pyroderces rileyi* which lays its eggs in maize cobs. The genus *Ephestia* includes species that gradually diminish in importance as the harvested crop dries out, e.g., *E. figulilella* and *E. calidella,* and important storage pests that multiply considerably in stored seeds. Most moths can fly short distances into the field during favorable weather.

The larvae (caterpillars) of storage moths grow larger than most beetle larvae and are more readily able to attack sound seed. Almost any caterpillar thus presents a hazard, and in addition most of them spin silk which webs seeds together so that they are difficult to handle. Most moths will lay more eggs if they have access to water or nectar.

The great majority of stored products moths belong to the families Phycitidae and Tineidae, but there are a number of other families represented by only one or two species. The use of common names for moths is confusing, and correct identification of the battered specimens so often found in warehouses is difficult since it is necessary to study the chitinous parts of the genitalia. Common names will not be used here.

1. Sitotroga cerealella (Fig. 12a)

This species has been found infesting wild grasses of the genera *Setaria* and *Sorghum* in South Africa as well as the cultivated millets of these

FIG. 12. (a–c) Adults of some important moths found in stores; (a) *Sitotroga cerealella;* (b) *Ephestia cautella;* and (c) *E. elutella.* (d) Larva of *Corcyra cephalonica.*

genera (Joubert, 1966). It has spread throughout the warmer parts of the world attacking principally barley, wheat, maize, and sorghum. Eggs are laid both singly and in clumps and either on grains or fabric, and the newly hatched larvae tunnel into kernels. The rest of the larval stage and the pupal stage are spent within the kernel. The adult emerging to the free airspace has some difficulty if the infested kernels are buried deep in a bulk; the young larvae hatching from eggs also bore into kernels near the surface.

The life cycle can be completed in about 4 weeks at 28°C so that six or more generations a year are possible in the tropics. Development is possible down to 17°C. This species readily attacks the ripening seed on the growing crop and, therefore, often rapidly builds up a considerable population early in storage. The adult lives some 2 weeks and lays nearly 400 eggs.

2. PHYCITID MOTHS

The majority of the important warehouse moths belong to the family Phycitidae. The genus *Ephestia* has been split by some taxonomists into a number of genera of which *Cadra* and *Anagasta* have been widely used. All the phycitid moths have a wide range of food plants, but all eat the softer parts of the foodstuff and choose the softer food materials. Thus nuts and dried fruit and the embryos of wheat and some other seeds are especially vulnerable. Most of the important moths in this family can withstand dry climates.

a. *Plodia interpunctella* (Fig. 13a,b).   This is a cold-hardy species but one that does not grow very rapidly in temperate areas, seldom being able

FIG. 13. (a and b) *Plodia interpunctella;* (a) adults and a larva; (b) cocoon containing pupa in a groundnut. (c and d) Larva of (c) *Ephestia elutella* and (d) *Hofmannophila pseudospretella.*

to complete one generation a year in the United Kingdom. It thrives in the tropics even in semidesert areas but does best on freshly harvested seeds.

Most of the detailed laboratory work with this insect has been concerned with a very complex diapause which may be induced by short daylight periods and by low temperature. It is most often found with maize, groundnuts, and dried fruit. There is a pronounced larval migration in this species prior to spinning of cocoons in the upper parts of warehouses.

b. *Ephestia kuehniella.*   In recent years the generic name *Anagasta* has been widely used. For a short period, the trivial name *sericarium* was mistakenly adopted, but fortunately was not much used.

This species has been typically associated with flour mills since the introduction of roller mills provided suitable climate and food supply together, but it is able to damage wheat and frequently does so in the grain stores of mills. It is presumed to have originated in the Mediterranean region and does not extend its range into the Far East or the hotter parts of the tropics in spite of the exceptional powers of the larva to grow in very dry foods at very low humidity. It is cold-hardy and can develop at 17°C and possibly below. The most rapid growth is about a month for a generation. It is doubtful that this species would damage any seeds except cereals.

c. *Ephestia elutella* (Figs. 12c and 13c).   This species is also of Mediterranean origin and has a well-marked diapause which enables it to pass the winter in the cooler temperate parts of its range. It does not occur in northern latitudes such as the Scottish highland area nor in the tropics. It was first important as a pest of dried fruit, and is one of the few serious pests of stored tobacco, but also became important in cereals stored for long periods during wartime. The principal damage to cereals was the eating of the embryos from large numbers of kernels. This species also migrates to the ceilings of warehouses prior to spinning of cocoons. In the Mediterranean type of climate, it can pass three or four generations in the summer before spending the cooler season in diapause. This species should not be a serious pest of seeds.

d. *Ephestia cautella* (Fig. 12b).   This is one of the species often placed in the genus *Cadra*. It is the most common and widespread storage moth, exceeded in prevalence only by the beetle *Tribolium castaneum*. It extends its range into the tropics including the dry areas but is not very cold-hardy and persists in the cooler temperate areas only because of repeated importation. It is common in dried fruit and in most tropical produce especially maize, groundnuts, cocoa, and palm kernels. It is found early after harvest and, although it does not fly strongly, may sometimes attack seeds before harvest. It usually manages two or three generations, becoming increasingly abundant before exhausting the food available to it and then decreases in numbers. The larval migration of this species is less marked, possibly because it is most often found in produce stored in bags in which it pupates when it cannot reach the ceiling by moving only upward. This species is important because it is abundant and able to attack a wide variety of products.

e. OTHER PHYCITIDS.   There are two other Mediterranean species of *Ephestia* that are associated with the freshly harvested or ripening crop. *Ephestia figulilella* is probably associated with fallen fruit and is not often found in warehouses. *Ephestia* or *Cadra calidella* is more frequent in warehouses although it is most commonly associated with carobs especially in Cyprus, but in warehouses it is gradually superseded by *Ephestia cautella*. Two other species that are principally field pests occasionally appear in warehouses—*Spectrobates* (previously known as *Myelois* and *Ectomyelois*) *ceratoniae* is sometimes found on dried fruit and more commonly on carobs and *Mussidia nigrivenella* on maize.

3. GALLERIIDS

Two species of this family are capable of damaging stored seeds. *Corcyra cephalonica* (Fig. 12d) is especially associated with rice and groundnuts in the tropics. It develops much more slowly and reaches a larger size than *Ephestia* spp. and probably has a higher optimum temperature; when this species is present with *Ephestia cautella,* it becomes prominent much later in storage. *Corcyra cephalonica* has spread throughout the tropics but is most abundant in the Indian subcontinent.

Aphomia gularis also is best known from the Indian subcontinent and is principally a pest of edible nuts. This is also a tropical species and develops rather slowly but it is established in temperate areas in mills that use nuts for manufacture of blended foods especially animal feeds. This species has a diapause. Galleriids spin silk which contaminates seeds and binds them together, and the larval cocoon is much tougher than that spun by phycitid moth caterpillars.

4. PYRALIDS

The pyralid moths are not important except in so far as they warn of poor storage conditions. They are scavengers that appear when the store and produce are damp and the premises not properly cleaned. *Pyralis farinalis* is a temperate species that attacks cereals, completes about two generations a year, and spins a very flimsy cocoon. The pyralids are generally brighter colored (with some yellow or gold scales) than are other storage moths.

5. OECOPHORIDAE

There are two very common and widespread storage moths of temperate regions in this family, but, since both require high humidity and damp seeds, neither should have much economic importance. *Endrosis sarcitrella,* occasionally referred to as *Endrosis lactella,* is mostly found in beans and peas and on oat products, although not so commonly on oat seeds. It can complete development at both 10° and 25°C (Woodroffe, 1951b) and does not have a diapause. It can have several generations annually.

Hofmannophila (previously *Borkhausenia*) *pseudospretella* (Fig. 13d) is a larger moth, and the older caterpillars can be very conspicuous, especially as they enter a diapause. The diapause adjusts the life history to an annual cycle and enables the larvae to withstand low humidity. This species also attacks peas and beans but is able to damage wheat by eating out the germ. Both species are common household insects in the damper temperate countries. (See Woodroffe, 1951a.)

6. TINEIDS

There are a large number of species of tineid moths in storage and domestic habitats. Most have been mentioned in the literature under the generic name *Tinea* or *Tinaea* but this has been subdivided into many genera as taxonomic knowledge has accumulated. Most species are clothes moths or are associated with animals, especially birds, bats, and rodents as scavengers on animal and plant remains, but one, *Nemapogon granellus,* is a granary pest. It prefers offals but can attack cereals, especially rye and wheat. This species often pupates in crevices or food with half the body projecting from the surface, both in bulk or bagged produce and, when abundant, gives the impression of a petrified forest. It is probable that this species needs high humidity and it has a diapause. It is a temperate insect which has been abundant in continental Europe and also recorded in North America and Japan. It is said to be replaced by *Nemapogon infimellus* in more northerly regions of Germany.

VI. Conclusion

Exceptional care should be taken of seeds during storage because they are such a valuable commodity. Except in a few very warm and wet regions, it should be possible to store seeds free of insects and to avoid deterioration resulting from these insects. It is obviously better to prevent damage rather than spend money on control methods that may themselves harm the seed, and it is better to use capital to build adequate seed storage rather than spend money annually on repeated palliatives. The design of storage structures and of machinery for handling produce, especially seeds, ought to take storage pests more into account, because hiding places for the organisms and dead spaces where a food supply can accumulate are the major reasons that pests persist and spread.

ACKNOWLEDGMENT

The illustrations were prepared by Mr. J. H. Hammond from his collection of photographs. They are published by the courtesy of the Pest Infestation Control Laboratory, Ministry of Agriculture, Fisheries and Food, Slough, England and Crown Copyright is reserved by the Controller, Her Britannic Majesty's Stationery Office.

REFERENCES

Agrawal, N. S., Christensen, C. M., and Hodson, A. C. (1957). Grain storage fungi associated with the granary weevil. *J. Econ. Entomol.* **50,** 659.

Bailey, S. W. (1965). Air-tight storage of grain; its effect on insect pests. IV. *Rhyzopertha dominica* (F.) and some other Coleoptera that infest stored grain. *J. Stored Prod. Res.* **1,** 25.

Bailey, S. W. (1966). Review of "The Entomology of Radiation Disinfestation of Grain" (P. B. Cornwell, ed.). *J. Stored Prod. Res.* **2,** 171.

Birch, L. C. (1944). Two strains of *Calandra oryzae* L. (i.e. *Sitophilus oryzae* (L.) (Coleoptera). *Aust. J. Exp. Biol. Med. Sci.* **22,** 271.

Birch, L. C. (1954). Experiments on the relative abundance of two sibling species of grain weevils. *Aust. J. Zool.* **2,** 66.

Bond, E. J., and Monro, H. A. U. (1954). Rearing the cadelle, *Tenebroides mauretanicus* (L.) (Coleoptera Ostomidae) as a test insect for insecticidal research. *Can. Entomol.* **86,** 401.

Booker, R. H. (1967). Observations on three bruchids associated with cowpeas in Northern Nigeria. *J. Stored Prod. Res.* **3,** 1.

Boudreaux, H. B. (1969). The identity of *Sitophilus oryzae. Ann. Entomol. Soc. Amer.* **62,** 169.

Breese, M. H. (1960). The infestibility of stored paddy by *Sitophilus sasakii* (Tak.) and *Rhyzopertha dominica* (F.). *Bull. Entomol. Res.* **51,** 599.

Breese, M. H. (1964). The infestibility of paddy and rice. *Trop. Stored Prod. Inform.* No. 8, p. 289.

Brindley, T. A., Chamberlin, J. C., and Schopp, R. (1958). The pea weevil and methods for its control. *U.S., Dep. Agr., Farmers' Bull.* **1971,** 1–24.

Burges, H. D., and Burrell, N. J. (1964). Cooling bulk grain in the British climate to control storage insects and to improve keeping quality. *J. Sci. Food Agr.* **15,** 32.

Caswell, G. H. (1960). Observations on an abnormal form of *Callosobruchus maculatus* (F.). *Bull. Entomol. Res.* **50,** 671.

Davey, P. M. (1958). The groundnut bruchid, *Caryedon gonagra* (F.). *Bull. Entomol. Res.* **49,** 385.

Doggett, H. (1958). The breeding of sorghum in East Africa. II. The breeding of weevil-resistant varieties. *Emp. J. Exp. Agr.* **26,** 37.

El Sayed, M. T. (1935). On the biology of *Araecerus fasciculatus,* De Geer (Col., Anthribidae) with special reference to the effects of variations in the nature and water content of the food. *Ann. Appl. Biol.* **22**, 557.

Floyd, E. H., and Newsom, L. D. (1959). Biological study of the rice weevil complex. *Ann. Entomol. Soc. Amer.* **52**, 687.

Giles, P. H. (1969). Observations in Kenya on the flight activity of stored products insects, particularly *Sitophilus zeamais* Motsch. *J. Stored Prod. Res.* **4**, 317.

Giles, P. H., and Ashman, F. (1971). A study of preharvest infestation of maize by *Sitophilus zeamais* Motsch. in the Kenya highlands. *J. Stored Prod. Res.* **7**, 69.

Green, A. A., Kane, J., Heuser, S. G., and Scudamore, K. A. (1968). Control of *Ephestia elutella* (Hb) (Lepidoptera, Phycitidae) using dichlorvos in oil. *J. Stored Prod. Res.* **4**, 69–76.

Highland, H. A. (1967). Resistance to insect penetration of carbaryl-coated kraft bags. *J. Econ. Entomol.* **60**, 451.

Howe, R. W. (1952). Miscellaneous experiments with grain weevils. *Entomol. Mon. Mag.* **88**, 252.

Howe, R. W. (1953). The rapid determination of the intrinsic rate of increase of an insect population. *Ann. Appl. Biol.* **40**, 134.

Howe, R. W. (1956). The biology of two common storage species of *Oryzaephilus* (Coleoptera, Cucujidae). *Ann. Appl. Biol.* **44**, 341.

Howe, R. W. (1965a). Losses caused by insects and mites in stored foods and feedingstuffs. *Nutr. Abstr. Rev.* **35**, 285.

Howe, R. W. (1965b). *Sitophilus granarius* (L.) (Coleoptera, Curculionidae) breeding in acorns. *J. Stored Prod. Res.* **1**, 99.

Howe, R. W. (1965c). A summary of estimates of optimal and minimal conditions for population increase of some stored products insects. *J. Stored Prod. Fes.* **1**, 177.

Hyde, M. B. (1968). Successful storage of high moisture grain. *Esso Farmer* **20**, 11.

Jay, E. G., and Pearman, G. C. (1969). Protecting wheat stored in metal cans with carbon dioxide. *J. Ga. Entomol. Soc.* **4**, 181.

Jones, J. D. (1943). Intergranular spaces in some stored foods. *Food* **12**, 325.

Joubert, P. C. (1966). Field infestations of stored-product insects in South Africa. *J. Stored Prod. Res.* **2**, 159.

Kuschel, G. (1961). On problems of synonymy in the *Sitophilus oryzae* complex. (30th contribution, Col. Curculionidae.) *Ann. Mag. Natur. Hist.* [13] **4**, 241.

McFarlane, J. A., and Weaving, A. J. S. (1967). A means of differentiating between *Acanthoscelides obtectus* (Say) and *Zabrotes subfasciatus* (Boh.) (Coleoptera, Bruchidae) in white haricot beans at the pupal stage. *J. Stored Prod. Res.* **3**, 261.

Messenger, P. S., and Flitters, N. E. (1959). Effect of variable temperature environments on egg development of three species of fruit flies. *Ann. Entomol. Soc. Amer.* **52**, 191.

Navarro, S., Donahaye, E., and Calderon, M. (1969). Observations on prolonged grain storage with forced aeration in Israel. *J. Stored Prod. Res.* **5**, 73.

Oxley, T. A. (1948). A study of the water content of single kernels of wheat. *Cereal Chem.* **25**, 111.

Page, A. B. P., and Lubatti. O. F. (1963). Fumigation of insects. *Annu. Rev. Entomol.* **8**, 239.

Parkin, E. A. (1963). The protection of stored seeds from insects and rodents. *Proc. Int. Seed Test. Ass.* **28**, 893.

Peake, F. G. G. (1953). On a bruchid seed-borer in *Acacia arabica*. *Bull. Entomol. Res.* **43**, 317.

Phelps, R. J., and Oosthuizen, M. J. (1958). Insects injurious to cowpeas in the Natal region. *J. Entomol. Soc. S. Afr.* **21**, 286.

Pingale, S. V. (1953). Effect of damage by some insects on the viability and weight of stored grain. *Bull. Centr. Food Technol. Res. Inst., Mysore* **2**, 153.

Pixton, S. W. (1967). Moisture content—its significance and measurement in stored products. *J. Stored Prod. Res.* **3**, 35.

Pixton, S. W., and Hill, S. T. (1967). Long term storage of wheat. II. *J. Sci. Food Agr.* **18**, 94.

Powell, J. D., and Floyd, E. H. (1960). The effect of grain moisture upon development of the rice weevil in green corn. *J. Econ. Entomol.* **53**, 456.

Prevett, P. F. (1961). Field infestation of cowpea (*Vigna unguiculata*) pods by beetles of the families Bruchidae and Curculionidae in Northern Nigeria. *Bull. Entomol. Res.* **52**, 635.

Prevett, P. F. (1966a). Observations on the biology of six species of Bruchidae (Coleoptera) in Northern Nigeria. *Entomol. Mon. Mag.* **102**, 174.

Prevett, P. F. (1966b). Observations on biology in the genus *Caryedon* Schönherr (Coleoptera: Bruchidae) in Northern Nigeria, with a list of associated parasitic Hymenoptera. *Proc. Roy. Entomol. Soc. London, Ser. A* **41**, 9.

Prevett, P. F. (1966c). The identity of the palm kernel borer in Nigeria, with systematic notes on the genus *Pachymerus* Thunberg (Coleoptera, Bruchidae). *Bull. Entomol. Res.* **57**, 181.

Rees, A. R. (1963). Some factors affecting the germination of oil palm seeds under natural conditions. *J. West Afr. Inst. Oil Palm Res.* **4**, No. 14, 201.

Richards, O. W. (1944). The two strains of the rice weevil, *Calandra oryzae* (L.) (Coleopt., Cuculionidae). *Trans. Roy. Entomol. Soc. London* **94**, 187.

Sano, I. (1967). Density effect and environmental temperature as the factors producing the active form of *Callosobruchus maculatus* F. (Coleoptera, Bruchidae). *J. Stored Prod. Res.* **2**, 187.

*Strong, R. G., and Lindgren, D. L. (1959). Effect of methyl bromide and hydrocyanic acid fumigation on the germination of wheat. *J. Econ. Entomol.* **52**, 51.

Strong, R. G., and Lindgren, D. L. (1961). Effect of methyl bromide and hydrocyanic acid fumigation on the germination of corn seed. *J. Econ. Entomol.* **54**, 764.

Utida, S. (1953). "Phase" dimorphism observed in the laboratory population of the cowpea weevil *Callosobruchus quadrimaculatus*. *Jap. J. Appl. Zool.* **18**, 161.

Utida, S. (1965). "Phase" dimorphism observed in the laboratory population of the cowpea weevil, *Callosobruchus maculatus* IV. The mechanism of induction of the flight form. *Jap. J. Ecol.* **15**, 193.

Utida, S. (1968). The influence of the parental condition of the production of the winged form in the population of *Callosobruchus maculatus*. *Jap. J. Ecol.* **18**, 246.

Utida, S. (1969). Photoperiod as a factor inducing the flight form of the southern cowpea weevil, *Callosobruchus maculatus*. *Jap. J. Appl. Entomol. Zool.* **13**, 129.

VanDerSchaaf, P., Wilbur, D. A., and Painter, R. H. (1969). Resistance of corn to laboratory infestation of the larger rice weevil, *Sitophilus zeamais*. *J. Econ. Entomol.* **62**, 352.

Venkat Rao, S., Nuggehalli, R. N., Swaminathan, M., Pingale, S. V., and Subrahmanyan, V. (1958). Effect of insect infestation on stored grain. III. Studies on Kaffir corn (*Sorghum vulgare*). *J. Sci. Food Agr.* **9**, 837.

Waloff, N., and Richards, O. W. (1946). Observations on the behaviour of *Ephestia elutella* Hübner (Lep., Phycitidae) breeding on bulk grain. *Trans. Roy. Entomol. Soc. London* **97**, 299.

---

*There are several other papers dealing with different kinds of seed by these authors in this journal between 1959 and 1963 inclusive.

Watters, F. L. (1965). Physical methods of insect control. *Proc. Entomol. Soc. Manitoba* **21**, 18.

Woodroffe, G. E. (1951a). A life-history study of the brown house moth, *Hofmannophila pseudospretella* (Staint.) (Lep., Oecophoridae). *Bull. Entomol. Res.* **41**, 529.

Woodroffe, G. E. (1951b). A life-history study of *Endrosis lactella* (Schiff.) (Lep. Oecophoridae). *Bull. Entomol. Res.* **41**, 749.

Wright, R. H. (1970). Some alternatives to insecticides. *Pestic. Sci.* **1**, 24.

Zachariae, G. (1959). Das Verhalten des Speisebohnenkäfers *Acanthoscelides obtectus* Say (Coleoptera: Bruchidae) im Freien in Norddeutschland. *Z. Angew. Entomol.* **43**, 345.

Zachariae, G. (1960). Kann sich der Speisebohnenkäfer *Acanthoscelides obtectus* Say als Freilandschädling in Norddeutschland einbürgern? *Z. Angew. Entomol.* **45**, 225.

5

ESSENTIALS OF SEED TESTING

Oren L. Justice

I. Introduction

Agricultural and horticultural productivity depends, to a great extent, on the quality of seeds which are planted. The farm value of field and vegetable crops grown from seeds in 1969 in the United States was in excess of 21 billion dollars. The seed industry alone represents a billion dollar operation. Estimates of crop losses due to planting seeds of inferior quality range from 5% upward, and another 1–2% loss is incurred by not having needed information on seed quality at the time of planting. In considering quality, a seed can be compared with a precious gem. The lay person cannot determine from appearance or examination the value of either. Only the specialist who is well trained can be expected to assess the quality of either the seed or the gem and both must rely upon the results of tests before final judgment is passed. It follows that farmers, seedsmen, control officials, or others who wish to know the quality and, hence, the value of a seed lot must submit a representative sample for testing in accordance with established procedures.

The purpose of seed testing is to provide the interested person with factual information relating to the submitted sample. Ordinarily the seed technologist is not in a position to relate the sample he tests to the seed lot from which it was taken; however, other persons may be able to do so (Kjaer, 1961). Considerable discussion has ensued and differences of opinion have been expressed as to whether the results of tests represent the field-planting value of seed lots or whether, indeed, this is the purpose of seed testing (Brown and Toole, 1934; Clark, 1942; Heydecker, 1962). The current practice is to test seeds under controlled laboratory conditions or grow plants in the field for variety verification according to specified procedures. With available research data these procedures have been adjusted as much as possible to yield results that compare most favorably with field-planting value. Obviously the grower desires to know the planting value of the seeds he purchases. Planting value, as used here, refers to the various quality factors that will affect the crop, such as field stand, variety, total yield, and introduction of diseases and undesirable weeds. The merchant is very interested in test results which can be duplicated by other testing stations throughout the world. This is necessary if seed testing is to serve the merchant who may ship seeds in either domestic or foreign commerce. It follows that uniform methods for determining seed quality generally must be acceptable and prescribed, and when properly applied by trained personnel must yield results with a high degree of accuracy regardless of where or when the tests are made (Linehan, 1960; Clark, 1961).

Although the first hundred years of seed testing was recently celebrated (Kick *et al.,* 1970), only one comprehensive review of the literature has been made (Porter, 1949). Justice *et al.* (1952) published a comprehensive manual on testing agricultural and vegetable seeds, including identification of crop and weed seeds. Barton's "Bibliography of Seeds" (1967) has been a great asset in recent years for locating background information when planning research or preparing manuscripts. Much of the literature in the English language, directly applicable to seed technology, has been published in *(a)* the *Proceedings of the Association of Official Seed Analysts* (United States and Canada) and *(b)* the *Proceedings of the International Seed Testing Association.*

Research has been an integral part of the seed testing program in the U.S. Department of Agriculture and many state universities. Among the universities which have been most actively engaged in seed quality research along with their seed testing programs over the past 25 years are California, Iowa, Mississippi, New York, North Carolina, and Oregon. The various publications series of these universities and of the U.S. Department of Agriculture carry articles relating to research on seed quality.

II. Origin and Development of Seed Testing

Seed testing was developed as a necessary reaction to unscrupulous practices prevalent in the nineteenth century (Nobbe, 1876; Burchard, 1893). As early as 1816 the City and Republic of Bern, Switzerland, enacted a law prohibiting the sale of adulterated clover seeds (Brown, 1941). Some of the contaminants intentionally used in Europe during the nineteenth century for adulterating seeds were *(a)* devitalized seeds of similar, less expensive kinds, *(b)* quartz sand sieved and stained to resemble the seeds to be adulterated, *(c)* screenings containing high percentages of weed seeds, and *(d)* viable seeds of similar, less expensive species which could not be differentiated from the kind offered for sale

(Kjaer, 1961). The British Parliament adopted the Adulterated Seeds Act in 1869 to stop such practices (Brown, 1941).

Large amounts of adulterated seeds were offered for sale in the United States in the latter part of the nineteenth century and early part of the twentieth century (Galloway, 1909; Woods, 1910). Common examples of seed adulteration in the United States during 1890–1915 include: sweet clover and black medic in alfalfa and red clover; Canada bluegrass in Kentucky bluegrass; and perennial ryegrass in meadow fescue or vice versa, depending on the difference in prices. Seeds of dodder, a parasite, constituted a common impurity in forage seeds. An examination of 873 samples of red clover and alfalfa seeds made by the Federal Seed Laboratory (Washington, D.C.) in 1906 showed that 30.6% of the samples contained dodder. Analyses of 61 samples of low-quality red clover seeds imported into the United States in 1905 and 1906 revealed averages of thirty kinds of weed seeds per sample, 3088 weed seeds/oz, 74% pure seed, and dodder in 75% of the samples (Justice, 1961).

The sale in Europe and America of low-germinating seeds or dead seeds added to the uncertainties of crop production. The average germination of 12,454 packets of vegetable seeds collected from commission boxes and tested by the Federal Seed Laboratory (Washington, D.C.) from 1907 to 1910 was 60.5%. Mail-order seeds were somewhat better: 6117 samples purchased in 1911 gave an average germination of 77.5% (Brown and Goss, 1912). These and other unscrupulous practices stimulated the study of seeds and trade practices in many countries and led to the establishment of laboratories where seeds could be tested.

The world's first seed testing laboratory was established by F. Nobbe at Tharandt, Saxony, Germany, in 1869 (Burchard, 1893). Seven years later he published his famous book, "Handbuch der Samenkunde," which was a standard reference for more than 50 years (Nobbe, 1876). At the same time, but independently, E. Möller-Holst was planning a private laboratory in Copenhagen, Denmark. This laboratory was not functional until 1871 but it was later converted into a government facility and is the forerunner of the present Danish Seed Testing Station, recently employing a staff up to seventy persons (Kjaer, 1961). Seed testing spread rapidly in Europe during the next 20–30 years. According to Burchard (1893), forty-two seed testing stations had been established by 1893 in Germany alone. The first seed testing laboratory in the United States was opened at the Connecticut Agricultural Experiment Station in 1876 under the direction of E. H. Jenkins who had studied under F. Nobbe (Brown, 1941). Other laboratories were established in Massachusetts (1889), Vermont (1895), Maine (1897), and in the U.S. Department of Agriculture (1894), the latter under G. H. Hicks. Educational work, including the training of technicians was being carried out at nine or more colleges or universities by 1900 (Higgins *et al.,* 1961).

Outside the United States at least 130 laboratories were testing seeds by 1905 (Justice, 1961). This dramatic growth has continued to the present time. In 1970, seventy official government seed testing laboratories in the United States and Canada are members of the Association of Official Seed Analysts. Approximately fifty countries with about 150 participating laboratories are members of the International Seed Testing Association. Laboratories of private seed firms in the United States and Canada total at least sixty-five and there are about thirty-five commercial laboratories in the two countries testing seeds on a fee or contract basis. Workers in these private and commercial laboratories comprise the membership of the Society of Commercial Seed Technologists.

In recent years the International Seed Testing Association has sponsored courses in seed technology for training personnel in the developing countries and to promote uniform application of rules for seed testing. Such training conferences were held recently in Kenya, Brazil, New Zealand, and England. The International Seed Testing Association held its sixteenth conference in New Zealand in 1968, demonstrating its sincerity to reach and assist the developing nations.

III. Development of Standard Procedures for Determining Seed Quality

A. Development of Rules for Testing Seeds

Nobbe (1876) was the first to publish on methods of testing seeds. Actually, relatively few pages of his 631-page book are given over to methodology. Burchard (1893) expanded on the German methods considerably and refers to rules adopted by the German stations; however, the matter of adoption of rules or procedures by the German stations is not clear. Burchard describes and illustrates a temperature-controlled germination chamber which he designed and an earthenware dish for germination, designed by Nobbe.

The first formalized procedures for testing seeds were developed in the United States and published in 1897 (Jenkins *et al.,* 1897). The Association of American Agricultural Colleges and Experiment Stations appointed a committee in 1896 to devise and adopt a standard form of seed testing apparatus and methods of procedure for use in the experiment stations. The report of this committee is presented in three parts: (*a*) Rules for Seed Testing, (*b*) Germination Chamber, and (*c*) Blanks for Record, Sampling and Report. As would be expected, the three pages of rules for testing seeds are rather brief compared to present rules (Association of Official Seed Analysts, 111 pages in 1965 edition; International Seed Testing Association, 142 pages). The only instructions given for performing a purity analysis were that *(1)* the results were to be based on weight, and *(2)* the sizes of working samples for making the purity analysis were specified for a number of seed kinds. No methods were specified for making a noxious-weed seed examination; however, the section on purity analysis did contain a statement advising that, if the sample was suspected of containing any seeds of a pest such as dodder, Canada thistle, wild mustard, or plantain, then at least 50 gm should be examined for the impurity. Brief directions are given for making germination tests including instructions for taking the seeds for germination, germination substrata (blotting paper and sand), moisture content of substrata, duration of test, germination environment, necessity of duplicate tests and when to run supplementary tests in sand. A double-jacket seed germinator, including detailed construction plans, is described. Temperature was controlled by thermoregulators which either allowed water to enter for cooling or caused a micro-bunsen-burner to function for heating.

The Association of Official Seed Analysts, founded in 1908, adopted its first rules for testing seeds in 1917 (published in mimeograph form only). These rules were amended in 1924 to provide methods of testing 53 kinds of field seeds, 27 kinds of vegetable seeds, and 18 kinds of flower seeds (M. T. Munn, 1924). Since 1924 the Association has revised its rules at least 10 times (Anonymous, 1970).

Although the International Seed Testing Association dates back to a meeting held in Copenhagen, Denmark, in 1921, and was formally organized in 1924, it did not adopt a set of rules for testing seeds until 1931 (Anonymous, 1931). These rules were rather complete considering background of available information and the difficulty of obtaining agreement among a number of countries. Minor revisions have been made at most triennial congresses and major revisions were made in 1953 and 1965 (Anonymous, 1966). Since 1953, each section of the International Rules has been designated either as a prescription (mandatory application) or recommendation.

B. Criteria on Which Rules Are Based

In developing the rules for seed testing the following objectives have served as guidelines: *(a)* to provide methods by which the quality of seed samples can be determined accurately; *(b)* to prescribe methods by which seed analysts working in different laboratories in different countries throughout the world can obtain uniform results; *(c)* to relate the laboratory results, insofar as possible, to planting value; *(d)* to complete the tests within the shortest period of time possible, commensurate with the above-mentioned objectives; and *(e)* to perform the tests in the most economical manner. Much progress has been made with respect to accuracy and uniformity in testing but much remains to be done in relating laboratory test results to planting value. Information gained through routine testing and research has served as the basis for revising the rules for testing seeds. Until about 1915 most of the information was gained through records of routine testing. In the next 25–30 years information gained by this method and through research appears to have been used to approximately an equal extent. Since about 1940 to 1945, research has provided most of the data used in altering the rules for seed testing.

The early rules for testing seeds provided for three types of tests, namely: *(a)* the purity test whereby the working sample was divided into pure seeds, other crop seeds, weed seeds, and inert matter, the results being expressed in percent on a weight basis; *(b)* an examination for noxious-weed seeds, in which a much larger sample than that used for the purity test was tested and the result expressed as the number of seeds in an ounce or pound of seeds; and *(c)* a germination test including the determination of hard seeds, both results being expressed as percents by count. The present rules of the Association of Official Seed Analysts (Anonymous, 1970) include the three tests mentioned above. Also, these rules establish special procedures for distinguishing the crop species of *Lolium* and of *Melilotus* which cannot be differentiated by visual observation. Procedures are approved for detecting chemical treatments of seeds and for identification of seeds by seedlings and growing plants and by the chemical phenol test. Determination of viability by the embryo excision method and by the tetrazolium biochemical procedure is permitted for seeds of a few tree species.

The rules of the International Seed Testing Association contain methods for the three types of tests discussed above and for the following quality factors: viability of tree seeds by the tetrazolium procedure; seed health condition (i.e., seed-borne organisms); genuineness of species and cultivar; moisture content; provenance or locality of harvest; unit weight of seeds; and homogeneity of seed lots.

C. Some Major Problems

1. The Quicker and Stronger Methods of Purity Analysis

Purity analysis of grass seeds with subtending chaffy structures is tedious, time-consuming, and difficult. In 1931, European seed technologists believed firmly that all grass seed units with broken, shriveled, decayed, or otherwise damaged caryopses would not produce a seedling and should not be regarded as "pure seed." In other words, the analyst had to decide by observation, dissection, and use of optical aids whether, in his opinion, the caryopsis was germinable. This method of purity determination came to be known as the *Stronger Method*. However, the Americans and Canadians believed that each floret with a caryopsis, regardless of its condition, should be regarded as pure seed (Lafferty, 1932; M. T. Munn, 1936). The method used to make this determination was faster than the Stronger Method and was known as the *Quicker Method*. Because agreement on this issue could not be reached the International Seed Testing Association rules provided for both procedures until 1953. This was an unfortunate development as it tended to split the seed technologist into two groups and was conducive to yielding different results when a seed lot was sampled and tested in different countries. In 1953, a single procedure was adopted which was in keeping with the so-called Quicker Method.

2. SEEDLING EVALUATION

In the early years of seed testing any seed of which the radicle had penetrated the seed coat was regarded as germinated. Analysts soon realized that many sprouts which were cracked, broken, diseased, or weak could not possibly produce a plant, especially when planted under field conditions. Research conducted at the U.S. Department of Agriculture between 1910 and 1915 confirmed this observation quite clearly. When results of this research were applied in routine testing, strong opposition developed in the United States. In general, greatest opposition came from those groups concerned with sale of seed lots, whereas consumers and control officials defended the requirement. The Association of Official Seed Analysts carried on a vigorous referee testing program which eventually convinced most of the opposition. The present rules for seed testing contain specific guidelines for evaluating seedlings of most economic species. Surely the present procedure moves the germination test a step closer to the field-planting value of seed lots.

3. BIOCHEMICAL TEST FOR VIABILITY

Lakon (1942) introduced the topographical tetrazolium biochemical test for seed viability after rejecting selenium as an indicator due to its toxicity (Lakon, 1940). The tetrazolium test has been refined to the point that it is a useful tool for determining whether a seed is alive or dead (more properly, what parts of the seed are alive). Basically, the test distinguishes between live and dead tissue. A very important question is, to what extent can the tetrazolium test be used in routine seed testing and in seed research—can it be used as a substitute for the germination test or, perhaps, as a supplementary test? Although this question has been considered over the past 20–25 years, there still is divided opinion as to what extent tetrazolium should be used in routinely determining seed quality (Moore, 1969). A more detailed treatment of the tetrazolium test is found in Section VII,D,1.

IV. Botanical Aspects

A. Seed and Seed Units

Botanically a seed is a ripened ovule, consisting of the embryo, integuments, and usually an endosperm. In many botanical groups the seed may be accompanied or enclosed by accessory structures such as the pericarp of achenes, nutlets and caryopses, lemma and/or palea of grasses, or fused flower and fruit parts in the case of *Beta* and *Tetragonia*. These structures are also referred to as seeds by lay persons and seedsmen. This popular concept of seed will be followed in this chapter. The term *seed unit* will also be used sometimes in referring to structures regarded as a seed for testing purposes. The fruit of the cereals and grasses is a caryopsis or grain. This structure, singularly or surrounded by lemma and/or palea, is regarded as a seed or seed unit. In some genera of grasses, a floret or spikelet with caryopsis is also regarded as a seed unit. The grass genera, with important crop species in the United States, were classified by Justice *et al.* (1952) into the three following categories. *(a)* Seed unit consists of caryopsis or floret—*Agropyron, Cynadon, Cynasurus, Dactylis, Elymus, Eragrostis, Festuca, Hordeum, Lolium, Oryza, Pennisetum, Phalaris, Poa* (also bulblets in *Poa bulbosa*), *Secale, Sorghum, Sporobulus, Triticum,* and *Zea. (b)* Seed unit consists of caryopsis, floret or spikelet with at least one caryopsis—*Agrostis, Alopecurus, Andropogon, Arrhenatherum, Avena, Anthoxanthum, Axonopus, Bouteloua, Bromus, Chloris, Echinochloa, Holcus, Melinis, Oryzopsis, Panicum, Paspalum, Setaria, Sorgastrum,* and *Zoysia. (c)* Seed unit consists of caryopsis, bur or floret—*Buchloë.*

Typically, the seed unit in the following genera can be either a seed or a pod with a seed: *Alysicarpus, Desmodium, Lespedeza, Medicago, Melilotus, Onobrychis,* and *Trifolium.* Also, the burs of *Medicago arabica* and *Medicago hispida* are treated as seed units provided they contain at least one seed. The seed unit of peanut is either a seed or pod with at least one seed (Justice *et al.*, 1952).

In most other plant families the seed unit is a seed or a seedlike fruit (achene, nutlet, mericarp, or utricle) which is treated as seed provided the structure contains a true seed (Porter, 1959). The natural seed unit of beet is a seed ball consisting of fused flower parts and one to many seeds. A different definition must be accepted for milled seed of beet, commonly known as "sheared seed" and "rubbed seed." The seed unit of New Zealand spinach is a "fruit" with enclosing bony structure and one to many seeds, the different structures being fused and compacted (Martin, 1946).

B. Morphological and Physiological Variability

A wide range of variation with respect to the morphological development and physiological maturity of seeds of a given crop can be found within the same seed lot. Seeds are harvested at such a time as to obtain maximum yield, yet maintaining optimum quality. Regardless of the harvest date, the unclean seeds will probably contain individual seed structures varying from mere fertilized ovules to mature seeds. These seeds are not only in different stages of development morphologically but are in different physiological stages (Evenari, 1965). Cleaning equipment of various types has been designed to remove immature seeds but none has been completely successful whether used alone or in combination with other devices. In fact, some seeds which apparently are mature morphologically, are not mature physiologically (i.e., have not after-ripened).

It is evident, therefore, that no single testing method can be expected to reveal the correct germination capacity of all seed lots for a given species. Thus, alternative germination methods or special treatments for overcoming dormancy have been provided in the rules for seed testing (Anonymous, 1970). However, as the number of alternative methods for testing a given species is increased, a high probability exists that variability of test results among different seed testing stations will increase. For this reason the number of alternative methods has been kept to a minimum, commensurate with requirements.

C. Association of Weed and Crop Seeds

Granted that a seed sample (or lot) consists principally of pure seeds (i.e., the species or cultivar under consideration), what are the characteristics of the impurities? One would expect the foreign seeds to have been produced from plants which grow under the same or similar conditions as the crop species grew. This is a good assumption but caution must be exercised as the foreign seeds may have been introduced subsequent to harvesting, either accidentally or deliberately. Ordinarily, the foreign seeds, and indeed some inert particles, have characteristics similar to the pure-seed component. Engineers have designed commercial cleaning equipment which takes advantage of all known differences useful in removing unwanted seeds from desired crop seeds. These cleaning machines work on various physical principles. Although seed lots are usually passed through several cleaning machines, some weed and other crop seeds may not be removed. It can be expected therefore, that any weed seeds, other crop seeds, and to some extent inert matter not removed, would possess characteristics of the desirable crop seeds being cleaned.

Seeds of some crop species are so similar in appearance that identification is difficult or impossible. For example, the lay person would not be able to make a separation of alfalfa seed and yellow-flowered sweet clover seed, yet a trained seed technologist can. The same holds true for a number of other species in the same genus or in different genera. Nevertheless, the seed technologist cannot distinguish all the seeds in a mixture of annual and perennial ryegrass or between white-flowered and yellow-flowered sweet clover seeds. Similar problems exist in distinguishing seeds of certain closely related crop and weed species. Examples are Johnson grass seeds in Sudan grass seeds, quackgrass seeds in western wheatgrass seeds, and wild mustard seeds in cultivated mustard, kale, rape, etc., seeds. Special problems such as these have been treated by Justice *et al.* (1952).

V. The All-Important Sample

A. Basis for Sampling Seeds

The basic objective in sampling a seed lot is to draw a portion that is representative of the entire lot. Such a sample is the basis for analysis and tests to determine the purity, germination, noxious-weed seed content, variety, and other quality factors (Justice *et al.,* 1952). Ordinarily, the seed technologist does not actively participate in the sampling of seed lots. However, because he may be called upon to make recommendations, a brief description of sampling is given. After the sample arrives at the laboratory, the technologist must subdivide it to get a specified amount of seeds for the working sample. If the results of tests are to be meaningful the test must be made on a sample or subsample representative of the seed lot from which it was drawn. Obtaining a small sample that will correctly represent a carload lot of chaffy seeds, such as Kentucky bluegrass is not an easy task. If seed lots were homogeneous, as some laws assume or require, sampling would be a simple task. The fact is that seed lots are not completely uniform (Carter, 1961). Thus, rather precise procedures have been established to insure the taking of representative samples (Anonymous, 1970). These procedures outline practices to be followed, and require that sampling instruments must be capable of reaching all parts of the seed container.

B. Sampling Equipment and Procedures

1. SEED TRIERS

Triers are commonly used in sampling seeds in bags and bins. The most commonly used instrument is a sleeve-type trier which consists of a hollow brass tube inside a closely fitting outer shell or sleeve. The tube and sleeve have open slots in their walls so that when the tube is turned until the slots in the tube and sleeve are in the same position, seeds can flow into the cavity of the tube; when the tube is given a half-turn the openings are closed. The tubes vary in length and diameter, having been designed for different kinds of seed and sizes of containers. Following are some examples: (a) 30-in. trier with outside diameter of 0.5 in. and 9 slots; (b) 30-in. trier with outer diameter of 1 in. and 6 slots; (c) 11-in. trier with outside diameter of 0.375 in. and 1 slot. Seeds are admitted into the tube of the 11-in. trier by withdrawing the tube past the slot. This trier makes a very small hole which is important in sampling seeds in cotton bags; however, it should not be used in sampling large bags as the trier is not long enough to reach the most distant parts.

Bin samplers are constructed on the same principle as bag triers but are much larger, ranging up to 63 in. in length and 1.5 in. in diameter with 6 to 9 slots. The 6-, 9-, or 12-in. thief-type trier is not recommended because its construction and size do not permit sampling in accordance with the rules for seed testing. Different types of triers are illustrated in Fig. 1.

2. DRAWING THE SAMPLE

The sample drawn should be made up of equal portions taken from evenly distributed parts of the quantity sampled. In quantities of six bags or less, each bag should be sampled, and in quantities of more than six bags the number to be sampled is 10% of the number of bags in the lot plus 5 (Anonymous, 1970).

If the sleeve-type trier is used, seed bags should be in a horizontal position to insure that as the tube of the trier is opened, seeds will drop in along the entire length of the bag. Perhaps the best sample coverage comes from following a diagonal path through the seed bag to an opposite corner. The trier should be inserted with the slots facing downward so that as it is inverted, or placed upright for collecting the sample, any seeds "dragged along" by the cross-ribs will be dislodged and will not be added to the portion drawn. Sampling bags of seeds that are standing upright results in obtaining unequal portions from various parts of the bag (Shenberger et al., 1946).

Certain chaffy grass seeds and non-free-flowing seeds, such as those of *Bromus* spp., *Andropogon* spp., *Paspalum* spp., or lint cottonseed, which may not be easily sampled with a trier can be sampled by hand. Hand sampling necessitates opening the seed bag, thrusting the hand into

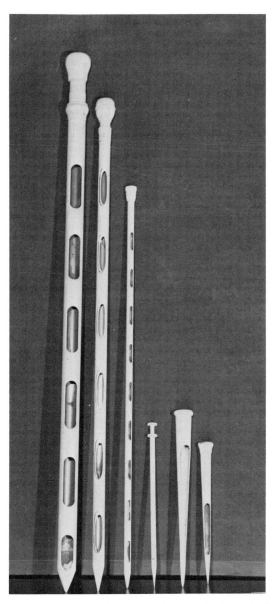

FIG. 1. Seed and grain triers. From left to right: bin sampler for grain; bag trier for grain; bag trier for clover and other small seeds; 11-in. bag trier for small seeds; two "thief-type" triers that are not recommended. (Courtesy of U. S. Department of Agriculture.)

different parts of the bag, and removing small portions of seed from the bottom, center, and top.

3. CAUTIONS IN SAMPLING

Sometimes laboratories receive samples of insufficient size for testing. The inspector or sampler should attempt to obtain samples of at least the minimum quantities required for noxious-weed seed examinations. An inspector or sampler may hesitate to draw a large quantity of seeds having a high retail value, but it is necessary to submit samples of required weights in order to provide adequate amounts of seeds for standard tests and retests if necessary.

Other things being equal, a large sample is more representative of a seed lot than is a small sample. Moreover, if there is a choice as to whether to reduce the sample before sending it to the laboratory, the larger quantity should be submitted. The laboratory can be expected to subdivide the sample more accurately than a sampler working under field or warehouse conditions. The following amounts of seeds are usually sufficient: 2 oz of the fine grasses, white clover, and seeds of similar size or smaller; 5 oz of red clover, bromegrass, flax, and seeds of similar size; 1 lb of Sudan grass, sorghum, cotton, beet, and seeds of similar size; 2 lb of cereals, beans, peas, vetch, and seeds of similar size; and 800 seeds of vegetable, flower, and tree species which require a germination test only. Larger samples may be necessary if either field or greenhouse variety trials are to be made (Carter, 1961).

Whether the seed sample is drawn by a seedsman or farmer for his own guidance, or by an official seed inspector, a number of precautions need to be taken to insure an accurate sample. The sampler must first determine that all seed bags sampled are identified as belonging to a single lot, either by a label or stencil mark on the bag. He must sample the prescribed number of bags for the size of lot at hand. If, in sampling a large lot, more seeds are obtained than is needed by the laboratory, care must be exercised in reducing the quantity. As a mechanical divider is usually not available, the sample reduction might best be done by placing the entire quantity on a sheet of paper or canvas, thoroughly mixing the seed by hand and halving the sample until the desired quantity is obtained. Careless splitting of the sample cannot be expected to produce two similar portions.

Avoidance of damage to seed bags is of prime consideration to seed merchants. The thief-type trier does little damage to cotton bags, but, as indicated above, its accuracy in drawing a representative sample is questionable. The longer trier can safely be thrust into burlap bags which are easily "scratched over." A few stitches at one of the top corners of machine-sewed cotton bags can be broken and then this break can be closed with a hand-stapling device, after the contents of the bag have been sampled. Any seeds known to have been treated with a poisonous fungicide should be so identified so the person who subsequently may handle the sample will be informed of the potential hazard.

4. CARE OF SAMPLES AWAITING TEST

Seed samples awaiting test in a laboratory retained for future retests or research should be stored under conditions that will, as nearly as possible, be conducive to maintenance of their original quality character-istics. Samples should be given adequate protection against possible damage by rodents, particularly if stored in attics or basements. It is essential that samples infested with weevils, moths, or chalcid flies be kept intact, preferably in sealed containers. A small amount of p-dichloro-benzene added to the seed container will kill the common insects that in-fest seeds. Caution should be exercised to insure that seed samples are not exposed to extremes of temperature or moisture while awaiting tests. Only samples that are received in the laboratory in airtight containers should be tested for moisture content. There is a strong possibility that the hard seed content of certain samples of legume seeds left in a labora-tory with a dry atmosphere for a few weeks may change. Caution should be exercised not to expose seeds to fumes from some of the more volatile herbicides as abnormal seedlings may be produced by accidental exposure to fumes of some of these compounds.

C. Reducing the Sample in the Laboratory

1. METHODS

Samples received by a laboratory generally need to be reduced to a working sample of standard weight. These weights are tabulated in the rules for seed testing, for the various kinds of seeds, and have been determined to provide an adequate number of seeds for purity analyses and noxious-weed seed determinations. Usually the working sample is a very small portion of a large bulk of seed. Whenever possible, the sample should be subdivided by a mechanical divider; their use eliminates the personal element in reducing the bulk to a working sample. Most mechani-cal dividers in common use are designed to split the sample into two approximately equal parts. The working sample is obtained by repeatedly dividing the sample until a quantity of approximately that stated in the rules for testing seeds is obtained. It is frequently necessary to run the "unused" part of the sample through a series of divisions to get a few seeds to bring the sample up to the necessary weight. Any attempt to correct the sample weight by personally adding or removing seeds to obtain a precise amount defeats the objective of using a mechanical divider.

Another method of subdividing the sample is by the halving method. The sample is placed in a pile on a clean surface and thoroughly mixing by hand; then, it may be successively halved by use of a sharp-edged instrument until an approximate weight is obtained. This is the least desirable of the methods discussed herein. Still another procedure for subdividing the sample is by the random cups method (Thomson and Doyle, 1955). A series of small cups or thimbles of known capacity are arranged on a tray or pan, in definite pattern, and the sample is poured systematically over this area. The working sample is obtained from randomly selected thimbles or cups. The method may be modified by using a tray, divided into an equal number of square compartments, every alternate one of which has no bottom.

2. MECHANICAL SEED DIVIDERS

One of the most commonly used mechanical seed dividers is the so-called Boerner divider named after its inventor (Boerner, 1915). This divider is made in different sizes, all according to the original plan. The essential parts consist of a hopper, inverted cone, and a series of baffles directing the seeds into two spouts. The baffles form alternate channels and spaces of equal width. They are arranged in a circle at their summit and are directed inward and downward, the channels leading to one spout and the spaces to an opposite spout. A valve or gate at the base of the hopper controls the seed flow. When the valve is opened the seeds fall by gravity over the inverted cone where they are evenly distributed to the channels and spaces; then, they pass through the spouts into the seed pans below.

The Gamet divider makes use of centrifugal force to mix and scatter the seeds over the dividing surface. The seeds flow downward from a hopper onto a shallow rubber cup below. Upon rotation of the rubber cup by an electric motor the seeds are thrown out by centrifugal force and fall downward. The circle or area where the seeds fall is equally divided into two parts by a sharp-edged, stationary baffle so that one-half of the seeds fall in one spout and the other half in the other spout. In using this divider care must be exercised in dividing very small samples as it is possible that a majority of the seeds may be thrown out in one spout. Hardin *et al.* (1965) compared the effectiveness of the Boerner and Gamet dividers and concluded that if used correctly, the Gamet divider yields a slightly more accurate subsample.

The Ottawa divider (Fig. 2) is a precision divider intended for use with

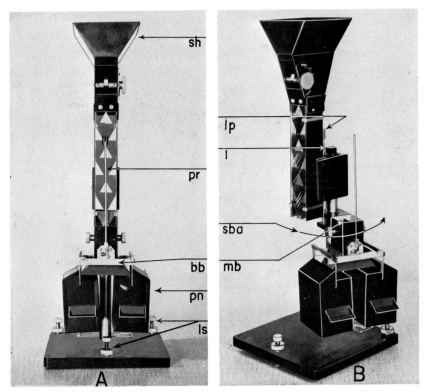

Fig. 2. Ottawa seed divider. (A) front view; (B) dividing column swung around. Key to labels: sh, seed hopper; pr, pointer; bb, balance beam; pn, pan; ls, leveling screws; lp, lock pin; l, level; sba, spout and balance mechanism swing away; mb, movable baffle. (Courtesy of Canada Department of Agriculture.)

free-flowing seeds the size of flax or smaller. The sample is introduced into a hopper and then falls down a shaftlike column during which time it undergoes repeated divisions and recombinations to insure proper mixing. Near the bottom of the column is a movable baffle resting on knife edges of regular balance construction. The tilt of the baffle determines the proportion of seeds that will flow into each of the two seed pans. To divide the sample into two parts of equal weight the seeds are allowed to run gently through the leveled divider without any further adjustment. Unequal divisions can be made as desired by adding weights to one of the pans.

The Oregon divider developed by Oregon State University under contract with the U. S. Department of Agriculture was designed to extract a secondary sample of predetermined weight from a primary sample, with a single pass through the machine. The prototype has been developed, but due to complexity and estimated cost of commercial models, improvement will be necessary before the divider is generally accepted.

Unfortunately due to the construction of some dividers, seeds can become lodged in sharp, rough angles or unfinished joints. Some seeds will adhere to smooth surfaces due to hairs, gums, and resins, whereas others are attracted by static electricity. The chances are high that with use these seeds will become dislodged and appear in another sample passed through the divider. Thus, cleaning after each use by directing a strong air blast through the divider is a good practice.

VI. Testing for Seed Purity and Noxious-Weed Seeds

A. Some Basic Considerations

The object in making the purity analysis is to deterime the identity of the important kinds of seed present and the percent by weight of each component; namely, pure seed, other crop seed, weed seed, and inert matter. Since germination tests are based on pure seed components, it can readily be seen that the purity analysis and germination tests complement each other. Thus, the actual plant-producing power of a seed lot can be determined only when the purity analysis and the germination tests are considered together.

Testing seeds for purity is a meticulous, painstaking operation, requiring constant use of the eyes. To avoid serious eyestrain, the purity-testing laboratory should be located with northern exposure. If artificial light is used, an adjustable, multiple-tube, fluorescent lamp of the daylight type should be provided for each technologist. Purity analysts usually examine and separate seed samples on "workboards" placed on top of desks or table. These workboards have a surface of 150 to 225 in.^{2} and enable analysts to adjust the work to a desired height, usually 3–6 in. above table height.

B. Required Equipment

Equipment required for making purity analysis is as follows: forceps and spatula for handling, separating, and pushing seeds over a surface; wide-field hand lens of 5, 6, or 7× magnification; reading magnifiers relatively free of curvature and distortion, wide-field stereoscopic microscope with a range of magnifications from 10 to 75×; scale with capacity up to 1000 gm and accurate to 0.5 gm; torsion balance with capacity of 120 gm and accurate to 0.01 gm; rapid-acting chemical balance accurate to 1 mg;

small seed containers to hold separations; vials with corks; 2-quart seed pans with pouring spouts; and a seedblower.

A seedblower that has proved successful is illustrated in Fig. 3. The base of this seedblower consists of a metal box containing an air impeller or fan, a low-pressure space, and a high-pressure space. The impeller is operated by a uniform speed motor. The seed cup and blowing tube are mounted over the high-pressure area. The metal seed cup, with screen bottom, supports the glass blowing tube which leads to a glass cylinder designed to retain the light material. Air pressure in the system is con-

FIG. 3. General seedblower. Pull-out seed cup at left; curved glass tube leading to collector of lightweight material at top; air adjustment controls at right; air valve and indicator front top; timer front bottom. (Courtesy of New Brunswick General Sheet Metal Works, New Brunswick, New Jersey.)

trolled by a sensite valve. This blower is designed specifically for small-seeded grasses and will not efficiently accommodate samples larger than 5 gm (Leggatt, 1938). Other blowers which will accommodate larger samples are available but these are less accurate (Porter, 1938).

A diaphanoscope is used to determine whether certain grass florets acutally contain caryopses. This is done by directing a strong beam of light through the lemma and palea, and other structures, if present. Custom-made diaphanoscopes usually are constructed by starting with a purity workboard. A hole, 1.25–1.5 in. diameter, is made in the center of the working surface of the workboard. A microscope lamp emitting a strong beam is mounted in the cavity below the working surface. The beam is directed diagonally downward against an adjustable concave mirror. By adjusting the mirror the beam is directed upward through the hole in the workboard. The workboard is covered with a pane or slab of frosted glass. Seeds are examined by drawing them across the opening and observing their relative density and silhouette to determine whether they contain caryopses (Justice *et al.*, 1952).

C. The Purity Separation

1. MANUAL SEPARATION

The sample is reduced in size (preferably with a mechanical divider) and the weight of the working sample is determined. The separation is then made by placing the seeds on the clean surface of a workboard and examining them to determine whether the sample (*a*) conforms to the name under which it was submitted, (*b*) contains small particles of inert matter which should be removed by sieving, and (*c*) contains lightweight material which may be removed by blowing. If inert matter which can be removed by sieving or by preliminary blowing is present, appropriate sieves or a blower are used for a partial separation. With the aid of forceps, scalpel, or spatula having straight smooth edges, seeds are drawn, a few at a time, from the pile, spreading them apart by pulling them toward the front of the board. This operation is continued until the entire sample is separated. After the preliminary separation is made the sample is returned to the workboard for further checking. In samples of less than 50 gm, all components, including pure seed, should be checked under magnification. After all the pure seeds are checked, the foreign material is separated into other crop seed, weed seed, and inert matter.

2. USE OF SEEDBLOWER

Analyses of many samples, especially grasses, can be facilitated by removing the lightweight material with the aid of a seedblower. If the sample to be analyzed contains lightweight material, an appropriate blowing schedule should be used. Generally, several blowings are made at successively wider valve openings. The light material from each blowing is examined at the workboard under magnification, and the different components present are separated and placed in dishes or vials. Many samples will require several blowings beyond the first which removes only the empty florets. The next blowing should remove some empty and some filled florets. The sample should be blown until the heavy portion or residue contains no inert florets of the seeds under consideration. The portion removed at each successive blowing should be separated into pure seed, other crop seed, weed seed, and inert matter (Isely, 1954).

3. WEIGHING COMPONENTS

Each of the four components must be weighed. If there is a gain or loss between the weight of the original sample and the sum of the four components in excess of 1%, another analysis should be made. Percentages of the components are determined on the basis of the sum of weights of the four components and not on the weight of the original sample.

The rules for testing seeds (Anonymous, 1970) provide that the working sample shall be weighed to four significant places and each component of the separation shall be weighed to the same number of decimal places. This causes a slight conflict because working samples, weighing 1, 10, and 100 gm, or slightly more, usually yield components with a maximum weight of 1 digit less than the weight of the original samples. Thus, in samples of 1.015, 10.15, and 101.5 gm, the heaviest component (pure seeds) might be 0.987, 9.87, and 98.7 gm, respectively. Hence, if the rules are followed strictly in these cases, the original sample is weighed to four significant places but the components to no more than three significant places.

The number of significant figures is determined by the weight of the greatest component, zeros of the same order in the lesser weights being considered significant. Digits are significant both before and after the decimal point. Table I shows the number of decimal places to which samples of different groups must be weighed in order to insure four significant figures for the components.

4. SEPARATING SEEDS OF SIMILAR KINDS AND VARIETIES

It is permissible to use as few as 400 seeds if the working sample consists of two similar species or a species and its variety which would require an excessive amount of time, eyestrain, and effort for separation of the entire working sample. Since tolerance tables are available for 400, 800, and 1000 seeds, it is recommended that one of these amounts be used

(Anonymous, 1970). Frequently, when the separation is so difficult that a reduced sample is used, a stereoscopic microscope will often be helpful or even essential. In testing mixtures of creeping and colonial bentgrass seeds, most of the separation can be made with a hand lens by an experi-

TABLE I

GUIDE TO NUMBER OF DECIMAL PLACES FOR WEIGHING WORKING SAMPLES AND COMPONENTS TO INSURE FOUR SIGNIFICANT FIGURES[a]

Working sample		Components of separation[b]	
Weight specified in rules (gm)	Example (gm)	Number of decimal places	Example (gm)
0.5	0.5108	4	P 0.9876
1.0	1.012		C 0.0014
			I 0.0102
			W 0.0031
			1.0023
2.0	2.025	3	P 9.876
5.0	5.108		C 0.014
10.0	10.02		I 0.102
			W 0.031
			10.023
25.0	25.59	2	P 98.76
50.0	51.63		C 00.14
100.0	102.8		I 01.02
			W 00.31
			100.23
500.0	504.2	1	P 487.6
			C 001.4
			I 010.2
			W 003.1
			502.3

[a] Courtesy of U. S. Department of Agriculture.
[b] Components of the separation are indicated as follows: P = pure seed; C = other crop seed; I = inert matter; W = weed seed.

enced analyst, but the microscope is essential for identifying those few nontypical seeds which are present in practically all samples. A microscope is essential in making positive identifications of some weed seeds.

5. INTERPRETATION OF CROP SEEDS

Due to the difficulties incident to identifying inert matter in some groups, adoption of specific rules for interpretation of pure seeds and other crop seeds has been necessary to promote uniformity. These special provisions are discussed below.

a. BROKEN AND CRACKED SEEDS. Legume seeds usually break along the line of cleavage between the cotyledons. Cotyledons of split peanut, cowpea, and soybean are classed as inert matter, regardless of whether or not the young root-shoot axis is present. For other Leguminosae, if at least half the seed and the radicle are present, the structure is classed as pure seed; all pieces less than these are classed as inert matter. Rye and barley seeds frequently break transversely. In questionable cases the broken fragments are lined up alongside several normal, unbroken seeds of various sizes for comparison. It is not necessary that the fragment have an embryo in order to be regarded as pure seed. A similar procedure is followed in determining the relative proportion of the original size represented by seed fragments in different groups. All cracked seeds are regarded as pure seed if the seed or fragment consists of more than half the original structure.

b. IMMATURE AND SHRIVELED SEEDS. Occasionally samples contain relatively high percentages of immature seeds which the analyst may think should be included with the inert material. Immature seeds are particularly prevalent in uncleaned seed lots. All these seeds should be regarded as pure seed if they can definitely be identified as the kind under consideration. So-called empty seeds (i.e., without embryo and endosperm) of the Cucurbitaceae and the Solanaceae are regarded as pure seed.

c. IMMATURE SEEDS IN GRASSES. This group differs from the preceding one in that the caryopses are enclosed within the lemma and palea. To apply the half-seed rule in determining purity in grasses the question arises as to just how well developed the ovary must be to fulfil the definition of a seed. Before attempting to separate inert matter from pure seeds in grasses the worker should aquaint himself thoroughly with all the structures of the normal pure seed unit. The caryopsis should be removed and studied for comparison with the immature, diseased, or empty florets. Most grass seeds received for testing are florets consisting of a caryopsis enclosed by the lemma and palea. The separation of pure seeds and inert matter is rather simple in grasses with thin, papery lemma and palea. The thick, horny lemmas and paleas of some grasses are opaque, requiring either the use of a diaphanoscope or hand pressure with forceps to determine the presence or absence of a caryopsis.

d. EMPTY FRUITS. The seed unit in several plant families (Compositae, Polygonaceae, Umbelliferae, Labiatae, and Valerianaceae) is a one-seeded, dry indehiscent fruit (mericarps in Umbelliferae). Although it appears logical that the half-seed rule should prevail in testing seeds of buckwheat, scarlet sage, cornsalad, lettuce, sunflower, etc., this procedure is not practical. The outer structures are comparable to the pod in the legumes. In chicory some thin-walled fruits contain heavy seeds; whereas, others with thick walls contain immature seeds or no seeds. The filled and empty fruits are not separated by blowing, and it would be impractical to open each fruit to determine its content. Moreover, opening the fruits would render the sample inadequate for germination testing. In the above families, only broken fruits known to contain no true seeds, half a seed, or less than half a seed are classed as inert matter.

e. SEEDS OF CERTAIN GROUPS LACKING SEED COATS. Crop seeds in the Leguminosaea, Cruciferae, and Pinaceae (legume, mustard, and conifer families) with seed coat entirely removed are classed as inert matter. Thus, seeds without seed coats must be identified as belonging to one of these three plant families. Normal, well-filled seeds of known samples with seed coats removed should be available for comparison with the seeds in question.

f. INSECT-DAMAGED SEEDS. Insect-damaged seeds are classed by the so-called half-seed rule. Generally if half or more of the seed has been consumed by the insect, it is considered to be inert matter (Cull and Justice, 1950). Vetch *(Vicia)* seeds damaged by weevils are classed as pure seed except that broken pieces less than half the original seed size are treated as inert matter. Alfalfa and red clover seeds are often infested with an insect known as chalcid fly. The chalcid fly deposits its egg in the ovary of the flower where the egg develops. When the seed is harvested, insects can be found in all stages of development from egg to mature fly. Infested seed is considered pure seed if half or more of the original seed is present. The infested seed can be removed by blowing, but in using enough air to remove all the infested seeds some pure seeds will also be removed. Thus, it is necessary to separate by hand the infested seeds from the pure seeds in this portion. The infested seeds are generally swollen and appear decayed; their color lacks the lifelike appearance of normal seeds. Emerging insects leave holes in the "seeds" but many infested seeds contain no evidence of this nature (Wheeler and Hill, 1957). If the major portion of the seed has been eaten away, the seed coat will collapse easily with pressure. Seeds that resist this pressure should remain with the pure seeds. This subject is treated in greater detail by Justice *et al.* (1952).

g. NEMATODE GALLS AND SCLEROTIA. In some instances nematodes have replaced most or all the contents of the seed, forming galls. Although these nematode galls alter the inner contents of the seed, from outer appearances there is a marked resemblance to true seeds. These structures should be considered inert matter. Nematode infestations are frequently found in fescue (Lewis, 1941) and bentgrass seed (Courtney and Howell, 1952) and occasionally in wheat and wheatgrass (D. C. Norton and Everson, 1963). If nematodes are suspected, a positive test can be made by crushing the seedlike structure, placing it in water for a few hours and then examining it under a microscope. The nematodes (eelworms) become active and can be identified easily by their wormlike motion.

Mycelia of fungi may develop to the extent that they replace the contents of the seed, forming sclerotia (Crosier and Patrick, 1952; Wheeler and Hill, 1957). Frequently, ergot replaces only a portion of the seed, especially in redtop. In such instances each seed must be considered on its own merits in applying the half-seed rule. Ryegrass seed infested with the blindseed fungus *(Phialea temulenta)* is not a sclerotium and thus should be classified as pure seed. A diaphanoscope is often helpful in detecting the presence of ergot and other sclerotia (Justice *et al.,* 1952).

h. PULPY SEED BALLS AND FRUITS. Seed units of beets *(Beta vulgaris)* and New Zealand spinach *(Tetragonia expansa)* are regarded as pure seed without reference to the presence or absence of true seeds. Segmented beet balls are classified likewise except that small fragments which obviously do not contain a seed are placed with the inert matter.

i. WINGS OF TREE AND SHRUB SEEDS. Broken and unattached wings of tree and shrub seeds are classed as inert matter. The following procedures apply when the wings are attached. Detach and classify as inert matter the wings, except that part which encloses the seed and is not easily removable of the following: *Abies, Larix, Libocedrus, Pseudotsuga, Pinus echinata, Pinus elliottii, Pinus palustris, Pinus rigida,* and *Pinus taeda.* The wings on seeds of *Acer, Betula, Catalpa, Chamaecyparis, Cupressus, Fraxinus, Liquidambar, Liriodendron, Platanus, Thuja,* and *Ulmus* are not detached. Remove and classify as inert matter the entire wings of *Cedrus, Picea, Tsuga,* and *Pinus* except the five species of *Pinus* indicated above.

6. INTERPRETATION OF WEED SEEDS

Procedures and interpretations applied in distinguishing between weed seeds and their counterpart, inert seedlike structures, differ from those used for crop seeds. In general, more seedlike structures are regarded as weed seeds than crop seeds, other factors being equal. This is because an error of classifying a questionable weed seed as inert matter is of much greater importance than a similar mistake when classifying a questionable crop seed. Moreover, the seriousness of the error is greater if noxious-weed seeds are involved.

a. DAMAGED SEEDS AND CARYOPSES. Damaged seeds with over half the embryo present are classed as weed seeds; structures with less than half embryos are regarded as inert matter. In the case of caryopses (grasses) the same interpretation prevails but the decision is based on the root-shoot axis (the scutellum excluded), rather than on the embryo. Undeveloped glumes and florets without embryo and endosperm are classed as inert matter. If the caryopsis of *Agropyron repens* is one-third the length of the palea, or longer, the floret is treated as a weed seed (Everson, 1954). Free caryopses devoid of embryos are classed as inert matter. Magnification is necessary in making determinations on small seeds such as *Poa, Agrostis, Panicum,* and *Eragrostis.* Desiccated stamens should not be confused with immature caryopses.

b. BULBLETS OF WILD GARLIC AND WILD ONION. Not all the bulblets of *Allium vineale* and *Allium canadense* are viable. Bulblets devoid of husks and which pass through a 0.077-in. round hole sieve (Justice, 1942) and bulblets which show evident damage to the basal end, whether the husk is present or absent (Hooker, 1942), are regarded as inert material. All other bulblets are classed as weed seeds.

c. BUCKHORN OR NARROW-LEAVED PLANTAIN. Black seeds with no color evident are classed as inert matter (Thompson, 1941; Everson, 1952). The following standardized conditions are recommended for classifying questionable seeds: a 10× stereoscopic microscope and fluorescent light provided by two 15-W daylight tubes (Anonymous, 1970).

d. DODDER SEEDS. Seeds of dodder which are ashen gray to creamy white in color and those devoid of embryos are included with inert matter. The embryo lies coiled in the center of the seed, surrounded by the endosperm. The seeds should be judged from outward appearance and the questionable ones dissected to determine whether an embryo is present. Questionable seeds include those with normal color or nearly normal color, but slightly swollen, dimpled, or with small pin-point holes. The comprehensive paper by Gaertner (1950), although not covering the above points, merits review by the serious student.

e. NAKED RAGWEED SEEDS. Seeds devoid of the involucre and peri-carp are classed as inert matter. Although the presence or absence of the pericarp and involucre can be determined readily, it is important that these naked seeds be correctly identified as ragweed.

f. EMPTY ACHENES. Empty seeds or those without embryos or endosperm, such as occur in the buckwheat, sedge, and composite families, can often be detected either with a diaphanoscope or by dis-section. It may be necessary to dissect or examine a few questionable seeds, especially burs of cocklebur *(Xanthium)*.

g. WEED SEEDS DEVOID OF SEED COATS. Weed seeds devoid of seed coats in the legume family and the genus *Brassica (Cruciferae)* only are to be regarded as inert matter. This differs from the classification of crop seeds in that all crop seeds in the entire mustard family, without seed coats, are classified as inert matter.

7. TESTING COATED AND PELLETED SEEDS

Seeds encased with an inert covering have appeared in trade channels under designations such as "coated" and "pelleted" seed (Heit and Munn, 1952). Two processes have been used in covering the seed. In one process individual seeds are covered, and in the other process the num-ber of seeds covered may range from one to several. Occasionally, an inert fragment is covered. In that case the pellet would be devoid of a seed. No special provisions have been developed for testing coated or pelleted seeds for purity and germination. The present rules for testing seeds are interpreted to mean that the true seed is the unit for test. The principal problem in the purity analysis appears to be that of removing the covering in order to determine the component parts of the sample and whether weed seeds, including noxious-weed seeds, are present (Allen, 1949). Until an acceptable procedure is developed the following pro-cedure might be used for testing. For purity analyses the sample sizes would be those set forth in the rules, after the inert coating is removed. The coating can be removed by soaking the structures in water and strain-ing off the inert material after which the seed is thoroughly dried. The weights of the pure seed, other crop seed, and weed seed would be de-termined as in a regular test. The difference between the total of these weights and the original sample weight would represent the weight of the inert component.

8. SPECIAL PROCEDURES

A number of special procedures are used to facilitate the determination of pure seed in certain species. For example, mottled seeds in sweet clover provide an estimate of the percentages of yellow-flower and white-flower sweet clover (Caldwell, 1941). By Caldwell's procedure the percent of mottled seed multiplied by 4 represents the percent of yellow-flower sweet clover in a sample.

Another test which has served a very useful purpose is the fluorescence or ultraviolet test of ryegrass (Gentner, 1929; Justice, 1946). When germination tests are made on white filter paper the roots of annual ryegrass secrete a substance which fluoresces under long-wave ultraviolet light; whereas, the roots of perennial ryegrass do not. A 5% allowance is made for uncontrolled hybridization and a correction for percentage germination is made. The introduction of new cultivars within the past 10 years has greatly weakened the value of the fluorescence test. Also, Nitzsche (1960) has shown that fluorescence is affected by a number of factors including temperature, period of storage of the seeds, and nitrate ions.

Perhaps the greatest progress made in the past 25 years in the area of purity analysis has been the introduction and acceptance of a standard, uniform blowing method for testing Kentucky bluegrass *(Poa pratensis)* seed. The method was first proposed after independent research by Leggatt (1938) and Porter (1938). Obviously neither seed technologists nor seed merchants were ready to accept such an unorthodox method at that time. Subsequently, additional research and comparative testing were done on a national and international basis primarily by Porter and Leggatt (1942) and Everson *et al.* (1962). The method was accepted by the Association of Official Seed Analysts in 1960 (Anonymous, 1970) and by the International Seed Testing Association in 1965 (Anonymous, 1966). By this procedure seedblowers are standardized and calibrated according to specially prepared samples consisting of (*a*) the lighter seeds that have been stained one color and should blow out and (*b*) heavier seeds that have been stained a contrasting color and should remain in the seed cup. The only adjustment made after blowing is to separate other crop seeds, weed seeds, and inert particles other than florets. In recent years the blowing procedure has been extended to include orchard grass (West, 1952), Pensacola Bahia grass (Morgan, 1965), and the various cultivars of Kentucky bluegrass (Anonymous, 1970).

Commercial seeds of a number of grasses contain spikelets with more than one floret. Until recently analysts were required to separate these florets from the spikelet and determine on a seed-by-seed basis whether the floret was fertile or sterile. Multiple florets were classed as pure seed or inert matter accordingly. Ching and Jensen (1957), Niffenegger and Davis (1958), and Madsen (1960) published information on the relationship of sterile and fertile florets in multiple units of Chewings fescue, creeping red fescue, orchard grass, crested wheatgrass, and intermediate wheatgrass. Conversion tables were developed for calculating the percentages of pure seed and inert matter of multiple florets by merely determining the percentages of multiple florets and single florets in the sample. This method is not approved for use when the proportion of multiple florets is less than 5.0%.

A chemical procedure known as the phenol test is used for distinguishing varieties of common wheat and other species of *Triticum*. Seeds are moistened by placing them on germination blotters saturated with a 1% solution of phenol. After 4 hours of soaking they are classified into five color groups and compared with color standards or the known color of the variety tested (Walls, 1965).

D.   Examination for Noxious-Weed Seeds

1.   OBJECT AND NATURE OF TEST

The object of a noxious-weed seed examination is to identify and determine correctly the rate of occurrence of those weed seeds defined by laws and regulations as noxious. The rate of occurrence refers to the number of seeds in a unit weight, based on the actual number of seeds found in the sample examined. The noxious-weed seed examination is often regarded and conducted as a part of the purity analysis, primarily because the methods used in conducting the two tests are similar and both tests are made by the same analyst. However, it is common practice among seed testing laboratories not to make noxious-weed seed examinations except upon request. Each state as well as the federal government has established a legal list of noxious-weed seeds.

2.   SIZE OF SAMPLE

A particular seed lot may contain many seeds of a certain kind of noxious weed or it may contain only an occasional one. In testing a lot of red clover seeds containing 18 dodder seeds/lb, the 5-gm sample used for purity analysis might contain no dodder seeds. Thus, if only this sample were examined for noxious-weed seeds the lot might be sold as "dodder-free." To avoid such errors it is necessary to use considerably larger samples for noxious-weed seed examinations than for purity analyses. Statistically, the size of sample for a noxious-weed seed examination should be based on the rate of occurrence of the noxious seeds. Obviously, this is not feasible; therefore, sample size has been established more or less on the basis of the size of seeds or number of seeds per unit weight of the crop species being tested. In general, the Association of Official Seed Analysts has adjusted the size of samples for noxious-weed seed examinations to 10 times the sample size for purity analysis; however, some exceptions exist. Whereas sample size for purity analysis ranges from 0.25 to 500 gm, sample size for noxious-weed seed examinations ranges from 2.5 to 500 gm. The exact sample size is indicated in the rules for testing seeds for each crop species listed (Anonymous, 1970).

When noxious-weed examination and purity analysis are made at the same time, the amount of seeds used for the purity analysis can be subtracted from that required for the noxious-weed seed examination. This figure represents the additional amount that must be tested for noxious-weed seeds. For example, since a 10-gm sample is used for the purity analysis of crimson clover seed, only an additional 90 gm need be examined for noxious-weed seeds. The number of each kind of noxious-weed seed found in the 10-gm analysis and in the 90-gm examination are added, and the total represents the number found in 100 gm, i.e., the required amount.

3. CONDUCTING THE EXAMINATION

The sample is reduced to the proper size by a mechanical divider or other acceptable method described under Section V,C. The procedure for making the examination is essentially the hand separation procedure described for the purity analysis (Section VI,C). It is not essential, that a magnifier be used in the examination of large seeds, but one should always be available for the critical examination of questionable seeds. Magnification is essential in making noxious-weed seed examinations of the kinds of seeds requiring a sample size of 25 or 50 gm.

The Rules for Testing Seeds of the Association of Official Seed Analysts (Anonymous, 1970) provide that if 30 or more noxious-weed seeds of a single kind are found in the purity analysis (or in an amount equal to the sample size for purity analysis, when only a noxious-weed seed examination is made), the sample need not be examined further for that kind of seed. At this point, the amount of the crop seeds in which the 30 weed seeds were found is weighed and the rate of occurrence is determined on that basis. However, the analysis must be continued for other noxious-weed seeds. Should 30 seeds of a second or third kind be found before the entire sample is analyzed, examination for each of these kinds of seed may be discontinued in accordance with the above procedure (Justice *et al.*, 1952).

4. INTERPRETATION OF SEEDS

Any sample may contain weed seeds about which there is question regarding either identification or interpretation as to whether the structures are weed seeds or inert matter. In each instance the structure must be examined critically, usually under magnification to make the determination. Interpretation of noxious-weed seeds is the same as for common weed seeds. Because a high degree of accuracy is essential, more seeds should be dissected, caryopses of grasses removed from glumes, and comparisons made with known seeds than when common weed seeds are involved. The number of each kind of noxious-weed seed found is converted to an appropriate unit weight basis.

VII.   Testing for Germination and Viability

In seed laboratory practice, germination is defined as the emergence and development from the seed embryo of those essential structures which, for the kind of seed in question, are indicative of the seed's capacity to produce a normal plant under favorable conditions. Germination is expressed as the percentage of pure seeds of the kind under consideration which produces normal seedlings. The environmental conditions for laboratory germination must not only be specific enough to initiate growth, but also must be favorable for development of the resultant seedlings to a stage whereby interpretation into normal and abnormal types may be made. Definite methods for laboratory germination of the more common agricultural, vegetable, flower, and forest tree seeds have been set out in the rules for seed testing. These methods have been developed from experience of seed technologists over many years and from research results gleaned from various disciplines of the plant sciences. As knowledge of the germination behavior of seeds increases, changes in these methods will no doubt take place. The 1970 edition of the Rules for Seed Testing of the Association of Official Seed Analysts includes specific methods for testing the germination of a total of 543 kinds of seeds. By groups they are distributed as follows: field or agricultural seeds, 179; vegetable and herb seeds, 79; flower seeds, 181; and tree and shrub seeds, 104. Most of these entries are for species, but a few are botanical varieties and several are genera.

A.   Equipment and Germination Substrata

1.   MISCELLANEOUS EQUIPMENT

A number of instruments and supplies commonly used in laboratories of various types are required for seed germination. These include forceps for handling seeds and seedlings, hand magnifiers, sample pans, miscellaneous glassware, florist sprinkling bulbs, and thermometers. In addition to good overhead light, each germination analyst should have an individual, adjustable fluorescent lamp. Major equipment is discussed below.

2. SEED-COUNTING DEVICES

Counting devices are regarded as standard equipment in seed testing laboratories. Two types of seed counters commonly used are (*a*) counting boards and (*b*) vacuum counters. Counting boards are frequently used for large seeds such as corn, beans, and peas. A typical counting board has a stationary bottom, approximately the size of the substratum to be used, perforated with 50 or 100 holes of the general shape and size of the seeds to be counted. Below the perforated base is a thin solid board which serves as a false bottom and can be moved forward and backward. In operation, the counting board is placed over the substratum, the seeds scattered over the board, and the excess seeds removed by tilting the board slightly. After checking to see that there is only one seed per hole, the movable bottom is withdrawn and the seeds fall in place upon the substratum.

A vacuum counting device has three essential parts: (*a*) a vacuum system including lines, (*b*) counting plates or heads, and (*c*) a valve. A laboratory in which corn and large-seeded beans are to be counted by vacuum will need a system that will support from 20 to 27 in. of mercury at the point where the counting head is attached. The capacity for suction must be great enough to hold the seeds with the unavoidable air leakage caused by irregular seed surfaces. The hose, valve, line, and all connections should have openings of sufficient size so as not to restrict the flow of air. A capacity of 5 to 8 ft^3/minute appears to be sufficient for clovers; whereas, larger seeds, such as corn and beans, may require a capacity of 15 to 18 ft^3/minute. Some vacuum counters are direct acting — the vacuum at the counting head is created only when the vacuum pump is in operation (i.e., the operator switches the motor on and off with each use). In other models a tank is placed between the counting head and pump; the pump operates automatically as the partial vacuum in the tank is reduced. Also, ordinary laboratory vacuum lines can be used, provided the vacuum is sufficient to hold the seeds to be counted.

Seed-counting heads are made in square, rectangular, or disc shapes to fit the substratum and size of seeds. The head has front and back faces between which there is a cavity for air passage. It is a completely airtight system except for the hole for attachment to the valve and the holes in the front plate for seeds. The size of head and number and arrangement of holes for seeds are determined by the size of seed to be counted and dimensions of the seed bed. Counting heads may be made of brass, aluminum, or other noncorrosive metals and plastic. Plastic is preferred for counter heads larger than approximately 50 in.^{2}, because of its lighter weight. The valve should (*a*) be easily adjusted to different settings between "on" and "off" positions, (*b*) remain in fixed position at any point of adjustment, and (c) be free of leakage at all times. Adjustment to different positions is necessary as the tenacity with which seeds cling to the openings determines, to a large extent, the ease and success of counting small and irregularly shaped seeds. A vacuum counting head is shown in Fig. 4.

FIG. 4. Vacuum seed-counter head for counting and placement of seeds. Fifty holes for seeds the size of wheat; cutoff valve at top. (Courtesy of U. S. Department of Agriculture.)

3. GERMINATORS

The kind of germinator used is not particularly important so long as the required temperature, moisture, and light conditions are met. The type of germinator most commonly used in the United States consists of a chamber for seed trays with provision for cooling, heating, and maintenance of high relative humidity. A chamber of approximately 9.5 ft^3 capacity with inside dimensions of 26 in. wide, 24 in. deep, and 27 in. high will accommodate thirteen trays 18.5 by 19.5 in. spaced at 2-in. intervals and will provide ample additional space between trays and around the edges for movement of air. Excellent germinator chambers of larger dimensions and capacity are also available. One of the most reliable germination chambers in use is the type which is cooled by refrigerated water passing through coils that line the inside of the chamber or through a water jacket between the walls of the chamber. Approximately 0.25 in. of water is held in the bottom of the chamber to assist in maintaining high humidity. There are various types of electric heaters that can be used inside the chamber, either submerged in water or suspended in the air. Valves are adjusted to allow a flow of water that will cool the chamber slightly below the desired temperature. The heater, which is controlled by a temperature regulator, holds the temperature at the preestablished setting. Four of these units are commonly assembled to give the appearance of a single cabinet.

Stermer (1968) designed a germinator based on the principles described above, which incorporates a number of improvements. According to Stermer, heating and cooling equipment "is designed to provide uniform temperatures without stratification, to prevent overshoot, and to maintain high relative humidity." A specially designed, inexpensive, condensate evaporator was designed to increase relative humidity at "stress" times such as the warm-to-cool cycle of a temperature alternation. Panel fluorescent lamps provide more uniform illumination than fluorescent tubes. Temperature is regulated by thermistor controls which are sensitive to a temperature change of $\pm 0.1°C$. Details of this germinator are available from the Agricultural Research Service, U. S. Department of Agriculture.

Another germination chamber which does not use chilled water for cooling is designed for rapid, automatic alternation of temperatures at preset times and for operation either as a dark or daylight germinator (Fig. 5). Cooling is accomplished by a refrigeration unit mounted below the chamber. The cool air is forced upward through a duct between the inner and outer walls of the chamber and is released inside the chamber at its top. Within the chamber the cool air moves downward through specially designed plenums from which it is distributed for cooling. For heating, electric heating coils are placed in water maintained at the bottom of the chamber. The special arrangement of plenums, a slow-running fan, and proper adjustment of heating and cooling controls permit a temperature change of $15°C$ within 15 to 20 minutes. Moisture is provided by a separately operable humidity spray system. Daylight fluorescent lamps provide light; however, controls permit adjustment for various intensities between zero and maximum intensity.

Illumination for light-sensitive seeds may be provided by natural daylight, fluorescent lights, or both. In one "daylight" germinator, 4-ft fluorescent tubes are placed 3 in. from the glass side walls of the chamber. Refrigerated water is released at the inside, top, and front of the germinator, allowed to run over a sloping glass retainer to the rear glass wall, and trickle down a vertical screen into a reservoir at the bottom of the chamber from which it is returned to the cooling system for continuous

Fɪɢ. 5. Alternating temperature germination chamber. Chilled air is released at top; heating is by electric coils in water reservoir at bottom; humidity is supplemented by pressure spray; 48-in. daylight fluorescent light tubes are mounted along sides. (Courtesy of J. P. Pfeiffer and Son, Baltimore, Maryland.)

use. An immersion-type heating coil is located in the water reservoir. Operation of the heater and flow of water are controlled by temperature regulators. A germinator having some of these features, designed to provide either a light or dark environment, is shown in Fig. 6.

The room-type germinator, sometimes referred to as a walk-in germinator, is used in large laboratories. Room germinators should be large enough to allow a row of trays along each wall with passage between. Like the germination chambers previously described, the inner walls, floor, and ceiling should be made of moistureproof, noncorrosive materials.

Fɪɢ. 6. Combination light or dark germinator. Cooling is by chilled water; heating by immersion–type heater in water reservoir; humidity is maintained by water gently falling over a water curtain into the reservoir; light is provided by flourescent tubes. (Courtesy of the Stults Scientific Engineering Corp., Springfield, Illinois.)

Provision must be made to avoid stratification of temperature, differential air currents, and areas of low humidity. Since the RH in germination rooms is somewhat less than 100% the amount of drying of the substrata is materially affected by the amount of moist substrata or number of tests in the room. In other words, no appreciable loss of moisture may occur if the room is used to capacity, but if only a few tests are being carried, serious drying of the substrata may take place. The room should (a) be adjacent to the regular germination laboratory, 6.5–7 ft high, 6.5–8 ft wide, and of desired length, (b) have walls and ceiling that are water-resistant and the floor waterproof, (c) have automatically controlled cooling and heating systems that will meet the needs for the desired constant or alternating temperatures, (d) possess a reliable spray system to maintain high atmospheric humidity, (e) provide slow circulation of the warm and cool air, and (f) be equipped with daylight fluorescent light for at least a section of the room.

In Europe, a germinator called the Jacobsen Apparatus or Copenhagen Tank is commonly used. One feature of the apparatus is a tank of water which serves to keep the tests moist. The top surface of the apparatus is approximately 44 × 50 in. Hollow stainless-steel strips are placed about 2 to 3 in. above the water tank with 5-mm spaces between strips. These strips are attached to a common header and from there to a source of (a) warm water and (b) cold water. Temperature on the surface of the strips is automatically controlled by the amount of warm and cold water allowed to circulate. Germination pads or substrata, with attached wicks, are placed on the metal strips, the wicks passing through the slots between heating–cooling strips to the water below. An inverted plastic funnel with a hole in the conical end is placed over each germination pad. Thus, temperature is controlled by automatically mixing warm and cool water that flows through the heating–cooling strips; moisture is provided through the wicks and the light from the room is sufficient for most seeds (Overaa, 1962). The equipment requires an enormous amount of floor space for the number of tests accommodated. Also, the handling of covers and substrata appears to be more cumbersome than moving a lightweight tray with paper substratum. The equipment has not gained general acceptance in the United States.

4. Minor Equipment

a. Ultraviolet Lamp. An ultraviolet lamp is required in laboratories testing the roots of ryegrass (*Lolium* spp.) for fluorescence. The lamp or unit should have maximum emission at approximately 3650 Å and radiation of sufficient intensity to activate the weak fluorescent lines.

b. Beet Seed Washer. Samples of the different varieties of beet (*Beta vulgaris*) seed may require washing with running water from 2 to 4 hours to remove the toxic principle causing darkening of the radicles (Stout and Tolman, 1941). Washing may be done by placing the seeds loosely in cheesecloth bags and the bags in running water. Uniform washing can be accomplished by constructing a metal pan having walls 2.5 in. high with a drain spout 0.75 in. from the top and a false bottom soldered to the walls 1 in. from the bottom. The false bottom has a 0.094-in. hole beneath each seed basket. The baskets, 1.5 by 1.5 in. in cross section and 2 in. high, are made of 20 by 6 mesh copper gauze. When the water enters the compartment between the bottom and false bottom, it squirts up through each hole, agitating the seeds while washing them.

c. SOIL AND SANDBOXES.   Different kinds of containers have been used to hold soil and sand in germination testing. A commonly used container consists of paraffined cardboard boxes, sizes 4.5 by 4.5 by 1.5 in. for seeds up to the size of clovers and 8.5 by 8.5 by 1.75 in. for larger seeds. The boxes are usually purchased unassembled but come creased and perforated. The sides of the large boxes tend to sag when filled, thus allowing the soil to dry along the edges. This can be avoided by using bands made of aluminum or other noncorrosive metal which fit around the cardboard boxes. Recently plastic boxes have been used as containers for soil and sand. A popular size now on the market is 4.5 by 4.5 by 1.125 in. The size and shape of plastic boxes should conform to the shape and dimensions of the appropriate seed-counting head.

5.   SUBSTRATA

a. REQUIREMENTS.   Most laboratory tests are made on filter paper, paper towels, or blotters. Other materials used include sand, granulated peat moss, and a form of mica called Vermiculite or Terralite. The substratum should (a) be nontoxic to the germinating seedlings, (b) be relatively free of molds, other microorganisms, and their spores, and (c) provide adequate aeration and moisture for the germinating seeds. Timothy, redtop, and Chewings fescue seeds can be used to check any substratum for toxicity as their roots will not develop normally in the presence of toxic materials (French, 1921). In making such a test some seeds from the same sample should also be planted on similar substratum known to be nontoxic, such as a high-quality filter paper. Toxicity in paper materials has, on occasion, been traced to sulfides that were not removed in the manufacturing process (Justice et al., 1952).

b. BLOTTERS.   Germination blotters should be blue or dark gray in color, weigh 275 lb/500 sheets of size 25 × 40 in., and free of toxic chemicals. For seeds the size of clovers, blotters are frequently cut to a size of 6 × 9.5 in. The blotter is folded at the center of the long dimension to form a "doublet" 4.5 × 6 in. In testing, the blotter is saturated with water, the excess water allowed to drain off and the seeds are placed between the folds. Only one such blotter should be placed on another, on a tray. The blotters are alternated each time a preliminary count is made to permit equal aeration. When blotters are used as a substratum for light-requiring seeds, the seeds are placed on two thicknesses of blotter without a cover. In this application large blotters (9.5 × 9.5 in.) may be placed directly on a tray, or blotter discs may be used in petri dishes.

c. PAPER TOWELS.   Paper towels provide an excellent substratum for many kinds of seeds, especially those the size of cereals or larger (M. T.

Munn, 1950). Towels are usually cut to a size of 11 × 14 in. and should meet the specifications mentioned for blotters. More specific specifications are listed in *U.S. Department of Agriculture Handbook No. 30* (Justice *et al.,* 1952). Towels are used primarily as a "roll test." Fifty or 100 seeds are placed on two or three thicknesses of wet towels, an equal number of thicknesses are placed over the seed, the bottom half inch is folded upward, and the entire unit is rolled in such a manner that the seeds will remain in place. The "rolls" are stood on ends in a germination chamber.

d. FILTER PAPER.   Filter paper meeting the above-mentioned specifications for blotters may be used. Frequently, filter paper is used in petri dish tests, 2 or 3 thicknesses per dish, and for germinating ryegrass seeds for the ultraviolet light test. Whatman No. 2 filter paper and others of equal quality have been satisfactory.

e. SAND.   Sand should be washed and free of organic material. The particles should be sharp and range in size between 0.05 and 0.8 mm in diameter. If used repeatedly, the organic matter must be removed after each use and the sand sterilized to kill any microorganisms which may be present. Quartz sand from the quarries at Ottawa, Illinois provides an excellent medium for check tests. The approximate amount of water to add to the sand can be calculated from the following formula:

$$\frac{118.3 \text{ ml (1 gill) sand}}{\text{its weight in grams}} \times 20.2 - 8.0 = \text{No. milliliters of water to add to each 100 gm of sand}$$

f. PEAT MOSS.   Granulated peat moss mixed with sand or soil is useful as a substratum for tests requiring a long period of treatment at a low temperature. Peat moss has a high water-holding capacity, is light and spongy, and tends to discourage growth of molds. A mixture of peat moss with soil and sand makes a good substratum for germinating New Zealand spinach seed.

g. MICA.   Expanded mica, known under names such as Vermiculite and Terralite, is widely used in greenhouses as a covering for seeds when grown in soil flats, and for propagation of cuttings. For laboratory use the advantages appear to be that these materials are *(a)* relatively inexpensive, *(b)* relatively sterile, *(c)* inert in nature, *(d)* light in weight, *(e)* spongy and porous, and have a tremendous water-holding capacity. Their disadvantages are twofold: *(a)* they are very "flaky" and will stick to the hands of workers and to the surfaces of tools when saturated, and *(b)* when saturated, they hold too much water for the proper development of normal seedlings of certain species.

h. Soil. In another section of this chapter soil is suggested as a substratum for check testing. Although soil for seed laboratory use has not been standardized, its occasional use for special tests should not be overlooked. Wellington (1965) suggests that soil of high organic content may adsorb toxic chemicals such as pesticides to the extent that normal growth takes place; whereas, little or no growth would occur on inert sand or paper substrata. Soil should contain a high proportion of organic matter, be devoid of clay, have a high water-holding capacity, and be relatively free of microorganisms when used. After water is added to the soil it should be forced through a sieve with 0.25-in. holes.

B. Laboratory Procedures

1. CONTROLLING MOISTURE AND AERATION

The substratum must be moist enough at all times to supply the needed moisture to the seeds and seedlings, but excessive moisture will restrict aeration. Except for a few kinds of seeds which require very moist substrata, the blotters, towels, or other substrata should never be so wet that a film of water is formed around the seeds. For most kinds of seeds, the blotters or other paper substrata should not be so wet that, by pressing, a film of water forms on the finger. The necessity of adding water to the substratum subsequent to placing the seeds in test will depend on the rate of evaporation from the substrata in the germination chambers. All tests should be examined daily to insure that the moisture content of the substratum is near optimum. Such mechanical aids as small rubber florist's bulbs and medicine droppers are useful for rewatering tests without danger of overwetting the substratum.

2. MAINTAINING TEMPERATURES IN GERMINATORS

Because temperature is one of the most critical factors in the laboratory germination of seeds, daily observations and records should be made of the temperature inside each germinator if a thermograph is not used. When alternating temperatures are required the test is held at the low temperature for approximately 16 hours and at the high temperature for approximately 8 hours per day. If temperatures cannot be conveniently alternated over weekends and on holidays, the seeds should be held at the lower temperature during such time. The daily alternation of temperature is either brought about by manually transferring the tests from one germination chamber to another or by using germinators equipped to change the temperature automatically at preset times.

3. PROVIDING LIGHT

Light is required for germination of seeds of most grasses and for some vegetable, flower, and tree seeds. Fluorescent light has been used successfully in lieu of natural daylight and is preferred in testing because the wavelength and intensity can be standardized, within limits. Available evidence indicates that daylight fluorescent light is as effective in stimulating germination as natural daylight However, when fluorescent lights are placed inside a germination chamber considerable heat is produced which must be counterbalanced by a cooling system which, in turn, tends to dry the substratum. Light should be evenly distributed over the tests, with a range of intensity from 75 to 150 ft-c (750–1250 lx). A variation of ±25 ft-c on the tests is not considered important at optimum intensity. Tests should be subjected to light for only a part of the test period. The usual procedure is to provide 8 hours of light during the day period. Although a somewhat shorter exposure period may be sufficient, not enough data are available to justify a definite statement on this point, especially as it relates to economic seeds (Bass, 1950; V. K. Toole, 1963).

4. COUNTING THE SEEDS

Counting the seeds for the germination test, without discrimination as to size or appearance, may be done by hand, with the aid of a counting board, or by use of a vacuum seed counter. Only free-flowing, smooth seeds lend themselves readily to counting by the vacuum method and only large seeds can be successfully counted with counting boards. However, a laboratory which tests a relatively large volume of crop species that can be successfully counted by the vacuum method, will find that the rapidity of placing the seeds in test and the even spacing of the seeds on the substratum more than justify the cost of a vacuum counter. The vacuum counter is designed to avoid personal bias in selecting seeds, but its misuse can defeat this purpose. The seeds should be spread well over the entire face of the counter head. If uneven seeds, such as immature flat seeds of alfalfa are placed on one edge and the counter tipped, the round seeds will roll across the face of the counter rapidly and be held on the holes out of proportion to their presence in the sample.

5. SPACING THE SEEDS

The proper spacing of seeds to reduce to a minimum the contact of seedlings with each other during germination cannot be overemphasized. This is especially important for large-seeded kinds and seed stocks infected with fungi. As a general rule the distance between seeds should be not less than 1.5–5 times the width or diameter of the seed to be tested. Since certain seeds, such as beans and peas, may double in size after absorption of water, and others, such as spinach and bluegrass, do not become appreciably larger, proportionate allowances for their expected increase in total area should be made when they are placed on the substratum.

6. Dates of Preliminary and Final Counts

The preliminary and final counts have been worked out for the different kinds of economic seeds. The date for the first count is approximate and a deviation of 1 to 3 days is permitted. Intermediate counts are made at the option of the analysts and will frequently vary with different samples. In general, tests on artificial substrata should be examined every 2 or 3 days, especially during the first 10 days of test. Grasses needing longer than 10 days for germination usually need counting only once a week after the fourteenth day of test. A convenient schedule for grasses having a test period of 21 or 28 days is to count on days 10, 14, 21, and 28. The prechilling period is added to the usual test period unless otherwise specified in the rules for seed testing. For example, a sample of Kentucky bluegrass seeds that requires prechilling could be held in test for a total of 33 days, the sum of a 5-day prechill period and a 28-day regular test period. Samples need not necessarily be held in test for the additional prechill period. The conditioning of the seed by prechilling frequently accelerates the speed of germination after the tests are placed at higher temperatures.

7. Controlling Fungi in Germination Tests

The control of saprophytic fungi which develop on seeds or substrata is a major problem to seed analysts, necessitating special attention in spacing, watering, selection of substrata, and care of the seeds while in test. All ungerminated seeds that are decayed, moldy, and obviously dead, should be removed from samples in which saprophytic molds are a problem at the time the initial and subsequent counts are made.

Saprophytic fungi and bacteria on seed surfaces may develop and spread easily in germination tests. Also, parasitic fungi associated with certain lots of seeds may occur on fresh, strong samples with the infection often being clear-cut and confined to individual seeds. Seedlings which decay as a result of infection from an adjacent decayed seed should be regarded as normal. Laboratory practices that will minimize the spread of molds include proper spacing of seeds, control of temperature, removing decayed seeds, proper aeration, and keeping the substratum on the "dry side," yet providing adequate moisture for germination.

8. Nongerminating Seeds

Samples suspected of being dormant should be subjected to specific conditions for breaking dormancy. During the test period dead seeds often decay, become covered with fungi and soft, or the embryos become soft and discolored, whereas dormant seeds are more likely to remain firm and relatively free of mold. If dormant seeds exhibit erratic sprouting, a retest made under conditions calculated to be more nearly optimum for overcoming dormancy may result in uniform germination. Samples with dead seeds are not likely to exhibit such erratic sprouting.

Seeds that remain hard in the test because they have not absorbed water, due to an impermeable seed coat, are to be regarded as hard seeds. Some seeds in samples of many leguminous crops, cotton, okra, and a few others are often impermeable and hard. The percentage of hard seeds occurring in the germination test will vary with age, kind, variety, and moisture content of the samples. The hard seed content of some freshly harvested legumes such as red clover, lespedeza, and field peas may decrease rapidly within the first few weeks or months of dry laboratory storage. Seeds of okra, vetch, and certain other legumes may increase in hard seed content during dry laboratory storage. Nutile and Nutile (1947) found that hard seededness in beans is inherited but the extent to which hard seeds are formed depends to a great extent on the relative humidity and temperature of the air in which they are stored,

Usually a hard seed can be distinguished easily from a swollen seed. In case of doubt, especially when testing large seeds such as beans or peas, swollen seeds when dropped on a hard table top or china plate, will produce a low-pitched thud; whereas, hard seeds will produce a higher-pitched ping.

If at the end of the test period there is doubt as to whether a seed is firm, swollen, or dead, it should not be discarded but left in test for 5 additional days. At the end of that period, the dead seeds have usually decayed and the swollen seeds will have germinated. Grasses such as Sudan grass often have firm, ungerminated seeds left on the substratum at the end of the test period (Weir, 1959). These dormant seeds must not be confused with or reported as hard seeds.

9. METHODS FOR OVERCOMING DORMANCY

Some seed lots of certain species contain seeds that do not germinate under conditions conducive to germination for most seeds of the species under test Seeds that do not germinate when exposed to conditions favorable for germination of the species concerned are referred to as *dormant* seeds. Dormant seeds may appear in most any seed kind, particularly in freshly harvested cereal, legume, and grass seeds and in tree seeds. In germination practice, alternative methods intended to induce these dormant seeds to germinate must be provided. The rules for seed testing of the Association of Official Seed Analysts and of the International Seed Testing Association provide such alternative methods. Some principles upon which these procedures are based are described below and in Chapter 3, Volume II, of this treatise.

a. LOW TEMPERATURES. Seeds are prechilled by placing them on a moist substratum at an indicated low temperature for a specified period of time before placing them at a higher temperature for the duration of the regular test period. The prechilling temperature most frequently used in seed laboratories is 10°C, although occasionally samples will be found which require a temperature as low as 5°C.

This is especially true for some of the native grasses and tree seeds (Forest Service, United States Department of Agriculture, 1948; Crocker and Barton, 1953). Also. certain cereals which may not exhibit deep dormancy can be germinated satisfactorily at 15°C without prechilling. The proponent of this method claims that the prechill period is eliminated by the 15°C temperature which promotes germination of most dormant samples (M. T. Munn, 1946).

b. HIGH TEMPERATURES. Exposure to high temperatures is necessary for overcoming dormancy of some freshly harvested seeds. For example, a constant temperature of 30°C will overcome dormancy in citron seeds. Fresh seeds of other cucurbits, peanuts, and tobacco have also responded favorably to high temperatures. A constant temperature of 35°C is specified in the rules for testing seeds of Alyce clover, although data show that as the seed ages it will respond to lower temperatures (Drake, 1947). Justice and Whitehead (1946) found that holding seeds of *Cyperus rotundus* on a moist substratum at 40°C overcame dormancy.

c. TEMPERATURE COMBINATIONS. Combinations of low and high temperature alternations, with and without light, have promoted prompt germination of some seeds (Crocker and Barton, 1953). Freshly harvested seeds of Canada bluegrass and Colonial bentgrass will germinate more promptly at low–high temperature alternations (such as 10°–30°C or 15°–30°C) than at the commonly used temperatures of 20° to 25°C without light. Combinations of temperatures such as 10°–25°C and 15°–25°C without light are optimum for dormant seeds of some species of fescue.

d. LIGHT. Light is essential for germination of seeds of most grasses and for some vegetable seeds. No distinction is made in the rules between the light requirements of dormant and nondormant seeds. E. H. Toole (1961) and V. K. Toole (1963) showed that the beneficial effects of light are mediated by the temperature at which the moist seeds are held before or after imbibition. In some cases the beneficial effect of light can be bypassed. Recent research indicates that lower light intensities and shorter exposures are more conducive to germination of *Poa pratensis* seed than longer and more intense exposures sometimes used in seed testing. V. K. Toole and Borthwick (1968) found that different ecotypes of *Eragrostis curvula* respond differently to light and that interruption of a 24-hour dark period by light satisfied the light requirement of some seeds of a given lot but inhibited the germination of other seeds. Temperature during the imbibition period markedly affected the germination response of seeds to light.

e. POTASSIUM NITRATE.   Potassium nitrate stimulates germination of certain dormant seeds (Mayer and Poljakoff-Mayber, 1963). A 0.2% solution (2 gm KNO_3 in 1000 ml H_2O) is used in moistening the substratum except that, when testing Kentucky bluegrass and Canada bluegrass seeds, a 0.1% solution is used. Either tap water or distilled water should be used when rewatering the tests.

f. MODIFICATION OF SEED COAT AND OTHER EXTERNAL STRUCTURES. The rules for testing seeds provide for the following practices of initiating germination in the kinds of seed indicated: clipping the seeds of alfilaria, basket flower, and gourds; piercing the seed coat of Alyce clover; degluming seeds of Bahia grass and fescue; removing the pulp from seeds (fruits) of New Zealand spinach; and removing the caryopses of buffel grass from the fascicles. These operations must be done without injury to the embryo and other essential organs. Other kinds of seeds may respond to similar treatments but none has been approved officially.

g. PREWASHING.   When germination is affected by a naturally occurring substance in the seeds which acts as an inhibitor, it may be removed by soaking and washing the seeds in running water before the test is made. Prewashing methods have been approved for several species including *Beta, Citrullus, Tetragonia,* and *Quercus.*

h. PREDRYING.   Drying prior to testing has been very effective in overcoming dormancy in grains with high moisture content. The rules for testing seed provide for drying of wheat and seeds of a few other species as an alternative method.

10. TESTING COATED OR PELLETED SEEDS

The method of pelleting and coating seeds and problems incident to testing such seeds were discussed in Section VI. Whether the inert covering must be removed from the seeds before a germination test is made has not been determined. Insofar as temperature is concerned probably such pellets can be germinated according to the present provisions of the rules, but it is questionable whether artificial substrata can be used successfully. Perhaps a more natural medium, such as sand or soil, will be necessary for proper germination and seedling evaluation. In laboratory tests the inert covering has appeared to retard germination of some seeds.

C. Seedling Evaluation

Germination is defined in the Rules and Regulations under the Federal Seed Act (Anonymous, 1940) as follows: "A seed shall be considered to have germinated when it has developed into a normal seedling. Broken seedlings and weak, malformed, and obviously abnormal seedlings shall not be considered to have germinated." This is in agreement with the definition of germination at the beginning of Section VII. Normal seedlings should have a well-balanced symmetrical growth pattern of all their essential parts. When one part shows stunting or weakness in respect to the growth of another part, some abnormality should be suspected. All seedlings should be allowed to develop to a stage at which the essential plant parts can be determined. Brown and Toole (1931) were among the early workers who showed that certain types of broken and weak seedlings did not produce plants. This work led to additional research (Andersen, 1957) and to inclusion in the rules for seed testing of definitions, examples, and illustrations aimed at assisting the seed technologist in classifying seedlings (Anonymous, 1966, 1970).

1. PROCEDURES FOR CLASSIFYING ABNORMAL SEEDLINGS

In current rules for testing seeds of the Association of Official Seed Analysts, 11 pages are used to describe normal and abnormal seedling types (Anonymous, 1965). In addition, more than 50 available photographs of normal and abnormal seedlings are listed. Some examples of normal and abnormal seedlings are illustrated in Figs. 7 and 8. The International Seed Testing Association establishes in its rules for testing seeds (Anonymous, 1966) four groups of abnormal seedlings, namely: "*(a)* damaged seedlings, *(b)* deformed seedlings, *(c)* decayed seedlings, and *(d)* seedlings with unusual hypocotyl development. *Damaged* seedlings are those without cotyledons; seedlings with constrictions, splits, cracks or lesions which affect the conducting tissues of the epicotyl, hypocotyl or root; seedlings with a primary root if the species is one of which the primary root is an essential structure." Exception to the latter requirement is made for *Pisum, Vicia, Lupinus, Glycine, Arachis, Gossypium,* and *Zea* and all species of Cucurbitaceae. If seedlings of species in these seven genera and of the Cucurbitaceae develop several vigorous secondary roots sufficient to support the seedling in soil, they shall be regarded as normal. According to the International Rules for Seed Testing, *deformed* seedlings include four types:

Those with weak or unbalanced development of the essential structures such as spirally-twisted or stunted plumules, hypocotyl, or epicotyl; swollen shoots and stunted roots; split plumules or split coleoptiles (of Gramineae) without a green leaf; watery and glassy seedlings, or seedlings which do not continue to develop after emergence of the cotyledons. Seedlings are regarded as *decayed* if any of the essential structures are so diseased or deteriorated that normal development is prevented. This interpretation does not apply if it is evident that the cause of infection is from some source other than the seed itself. Seedlings with *unusual hypocotyl development* are those which exhibit hypocotyl development from the micropyle, or radicle development from a seed structure other than the micropyle.

In another section of the International Rules for Seed Testing these four basic groups are explained in terms more meaningful to the seed technologist and categorized into six groups. Wellington (1970) has published detailed descriptions and illustrations of seedlings based on the International Rules for Seed Testing (Anonymous, 1966).

2. CAUSES OF ABNORMAL SEEDLINGS

a. DECLINING VITALITY. Seeds that are aged or have been subjected to unfavorable storage conditions are usually slow to germinate. One or more essential plant parts are frequently stunted or lacking and saprophytic fungi may interfere with the growth of the seedlings.

b. INFECTION WITH PATHOGENIC ORGANISMS. Although seeds infected with certain pathogenic organisms may initiate growth, one or more of the essential seedling structures frequently may be damaged or destroyed by fungi or bacteria. Since the manifestations of disease on the seedlings are largely dependent on environmental conditions during the test period, germination results may be erratic unless the conditions are carefully controlled. The following kinds of seeds frequently present problems in this respect: large-seeded legumes, cereals, sweet clover, alfalfa, cotton, rhubarb, beet, celery, radish, and flax.

c. MECHANICAL INJURY. Mechanical breakage of seeds may occur during the harvesting, threshing, handling, and processing operations, especially processing operations designed to scarify and remove certain accessory seed structures. Injuries of this type are found principally on the following kinds of seeds: legume, rye, flax, onion, sunflower, beets (injured by the shearing process), and seeds of many grasses.

d. INSECT INJURY. Seeds that have been infested with insects may produce seedlings which lack an essential part or structure, or the seedlings may be severely stunted or weakened. The principal problems are weevil damage to sorghum, field peas, and cowpeas; *Bruchus* damage to vetch; chalcid fly damage to alfalfa and red clover; and storage insect damage to cereal seeds (Wheeler and Hill, 1957).

e. CHEMICAL TREATMENTS. Seeds that have been overtreated with toxic fungicides, such as the organic mercurial compounds, commonly produce abnormal seedlings. The symptoms include stunting and thicken-

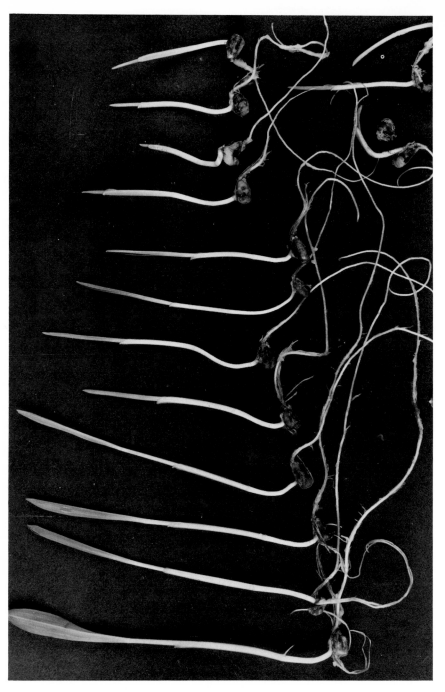

FIG. 7. Seedlings of corn infected with *Gibberella zeae*. Nine seedlings on left are regarded as normal; all others as abnormal and not used in determining percent germination.

FIG. 8. Seeds of garden bean. The three at left are abnormal and would not be regarded as having germinated. From left to right: broken seed with no hypocotyl development; fungal damage to cotyledons, epicotyl, and roots; broken epicotyl; normal seedling.

ing of the roots and hypocotyls when germinated on artificial substrata. Development may or may not be normal in soil, apparently depending on the dosage, types of soil, and possibly other factors. In some observed cases the treatment had permanently killed certain of the essential seed-ling organs. Seeds that have been treated or accidentally subjected to certain chemical pesticides, such as 2,4-dichlorophenoxyacetic acid and phenol compounds, may produce abnormal seedlings or prevent germination (Mayer and Poljakoff-Mayber, 1963). Check tests in soil or sand frequently aid in the evaluation of seedlings exhibiting this type of injury.

f. METAL GERMINATION TRAYS.   Seeds germinated on paper substrata placed directly on metal trays may produce seedlings with shortened, thickened, or discolored radicles. Such abnormalities are apparently caused by soluble zinc compounds from galvanized trays or from copper trays modified with acid during the soldering process (H. L. Munn and Staker, 1942). A sheet of waxed paper placed under the paper substratum will protect the seedlings from such injury. Also, galvanized trays may be protected with lead-free paints.

g. TOXICITY OF SUBSTRATA.   Seedlings grown on or in artificial substrata containing certain toxic materials, such as sulfuric acid or toxic pigments, may exhibit root inhibition or injury. Injury to the roots of timothy seedlings grown on blue blotting paper has been reported by French (1921). Similar symptoms have also been traced to ordinary paper towels from which the sulfur compounds used in processing had not been completely removed.

h. FROST DAMAGE.   Seedlings grown from frost-damaged seeds are especially difficult to evaluate. Growth is initiated in most of the frost-damaged seeds but the resultant seedlings are too weak and/or malformed to produce normal plants. Sand or soil tests are highly desirable, especially on frost-damaged cereals (Justice et al., 1952).

i. MINERAL DEFICIENCY.   Deficiencies of certain minerals in the soil on which seeds are produced will cause specific types of abnormal development of the cotyledons or plumules in some varieties of peas and beans. Manganese deficiency is characterized by appearance of sunken, brown, and slightly pithy areas in the center of the flat surfaces of the cotyledons, and sometimes with browning of the plumules in beans. Peas appear to be particularly susceptible to this type of injury which has also been reported as occurring on broad bean and runner bean (Cuddy, 1959). Boron deficiency has been reported for peas and is characterized by injury to the plumule which may be stunted, multiple branched, or undeveloped. The seed analyst can determine whether the injury is due to lack of boron by germinating the sample with and without added boron. Leggatt (1948) overcame boron deficiency by moistening the germination medium with a 0.01% borax solution or by the application of borax powder.

D. Viability Tests

Much research has been conducted to find an acceptable method by which the germination capacity of seeds can be estimated rapidly, without waiting days for the results of germination tests (Forest Service, United States Department of Agriculture, 1948). Baldwin (1942) reviewed seventeen types of rapid viability tests and classified them into three categories: physical tests, biochemical tests, and physiological tests. At that date (1942), the tetrazolium test was not known in the United States. Moore (1969) briefly reviewed research on biochemical viability tests, emphasizing the topographic tetrazolium test developed by Lakon (1942, 1949). It is evident that a rapid, reliable, viability test would obviate much of the cumbersome space-requiring equipment now used for germination testing, make results available much earlier, and determine the viability of both dormant and nondormant seeds. Only two methods, the topographic tetrazolium test and the embryo excision test have been accepted as official methods.

1. TOPOGRAPHICAL TETRAZOLIUM OR TZ TEST

In this biochemical test, living cells are made visible by reduction of an indicator dye. The indicator used in the TZ test is a colorless tetrazolium salt. Although a number of tetrazolium compounds can be used, 2,3,4-triphenyltetrazolium chloride is preferred by most workers. The colorless tetrazolium is reduced in living cells by action of dehydrogenase enzymes to form a red water-insoluble formazan product. The red color makes it possible to distinguish the living parts of seeds from the colorless dead parts. Treated seeds fall into three groups: *(a)* completely stained, viable seeds; *(b)* completely unstained, nonviable seeds, and *(c)* partially stained seeds. Varying proportions of necrotic tissues occur in different parts of partially stained seeds. Localization and spread of necrosis in the embryo and/or endosperm, not the intensity of color, determine whether such seeds are classified as viable or nonviable.

Theoretically the TZ procedure is good but its practical use in routine testing is mediated by many problems, including: difficulty in staining of some seeds; necessity of cutting or dissecting seeds to permit observation of stained parts; poor agreement with results of germination tests in some cases, especially for seed of low germination capacity; lack of uniform interpretation of staining and difficulty in interpreting the significance of different degrees of staining; and an increase in man-hours required to test 200 or 400 seeds compared to regular germination tests. According to the originator of the TZ test, Lakon (1949) and others (Delouche *et al.*, 1962; Moore, 1962; Grabe, 1970) many of these difficulties have been overcome. However, the test has not been accepted generally for routine testing of seeds. Its use has been limited to dormant seeds of woody plants which do not germinate by conventional methods. The Association of Official Seed Analysts has approved the TZ test for use on dormant seeds of *Fraxinus* spp., *Malus* spp., *Prunus* spp., *Pyrus* spp., and *Pinus cembra,* whereas the International Seed Testing Association has approved the test for use on *Chamaecyparis thyoides, Taxodium distichum,* nine species of *Pinus,* and all species of seventeen additional genera.

2. EMBRYO EXCISION METHOD

By this method, embryos are excised from the seeds, in accordance with the methods of Flemion (1938) and Heit (1955) and then placed on moist filter paper or blotter discs in petri dishes. The tests are placed at ordinary room temperature and light intensity; however, the temperature should not exceed 24°C. The embryos should germinate within a few days to 2 weeks. Viable nongerminated seeds can be easily distinguished from dead seeds after 2 weeks. This test has been adopted only by the Association of Official Seed Analysts and approved for seeds of eight species of *Pinus,* three other named species, and all species of five genera.

In a comprehensive study, Schubert (1965) compared the excised embryo method with the TZ method for determining the viability of dormant tree seeds. He concluded that the TZ method should receive preference over the embryo excision method but that improvements in the TZ test should be made by providing for the use of bacteriacides and stronger reducing solutions to resolve doubts in weakly stained tissues. Both the TZ test and the embryo excision test have served as useful tools for scientists working on seed research problems.

VIII. Special Tests

A. Seed Vigor Tests

A number of methods proposed for use in determining seed vigor have been reviewed in Chapter 6, Volume I, of this treatise. None of these methods has been accepted generally for routine use.

Some workers have claimed that seedling evaluation as described in Section VII,C,1 above constitutes vigor evaluation. When considered on a functional basis this viewpoint appears logical. However, seed technologists do not regard such seedling evaluation as seed or seedling vigor. The so-called cold test for corn (Isely, 1950; Clark, 1954; Heydecker, 1969) has received much attention and has proved useful for evaluating commercial seed lots. Briefly, the method consists in planting the seeds in unsterilized soil or a mixture of soil and sand, with high moisture content, at a temperature of approximately 10°C for 5 to 7 days, and then transferring the tests to a temperature of approximately 30°C for completion. Some laboratories use loam soil from a cornfield without further inoculation; whereas, in other laboratories the soil is inoculated with ground seeds which previously failed to germinate. The difficulty of standardizing such a test is apparent, and its future use as a laboratory method for routine testing is unpredictable.

No definition of seed vigor has yet been proposed which is generally acceptable to workers in this field, nor a practical and accurate method developed amenable to standardization. Merchants, seed technologists, and others agree that this is an area of first research priority (Woodstock and Niffenegger, 1971).

B. Detection of Seed Treatments

A bioassay for the detection of fungicides involves placing the seeds on a substrate previously inoculated with the spores of a sensitive fungus, such as *Glomerella cingulata,* or of a bacterium, such as *Bacillus subtilis.* The spores in the substrate germinate and the organism usually spreads rapidly. While the organism is growing the fungicide of the treated seeds diffuses into the substrate. After a suitable incubation time, usually 20–48 hours, a clear zone develops around each seed treated with an effective fungicide, indicating that the test organism has been inhibited. Generally, the greater the amount of fungicide on the seed, the larger the area of inhibition. Seed treated with various fungicides usually produce different amounts of inhibition, depending on the chemical involved.

Effectiveness of a seed treatment material containing captan, mercury, or thiram is determined by comparing a test sample with an untreated control sample of the same species for number of clear zones and total area of inhibition. A sample is considered effectively treated when tested by this method and found to produce at least 90% as many inhibition zones as the untreated control sample and a total area of inhibition at least 33% that of the untreated control (Kulik and Crosier, 1964).

Simon and Kulik (1971) modified an existing resorcinol colorimetric analytical method for use in detecting the fungicide captan on individual seeds. The procedure involves extraction with benzene, passage of the benzene solution through a Florisil column to remove interfering dyes, reacting the solution with a resorcinol solution for color development, and determining color intensity with a colorimeter.

C. Test for Seed-Borne Fungi

It can safely be stated that the average seed lot in commercial channels in the United States is sold without regard to the seed-borne disease organisms which it may carry. No doubt, agricultural production would be greatly increased if the farmer were aware of the sanitary conditions of the seed he plants. By sanitary condition is meant the kinds and relative occurrence of seed-borne fungi, bacteria, viruses, eelworms, and insects which cause disease and injury.

Some seed lots showing high germination when tested by official methods are practically valueless when planted under certain weather and soil conditions because they are infected with disease organisms which will eventually destroy the seedlings, and plants, or otherwise reduce yield. *Helminthosporium* on cereals is usually not detected in the regular germination test, and a high percentage of germination may be reported for infected samples. However, if the germination temperature is increased from 20° to 25° or 30°C and the seedlings are left in test for 14 or 15 days, infected samples may appear as worthless (Justice *et al.*, 1952).

Very few seed testing laboratories in the United States make any attempt to test for presence of injury or seed-borne diseases. Perhaps, this situation is due primarily to a paucity of methods applicable to routine seed testing by which the specific organisms can be identified. Although research in this phase of seed testing has lagged, it must be recognized that the development of practical methods will not be an easy task. This should not act as a deterrent to research but as a challenge, especially for those persons who recognize the importance of the problem.

In Europe many seed testing stations have facilities and personnel for carrying out seed health tests. The International Rules for Testing Seeds contain 4 pages on methods applicable to seed health testing; whereas, the rules of the Association of Official Seed Analysts do not mention seed health. The International Rules are used in Europe when testing for domestic purposes as well as for testing seed intended for foreign commerce. At present, a moderate research program on seed health testing is being carried out in the United States and in some European countries, especially The Netherlands. Some practical methods useful in seed health testing, assembled by A. Anderson, are given in *U.S. Department of Agriculture Handbook No. 30* (Justice *et al.*, 1952). For recent contributions the reader is referred to papers by de Tempe (1961), Malone and Muskett (1964), Limonard (1968), Noble and Richardson (1968), and A. N. Smith and Crosier (1965).

D. *Testing for Moisture Content*

Methods for determining moisture content of seeds may be roughly classified into *(a)* basic methods in which the moisture is driven out of the seeds by heat and measured by the loss of weight of the original material, or the weight or volume of the condensed moisture, and *(b)* practical methods designed for rapid routine work and standardized against one or more of the basic methods (Zeleny, 1961). Probably all the moisture cannot be driven out of seeds without driving out small amounts of other volatile constituents or causing chemical changes in the material which would result in weight changes. In applying any method, therefore, it is necessary to adhere closely to the prescribed procedure in order that the results of all tests made by that method will be comparable.

The Association of Official Seed Analysts has not published official methods for testing seeds for moisture content. However, the International Seed Testing Association has adopted rules covering most crop species commonly tested. These rules include only three procedures: the air oven 130°C method, the air oven 105°C method, and the toluene distillation method.

1. THE 130°C AIR OVEN METHOD

The air within the oven is at atmospheric pressure and circulated by convection or mechanical means. A temperature of 130°C and a heating time of 1 hour are specified for most kinds of seeds. The loss of weight that occurs during drying, calculated on a percentage basis, is taken to be the percentage of moisture in the seed before drying. Large seeds, such as grains, beans, and peas, must be ground in order to provide rapid penetration of heat and ready escape of moisture. A two-stage procedure is used on large seeds which are too wet to be ground easily without losing moisture in the grinding process. A weighed portion of the seeds is partly dried by exposing it to air in a warm place. The loss of weight in this preliminary drying is determined. The partly dried sample is then ground and its moisture content is determined in the usual manner. The moisture lost in both stages of the procedure must be considered in calculating the moisture content of the original seeds. Seeds of a high oil content usually should not be ground for oven moisture determinations because they are difficult to grind properly and because oxidation of the oil during drying may result in a gain in weight of the oil. Oxidation of oil is a particularly serious consideration in seeds that contain oils of high iodine number (i.e., "drying" oils) (Zeleny, 1954). Certain seeds contain constituents other than moisture that are volatile at 130°C; such seeds cannot be subjected to that temperature in the determination of moisture without introducing errors in the determination (Hart *et al.,* 1959).

2. THE 105°C AIR OVEN METHOD

This method differs from the above method in that a drying temperature of 105°C is used and the drying time is 16 hours. Seeds of the following crop species are to be tested by this method: shallot, onion, leek, garlic, carob, soybean, radish, peppers, eggplant, and fenugreek (Anonymous, 1966).

3. Toluene Distillation Method

A weighed portion of the finely ground seed is boiled in toluene in an apparatus that condenses the volatilized materials, collects the condensed water in a tube, and returns the condensed toluene to the boiling flask. The boiling is continued as long as any water accumulates in the tube provided for that purpose, and the moisture in the seed is calculated from the volume of water condensed (Zeleny, 1954). This method has the advantage that no water-insoluble volatile matter can be measured as moisture. Difficulty is sometimes encountered in reading accurately the volume of water distilled, because the separation between the toluene and water may not be sharp. The following tree seeds should be tested by this method: fir, cedar, beech, spruce, pine, and hemlock (Anonymous, 1966). Ordinarily the toluene method is regarded as a basic method along with the Karl Fischer method which is not described here. These methods obviate the measurement of nonwater volatiles as water but are not practical for routine work.

E. Testing for Trueness to Variety

According to Davidson and Clark (1961), 2640 names were used for 600 varieties of wheat, oats, barley, and rye in 1940. Confusion of varietal names in trade channels was especially bad following the introduction of hybrid corn. Only a forceful and cooperative law enforcement program between the U.S. Department of Agriculture and the various states avoided a similar condition in the United States when hybrid sorghum varieties were released. The goal in testing for variety is to determine if the seed in question possesses the characteristics of the variety represented. Determination of trueness to variety is complicated by the influence of environmental factors. Consequently, methods for field testing must be based on diagnostic characteristics which are not affected by environmental influences (Hawkins et al., 1964). In some kinds, such as beans, seed characteristics are stable, despite changes in environment.

Growth chambers, in which light, temperature, and humidity can be controlled, are used for testing trueness to variety of some forage crops and vegetables. When control of environment is not possible, as in field trials, seeds of known varieties are planted for comparison. If the seeds under test are of the same variety as those of the authentic samples, they will be influenced by the environment in the same way as the authentic samples and reliable determinations can be made in spite of environmental influences (Davidson and Clark, 1961). Three types of tests have proved useful for variety testing: (a) laboratory tests, (b) greenhouse and growth chamber tests, and (c) field tests.

1. LABORATORY TESTS

Relatively few kinds of seed can be tested for trueness to variety in the ordinary seed laboratory, but some determinations are possible. Offtypes in beans, soybeans, and peas can often be detected by the color or shape of the seed. Common Kentucky bluegrass seeds can be distinguished from seeds of the Merion variety by microscopic examination. Some offtypes in oats, wheat, and barley can be distinguished by the color and shape of the seeds (Radersma, 1964). Seedlings of red and white varieties of beets and of red and green varieties of cabbage can be distinguished readily in germination tests. The seedlings of the red varieties of beets and cabbage contain red pigment that is lacking in the white varieties of beets and green varieties of cabbage (Olsen, 1964). A 1% solution of phenol has been used to identify varieties of wheat and barley (Chmelar and Mostovoj, 1938). Results are based on color changes after 12 or more hours of soaking, followed by drying.

Some white and yellow oat varieties can be separated through the use of ultraviolet light. Yellow oats emit a dark bronze fluorescence. The difference can be observed even when the glumes have become so discolored through weathering that determination of their color under ordinary light is difficult (Finkner *et al.,* 1954). Ultraviolet light can be used also to detect certain offtypes of soybeans and peas and is useful in determining the percentage of seeds of annual and perennial ryegrass in mixtures (see Section VI,C). A few varieties of peas and soybeans produce seedlings the roots of which exude a fluorescent material. The fluorescent substance produced by the roots of these seedlings can be detected under ultraviolet light (Isely, 1956).

2. GROWTH CHAMBER AND GREENHOUSE TESTS

The development of better light conditions for growth chambers has opened a new field for testing. In addition, control over other environmental factors permits conditions that produce maximum differences in seedlings or plants of different varieties (Bass, 1959). The growing of seedlings under controlled environment in a greenhouse or a growth chamber offers special promise for testing forage crop seeds for varietal trueness. Usually, the plants lack diagnostic characteristics, but the seedlings and plants may be responsive to variations in temperature, day length, and quality of light (Clark, 1952; Isely, 1956). Seedlings of northern varieties of alfalfa, for instance, have short stems when grown at low temperatures with short photoperiods, whereas seedlings of southern varieties produce longer stems under the same conditions. This difference provides a way to distinguish between the two. Under similar conditions,

Empire birdsfoot trefoil has short stems, and European-type birdsfoot trefoil has long stems. The same type of technique can be applied to other kinds of plants (Davidson and Clark, 1961).

The controlled environment technique can be used to measure the resistance of seedlings to disease. A test of this type has been used satisfactorily for detecting varietal admixture in wheat and oats. Seedlings are inoculated with a specific race of rust and are observed to determine whether they show the resistant or susceptible type of reaction of the race being tested. A susceptible-type reaction in a seedling which supposedly is a resistant variety is evidence of varietal admixture (Nittler, 1958). Similar techniques can be used to determine wilt resistance in seedlings of alfalfa varieties, yellow resistance in cabbage, mildew resistance in soybean, and resistance to other diseases in various crop species.

3. FIELD TRIALS

The most common test of trueness to variety is the field trial. This test can be used for almost any kind of seed without the development of special techniques. Also, because plants are usually grown to maturity in the field trials, a full range of varietal characteristics can be observed. The cost of conducting field trials is a major disadvantage. Also, during seasons when the weather is unfavorable usual varietal characteristics may not be expressed or the trials may fail completely. Probably the most serious shortcoming of a field trial is that it cannot be conducted in advance of the normal planting season. Ordinarily this would not permit obtaining the test results before the seeds are placed on the market (Clark, 1952). Authentic seed samples are necessary for comparison with the samples under test. These samples should be planted at close intervals among the samples for which information is sought. It is necessary that enough plants be grown to permit detection of admixtures with a degree of precision consistent with the purpose of the trial. Large populations are required to detect offtypes with a low rate of occurrence (Davidson and Clark, 1961).

IX. Seed Identification

Seed testing is predicated on the assumption that the seeds in a sample can be correctly identified to the species level. Even the species under which a sample is submitted sometimes causes trouble due to the presence of atypical seeds. The same principle holds true for weed seeds, but perhaps to a greater extent. Quite frequently seed technologists are faced with the necessity of identifying crop and weed seeds from producing areas completely new to them. For example, vetch seeds from the Eastern

Mediterranean area have presented difficult identification problems in the United States. Frequently weed seeds are found of which the usual identification characteristics have been obliterated by harvesting, processing, and handling (Justice, 1960).

Notwithstanding the importance of seed identification the subject cannot be covered adequately in a treatment of this size and nature. Therefore, the reader will find it necessary to resort to references which treat the subject in detail. In laboratory practice the experienced seed technologist will readily recognize most of the seeds encountered, but how does he proceed to identify an unknown seed? First, an experienced worker would be expected to know the principal identification characteristics of the major plant groups such as families or orders. Thus, he should be able to place the seed in a family or group of families. The next logical procedure is to check the available keys, descriptions, and illustrations. Regardless of the success with references, he should then consult a seed collection and compare the unknown seed with correctly labeled samples. The library of the modern seed testing station should contain books on seed identification, seed production, agronomy, weeds, and botany from the major seed-producing areas which the station serves. Of these publications, those on seed identification are most useful to the seed technologist.

Of the different people in the United States who have contributed to our knowledge on seed identification, the name of F. H. Hillman stands out foremost. His early work in Nevada (Hillman, 1897) led to an appointment in the Federal Seed Laboratory, Washington, D.C. In 1918, he and H. H. Henry prepared a comprehensive treatise on crop and weed seeds under the title "The More Important Forage Crop Seeds and Incidental Seeds Commonly Found with Them" (not citable). This 15-plate set of illustrations was reissued in 1935 and still can be purchased from the U.S. Department of Agriculture (Office of Information, Washington, D.C.). Some or all of these plates have been reproduced and published in many countries. With respect to early contributions, "Seeds of Michigan Weeds" (Beal, 1910) and "Maryland Weeds and Other Harmful Plants" (J. B. S. Norton, 1911) should not be overlooked.

Two comprehensive and notable treatments of crop and weed seeds by A. F. Musil have appeared within the past 20 years. Both consist of keys, descriptions, and illustrations. One is a chapter in the *U.S. Department of Agriculture Handbook No. 30* (Justice *et al.*, 1952) and the other is *U.S. Department of Agriculture Handbook No. 219* (A. F. Musil, 1963). Seed publications of a comprehensive nature from countries other than the United States include the following: *(a)* Europe—Korsmo's (1935) colored plates of 306 species of weed seeds, Beijerinck's (1947) seed atlas with 140 plates, and Brouwer and Stählins's (1955) handbook; *(b)* New Zealand — Hyde's (no date) illustrations of weed seeds; *(c)* Asia — Fong's (1969) "Horticultural and Vegetable Seeds in Malaysia."

For further information on seed identification the reader is referred to Chapter 2 of this volume and to bibliographies in the *U.S. Department of Agriculture Handbook No. 30* (Justice *et al.,* 1952) and *No. 219* (A. F. Musil, 1963).

X. Tolerances for Testing Seeds

A. Nature of Tolerances

Ordinarily the exact percentages of germination, pure seed, or other quality factors cannot be determined. The application of appropriate statistical methods to test results permits estimation of the quality of seed lots within calculated ranges. The amount of allowable deviation from a standard or the allowable difference between test results is called tolerance. It represents the expected variation resulting from incomplete mixing of the seeds, variations in sampling, and uncontrolled differences in the application of testing procedures. A tolerance, or expected variation, is expressed in terms of a probability and the amount of tolerance. Probabilities are expressed as percentages (1% level) or as decimals (0.01). These expressions mean that the result of a subsequent test has a chance of about 1 out of 100 of exceeding the tolerance. Tolerances computed at the 1% level are greater than those computed at the 5% level. In some kinds of work, tolerances that may be exceeded as often as once in twenty trials may be satisfactory; whereas, other types of work require a greater degree of confidence (Justice and Houseman, 1961).

If a seed lot has been reasonably well mixed, a sample drawn at random, and a proper test made, a single test will indicate that in a specified number of cases, say 95 in 100 or 99 in 100, the true value of the lot is no more than one tolerance range removed from the results of the test. The probability statement allows for an occasional test result to exceed the tolerance range. The number of these exceptions is indicated in the probability statement. When test results do exceed the tolerance range, additional tests should be made in an attempt to determine the cause of the excessive variations. The magnitude of tolerances at any given probability level will depend on the percentage of seed component in the sample for which the tolerance is desired, the variations associated with testing procedures, the characteristics of the seed, and the size of sample tested. It also will depend on whether tolerances are intended to cover the differences between two tests or one test and a predetermined standard. Certain basic principles must prevail if the tolerances are to be used properly. The seed lot from which the sample is drawn should be relatively homogeneous. The sample must be drawn in a random manner from a sufficient number of containers or locations in the lot. Bias must be avoided insofar as possible in conducting tests. Random sampling assumes that each seed or particle in the seed lot has an equal chance of being drawn and that no selection of any type is exercised (Leggatt, 1939).

B. Tolerances for Specific Tests

1. PURITY TOLERANCES

Rodewald (1904) showed that variation in results of purity and germination tests agreed well with theoretical expectations. He recognized several sources of error that led to variations in test results. These included difference in technique, change in the material being tested, accidents, and personal factors. The errors were classified as systematic errors or accidental errors. In testing seeds of orchard grass for purity, he found the total error to be twice the accidental error; whereas, the total error was only one-quarter as great as the accidental error for red clover seed. Stevens (1918), using seeds of other crop species and working independently, reached essentially the same conclusions as Rodewald.

C. P. Smith (1916) proposed arbitrary tolerances for percentages of pure seed. His formula, adopted by the Association of Official Seed Analysts in 1917, was based on the premise that the sample was composed of the component under consideration and the sum of all other components. His original formula was simplified to T (tolerance) $= 0.2 + 20\%$ of the lesser part divided by 100, the lesser part meaning the component under consideration or the sum of all other components, whichever is smaller.

Collins (1929) proposed formulas for calculating purity and germination tolerances on the basis of the binomial distribution and for calculating noxious-weed seed tolerances on the basis of the Poisson distribution. Although Collins' formula for purity analysis was sounder than the one proposed by Smith, it was not adopted by the seed analysts' organizations.

From the 1930s to 1960 the Association of Official Seed Analysts and the International Seed Testing Association used the following formula for calculating pure seed tolerances:

$$T = 0.6 + 0.2 \left(\frac{a \times b}{100}\right)$$

Where a equals the percentage of the component under consideration and b equals $100 - a$. Tolerances for other crop seed, weed seed, and inert matter were calculated by the same formula, except that 0.2 was substituted for 0.6. Wider tolerances are required for chaffy grass seeds which do not blend so well. These tolerances were obtained by adding to the tolerances calculated by the above formulas, an additional tolerance obtained by multiplying the lesser of a and b by the regular tolerance, and dividing by 100.

Purity tolerances currently used were accepted by the Association of Official Seed Analysts in 1960 and by the International Seed Testing Association in 1965. Through cooperative research with twenty-one seed-testing stations, Miles *et al.* (1958) measured the variations between different seed bags, different probes from the same bag, different test samples from the submitted sample, different analysts, and day-to-day variation of the same analyst.

The formula derived from these studies is as follows:

$$T \text{ (tolerance)} = 1.414t[(B^2/n)(N-n)/N + C^2/n + W^2/n + A^2/n + I^2/n'^{1/2}$$

In the formula, $t = $ a factor corresponding to the desired probability level (the actual factors used were 5% probability level -1.65; 1% level -2.33; 0.1% level -3.09); $B = $ component of variation due to differences among bags; $C = $ component of variation due to differences among cores or probes within bags; $W = $ variations among working samples taken from the same submitted sample; $A = $ component of variation resulting from different analysts testing samples differently; $I = $ a component of variation arising from the fact that the same analyst may test the same sample differently from day to day; $N = $ number of bags in lot sampled; and $n = $ number of units of source of variation shown in same term of the equation.

Appropriate values for the different components of variance are inserted in the formula for computation of regular tolerances for either nonchaffy seeds or chaffy seeds. The same tolerance is applied to a given percentage, regardless of whether it refers to pure seeds, other crop seeds, weed seeds, or inert matter.

Mixtures consisting of kinds having unequal seed weights present additional problems. Miles and co-workers developed formulas and prepared tables of special tolerances for a number of different particle-weight ratios applicable to seed mixtures. Most of these tolerances are wider than those for unmixed seeds, although under certain circumstances they may be narrower. Miles (1963) published numerous tables, based on his formulas, for the convenience of workers.

2. GERMINATION TOLERANCES

Early germination tolerances were developed without the benefit of statistical theory or experimentation. The tolerance limits at the various percentage levels were determined through practice. This was done by comparing the results of tests on replicate samples from the same seed lot. The tolerances thus determined are somewhat greater than tolerances calculated from statistical theory at the 5% probability level. The tolerances used by the Association of Official Seed Analysts range between 5% for germination values of 95% and above to 10% for germination values below 60%. Other tolerances are 9, 8, 7, and 6%, respectively, for germination ranges of 60 to 69%, 70 to 79%, 80 to 89%, and 90 to 95%. Statistically, the tolerance of 10% for all germinations below 60% and the tolerance of 5% for germination above 95% are unrealistic. When the problems of testing are considered, there appears to be some justification for wider tolerances at the lower germination levels.

In 1965 the International Seed Testing Association adopted statistical tolerances for germination. These tolerances, calculated by Miles (1963), range from 2 to 11% for germination values between 2 and 50%, and tolerances of 2 to 11% for germination values of 99 to 51%, calculated at the 5% probability level.

3. TOLERANCES FOR NOXIOUS-WEED SEED EXAMINATION

Examinations for noxious-weed seeds, such as dodder and quackgrass, aim to determine the rate of occurrence of these weed seeds per ounce or per pound of crop seeds. Since the number of noxious-weed seeds usually is low in a sample, a statistical theory known as the Poisson distribution must be used (Collins, 1929; Przyborowski and Wilénski, 1935).

The reliability of a test for noxious-weed seeds is related primarily to the number of noxious seeds found in a sample. However, testing procedures based on the rate of occurrence of the noxious-weed seed would not be practical. Uniform sample sizes by weight, therefore, have been established for the various kinds of agricultural and vegetable seeds. Except for a few large-seeded kinds, the sample size for noxious-weed seed tests is at least 10 times that for purity analyses.

Noxious-weed seed tolerances are calculated from the formula:

$$Y = X + 1 + 1.96\sqrt{X}$$

Where X is the number of seeds labeled or represented and Y is the maximum number within tolerance of X (Anonymous, 1970).

Some examples of tolerances computed by this formula are listed in Table II.

4. OTHER TOLERANCES

Tolerances are available for a number of other types of tests including the following: fluorescence test, trueness-to-variety test, pure seed determination based on counted number of seeds, and pure live seeds. Statistical tables for all these situations are found in the Rules for Testing Seeds of the Association of Official Seed Analysts and in the International Rules for Seed Testing (Anonymous, 1966, 1970).

TABLE II

EXAMPLE OF COMPUTED TOLERANCES FOR NOXIOUS-WEED SEEDS

No. labeled or represented (X column)	Maximum No. within tolerance (Y column)
0	2
1	4
2	6
3	8
4	9
5	11
6	12
7	13
8	14
9	16
10	17

(Courtesy of U. S. Department of Agriculture.)

C. Discussion

Except for the germination tolerances used by the Association of Official Seed Analysts, all tolerances currently used are based on statistical theory. Purity tolerances applicable to a wide variety of situations are essential. Seed lots of chaffy grasses are more heterogeneous than lots of nonchaffy seeds. Hence, wider tolerances must be provided for chaffy seeds. Tolerances for seed mixtures, chaffy and nonchaffy, are different from those used for unmixed seed lots.

In some cases the results of a purity analysis or germination test may be compared with a *fixed standard*. In such cases a single test is subject to variation. In other cases a test is compared with a claim made on a label or business document. Since the claimed percent of pure seed or germination is assumed to have been based on a single test, two variables must be taken into account. Tolerances for two variables are greater than those used to compare a test against a fixed standard (Leggatt, 1939).

REFERENCES

Allen, C. E. (1949). A comparison of two types of crested wheatgrass pellets. *Proc. Ass. Off. Seed Anal.* **39**, 73.

Andersen, A. M. (1957). Evaluation of normal and questionable seedlings of species of Melilotus, Lotus, Trifolium and Medicago by greenhouse tests. *Proc. Int. Seed Test. Ass.* **22**, 1.

Anonymous. (1931). International rules for seed testing. *Proc. Int. Seed Test. Ass.* **3**, 314.

Anonymous. (1940). Rules and Regulations under the Federal Seed Act. Service and Regulatory Announcement No. 156, p. 9.

Anonymous. (1965). Rules for testing seeds. *Proc. Ass. Off. Seed. Anal.* **54,** No. 2, 1.

Anonymous. (1966). International rules for seed testing. *Proc. Int. Seed Test. Ass.* **31,** 1.

Anonymous. (1970). Rules for testing seeds. *Proc. Ass. Off. Seed. Anal.* **60,** No. 2, 1.

Baldwin, H. I. (1942). The determination of seed viability without germination. *In* "Forest Tree Seed," p. 169. Chronica Botanica, Waltham, Massachusetts.

Barton, L. V. (1967). "Bibliography of Seeds." Columbia Univ. Press, New York.

Bass, L. N. (1950). Effect of wave length bands of filtered light on germination of seeds of Kentucky bluegrass (*Poa pratensis*). *Iowa Acad. Sci.* **57,** 61.

Bass, L. N. (1959). Uniformity trials of a 15–30°C walk-in room type germinator. *Proc. Ass. Off. Seed Anal.* **49,** 119.

Beal, W. J. (1910). Seeds of Michigan weeds. *Mich., Agr. Exp. Sta., Bull.* **260,** 104.

Beijerinck, W. (1947). "Zadenatlas der Nederlandsche Flora." Veenman & Zonen. Wageningen.

Boerner, E. G. (1915). A device for sampling grain, seeds and other material. *U.S., Dep. Agr., Bull.* **287.**

Brouwer, W., and Stählin, A. (1955). "Handbuch der samenkunde für Landwirtschaft, Gartenbau und Gorstwirtschaft." DLG-Verlag-GMBH, Frankfurt am Main.

Brown, E. (1941). A brief history of seed testing in the United States Department of Agriculture. *In* "History of the Association of Official Seed Analysts," p. 32.

Brown, E., and Goss, W. L. (1912). The germination of packeted vegetable seeds. *U.S., Dep. Agr., Bur. Plant Ind., Circ.* **101.**

Brown, E., and Toole, E. H. (1931). The evaluation of seed tests. *Proc. Int. Seed Test. Ass.* **3,** 203.

Brown, E., and Toole, E. H. (1934). The purpose of seed testing. *Proc. Int. Seed Test. Ass.* **6,** 272.

Burchard, O. (1893). The object and methods of seed investigation and the establishment of seed-control stations. *Exp. Sta. Rec.* **4,** 793 and 882.

Caldwell, B. (1941). Mottled seed in sweetclover. *Ass. Off. Seed Anal. Newslett.* **15,** No. 1, 3.

Carter, A. S. (1961). In testing, the sample is all important. *Yearb. Agr. (U.S. Dep. Agr.)* p. 414.

Ching, T. M., and Jensen, L. A. (1957). Determination of inert matter in multiple florets of western grown fine fescues. *Proc. Ass. Off. Seed Anal.* **47,** 61.

Chmelar, F., and Mostovj, K. (1938). On the application of some old and on the introduction of new methods for testing genuineness of variety in the laboratory. *Proc. Int. Seed Test. Ass.* **10,** 68.

Clark, B. E. (1942). Comparative laboratory and field germination of onion seed. *Proc. Ass. Off. Seed Anal.* **34,** 90.

Clark, B. E. (1952). Testing forage crop seeds for trueness-to-type. *Proc. Ass. Off. Seed Anal.* **42,** 49.

Clark, B. E. (1954). Factors affecting the germination of sweet corn in low-temperature laboratory tests. *N.Y., Agr. Exp. Sta., Ithaca, Bull.* No. **769,** 1.

Clark, B. E. (1961). Ways and means of improving the uniformity of seed test results from one laboratory to another. *Proc. Ass. Off. Seed Anal.* **51,** 76.

Collins, G. N. (1929). The application of statistical methods to seed testing. *U.S., Dep. Agr., Circ.* **79,** 1.

Courtney, W. D., and Howell, H. B. (1952). Investigations on the bent grass nematode, *Anguina agrostis. Plant Dis. Rep.* **36,** 75.

Crocker, W., and Barton, L. V. (1953). Factors affecting germination, I. *In* "Physiology of Seeds," p. 87. Chronica Botanica, Waltham, Massachusetts.

Crosier, W., and Patrick, S. (1952). Some sclerotia-forming fungi in commercial seed stocks. *Proc. Ass. Off. Seed Anal.* **42,** 114.

Cuddy, T. F. (1959). Marsh spot of peas. *Proc. Ass. Off. Seed Anal.* **49**, 156.

Cull, H., and Justice, O. L. (1950). Classification of insect-infested seeds of legumes. *Ass. Off. Seed Anal. News Lett.* **24**, No. 3, 28.

Davidson, W. A., and Clark, B. E. (1961). How we try to measure trueness to variety. *Yearb. Agr. (U.S. Dep. Agr.)* p. 448.

Delouche, J. C., Still, T. W., Raspet, M., and Lienhard, M. (1962). The tetrazolium test for seed viability. *Miss., Agr. Exp. Sta., Tech. Bull.* **51**, 1.

de Tempe, J. (1961). Routine methods for determining the health condition of seeds in the seed testing station. *Proc. Int. Seed Test. Ass.* **26**, 27.

Drake, V. C. (1947). Effect of temperature and aging on the germination of Alyce clover (*Alysicarpus vaginalis* (L.) D. C.). *Proc. Ass. Off. Seed Anal.* **37**, 143.

Evenari, M. (1965). Physiology of seed dormancy, after-ripening and germination. *Proc. Int. Seed Test. Ass.* **30**, 49.

Everson, L. E. (1952). The germination of mature and immature seeds of buckhorn plantain. *Proc. Ass. Off. Seed Anal.* **42**, 83.

Everson, L. E. (1954). The germination of mature and immature seeds of quackgrass (*Agropyron repens*). *Proc. Ass. Off. Seed Anal.* **44**, 127.

Everson, L. E., Shih, C. S., and Cady, F. B. (1962). A comparison of the "hand" and "uniform" methods for the purity analysis of Kentucky bluegrass (*Poa pratensis*) seed. *Proc. Int. Seed Test. Ass.* **27**, 476.

Finkner, R. E., Murphy, H. C., Atkins, R. E., and West, D. W. (1954). Seed fluorescence in oats. *Proc. Ass. Off. Seed Anal.* **44**, 202.

Flemion, F. (1938). A rapid method for determining the viability of dormant seeds. *Boyce Thompson Inst. Plant Res.,* **9**, 339.

Fong, C. H. (1969). "Agricultural and Horticultural Seeds in Malaysia." College of Agriculture, Malaya.

Forest Service, United States Department of Agriculture. (1948). Woody plant seed manual. *U.S., Dep. Agr., Misc. Publ.* No. **654** (under revision, 1970).

French, G. T. (1921). The effect of commercial blue blotting paper on the germination of timothy seed. *Proc. Ass. Off. Seed Anal.* **12**, 55.

Gaertner, E. E. (1950). Studies of seed germination, seed identification, and host relationships in dodders, *Cuscuta* spp. *Cornell Univ., Agr. Exp. Sta., Mem.* **294**, 1.

Galloway, B. T. (1909). The adulteration and misbranding of the seeds of alfalfa, red clover, orchard grass, and Kentucky bluegrass. *U.S., Dep. Agr., Circ.* **28**.

Gentner, G. (1929). Über die Verwendbarkeit von ultravioletten Strahlen bei der Samenprufüng. *Prakt. Bl. Pflanzenbau Pflanzenschutz* [N.F.] **6**, 166.

Grabe, D. F. (1970). Tetrazolium testing handbook for agricultural seeds. *Ass. Off. Seed Anal., Handb. Contrib.* No. 29, p. 62.

Hardin, E. E., Copeland, L. O., and Knudson, L. A. (1965). A comparison of the relative effectiveness of the Boerner divider and several techniques of using the Gamet precision divider. *Proc. Ass. Off. Seed Anal.* **55**, 140.

Hart, J. R., Feinstein, L., and Golumbic, C. (1959). Oven methods for precise measurement of moisture in seeds. *U.S., Agr., Mkt. Serv.* [*AMS Ser.*] No. **304**, 4.

Hawkins, R. P., Horne, F. R., and Kelly, A. F. (1964). Identifying cultivars of grass and clover. *Proc. Int. Seed Test. Ass.* **29**, 837.

Heit, C. E. (1955). The excised embryo method for testing germination quality of dormant seed. *Proc. Ass. Off. Seed Anal.* **45**, 108.

Heit, C. E., and Munn, M. T. (1952). Observations on coated and pelleted flower and vegetable seeds. *Proc. Ass. Off. Seed Anal.* **42**, 141.

Heydecker, W. (1962). From seed to seedling: Factors affecting the establishment of vegetable crops. *Ann. Appl. Biol.* **50**, 662.

Heydecker, W. (1969). The "vigour" of seeds—a review. *Proc. Int. Seed Test. Ass.* **34,** 201.

Higgins, E. C., Elliott, G. A., and Stevens, O. A. (1961). Historical sketches of state laboratories. *In* "History of the Association of Official Seed Analysts. *Ass. Off. Seed Anal.* (1940–1959)," p. 11.

Hillman, F. H. (1897). Nevada and other weed seeds. *Nev., Agr. Exp. Sta., Bull.* No. **38,** 1.

Hooker, D. (1942). Wild onion bulblets. *Proc. Ass. Off. Seed Anal.* **34,** 49.

Hyde, E. O. C. (no date). Weed seeds in agricultural seed. *N. Z. Dep. Agr., Bull.* No. **316,** 1.

Isely, D. (1950). The cold test for corn. *Proc. Int. Seed Test. Ass.* **16,** 299.

Isely, D. (1954). Purity Analysis. *In* "Seed Analysis," pp. 3, 24. Iowa State Coll. Book store, Ames, Iowa.

Isely, D. (1956). Determination of variety or type in the laboratory and greenhouse—literature review. *Proc. Ass. Off. Seed Anal.* **46,** 75.

Jenkins, E. H., Hicks, G. H., McCarthy, G., Card, F. W., and Lazenby, W. R. (1897). Rules and apparatus for seed testing. *U. S., Dep. Agr., Circ.* **34,** 1.

Justice, O. L. (1942). Viability of bulblets of *Allium canadense* and *A. vineale* occurring in seed of cereals and crimson clover. *Proc. Ass. Off. Seed Anal.* **34,** 109.

Justice, O. L. (1946). A review of the literature on the use of the fluorescence test for the classification of *Lolium* species and hybrids. *Proc. Ass. Off. Seed. Anal.* **36,** 86.

Justice, O. L. (1960). Quality determination of forage seeds. *Advan. Agron.* **12,** 107.

Justice, O. L. (1961). The testing of seeds. *Yearb. Agr. (U. S. Dep. Agr.)* p. 407.

Justice, O. L., and Houseman, E. E. (1961). Tolerances in the testing of seeds. *Yearb. Agr. (U. S. Dep. Agr.)* p. 457.

Justice, O. L., and Whitehead, M. D. (1946). Seed production, viability and dormancy in the nutgrasses, *Cyperus rotundus* and *C. esculentus. J. Agr. Res.* **73,** 303.

Justice, O. L., Musil, A. F., Andersen, A. M., Cull, H., Drake, V. C., Wertman, F. L., Caldwell, B., Kent, C. A., and Zeleny, L. (1952). Testing agricultural and vegetable seeds. *U. S. Dep. Agr., Handb.* **30.** (Also in French and Spanish; FAO, Rome.)

Kick, H., Oslage, H. J., Ruge, U., Scheffer, F., Schlichting, E., Schmidt, L., and Wöhlbier, W. (1970). Internationales Symposium—Hundert Jahre Saatgutprufüng, 1869–1969. *Landwirt, Forsch., Sonderh.* **24,** 1.

Kjaer, A. (1961). Agricultural and horticultural seeds. Food and Agriculture Organization of the United Nations. *FAO Agr. Stud.* **55,** 95. (Also in Spanish; "Semillas–Manual para el Analisis de su Calidad." Agencia para el Desarrallo Internacional, Mexico City.)

Korsmo, E. (1935). "Weed Seeds; Ugressfro; Unkrautsamen." Glydendal Norsk Forlag. Oslo.

Kulik, M. M., and Crosier, W. F. (1964). Microbiological assay of fungicide-treated seeds. *Ass. Off. Seed Anal., Handb. Contrib.* No. 26, p. 9.

Lafferty, H. A. (1932). Purity determinations of Cocksfoot by the Continental and Irish methods, with special reference to the effect of "light" seeds on germination results. *Proc. Int. Seed Test. Ass.* **4,** 14.

Lakon, G. (1940). Die topographische Selenmethode, ein neues Verfahren zur Festellung der Keimfähigkeit der Getridefrüchte ohne Keimversuch. *Proc. Int. Seed Test. Ass.* **12,** 1.

Lakon, G. (1942). Topographischer Nachweis der Keimfähigkeit der getreidefrüchte durch Tetrazoliumsalze. *Ber. Deut. Bot. Ges.* **60,** 299.

Lakon, G. (1949). The topographical tetrazolium method for determining the germinating capacity of seeds. *Plant Physiol.* **24,** 389.

Leggatt, C. W. (1938). A new seed blower. *Proc. Ass. Off. Seed Anal.* **30,** 120.

Leggatt, C. W. (1939). Statistical aspects of seed analysis. *Bot. Rev.* **5,** 505.

Leggatt, C. W. (1948). Germination of boron deficient peas. *Sci. Agr.* **28,** 131.

Lewis, N. G. (1941). Nematode-infested Chewings fescue. *Proc. Ass. Off. Seed Anal.* **33**, 15.

Limonard, T. (1968). Ecological aspects of seed health testing. *Proc. Int. Seed Test. Ass.* **33**, 343.

Linehan, P. A. (1960). Nature of ISTA's service to the farmer and seed trader. *Proc. Int. Seed Test. Ass.* **25**, 59.

Madsen, S. B. (1960). Purity analysis of Cocksfoot seed. An investigation of the ratio between pure seed and inert matter in multiple florets. *Proc. Int. Seed Test. Ass.* **25**, 213.

Malone, J. P., and Muskett, A. E. (1964). Seed-borne fungi. *Proc. Int. Seed Test Ass.* **29**, 179.

Martin, A. C. (1946). The comparative internal morphology of seeds. *Amer. Midl. Natur.* **36**, 513.

Mayer, A. M., and Poljakoff-Mayber, A. (1963). Dormancy, germination, inhibition and stimulation. *In* "The Germination of Seeds," p. 61. Macmillan, New York.

Miles, S. R. (1963). Handbook of tolerances and of measures of precision for seed testing. *Proc. Int. Seed Test. Ass.* **28**, 525.

Miles, S. R., Carter, A. S., and Shenberger, L. C. (1958). Tolerances and sampling for purity analyses of seed. *Proc. Ass. Off. Seed Anal.* **48**, 152.

Moore, R. P. (1962). Tetrazolium as a universally accepted quality test of viable seed. *Proc. Int. Seed Test. Ass.* **27**, 795.

Moore, R. P. (1969). History supporting tetrazolium seed testing. *Proc. Int. Seed Test. Ass.* **34**, 233.

Morgan, P. W. (1965). A uniform blowing method for Pensacola bahiagrass (*Paspalum notatum* var. *saurae*) and a comparison of the uniform and hand methods. *Proc. Ass. off. Seed Anal.* **55**, 58.

Munn, H. L., and Staker, E. V. (1942). Toxicity to seedlings of zinc from germinator trays. *Proc. Ass. Off. Seed Anal.* **34**, 82.

Munn, M. T. (1924). Rules for seed testing. *N. Y., Agr. Exp. Sta., Circ.* No. **73**, 1.

Munn, M. T. (1936). Should our association be interested in the two purity analysis methods of the International Seed Testing Rules? *Proc. Ass. Off. Seed Anal.* **28**, 42.

Munn, M. T. (1946). Germinating freshly harvested winter barley and wheat. *Proc. Ass. Off. Seed Anal.* **36**, 151

Munn, M. T. (1950). A method for testing the germinability of large seeds. *N. Y., Agr. Exp. Sta., Bull.* 740.

Musil, A. F. (1963). Identification of crop and weed seeds. *U. S., Dep. Agr., Handb.* **219**.

Niffenegger, D., and Davis, D. J. (1958). A comparison of methods for testing crested wheatgrass seed for purity. *Proc. Ass. Off. Seed Anal.* **48**, 53.

Nittler, L. W. (1958). The use of different races of the stem rust organism to distinguish between varieties of oats. *Proc. Ass. Off. Seed Anal.* **48**, 73.

Nitzsche, W. (1960). Über die Inkonstanz der Fluoreszenz bei Weidelgrässern. *Z. Acker-u. Pflanzenbau* **110**, 267.

Nobbe, F. (1876). "Handbuch der Samenkunde." Weigandt, Hempel, Parey, Berlin.

Noble, M., and Richardson, M. J. (1968). An annotated list of seed-borne diseases. *Proc. Int. Seed Test. Ass.* **33**, 1.

Norton, D. C., and Everson, L. E. (1963). *Anguina* on western wheatgrass, *Agropyron smithii. Proc. Ass. Off. Seed Anal.* **53**, 208.

Norton, J. B. S. (1911). Maryland weeds and other harmful plants. *Md., Agr. Exp. Sta., Bull.* No. **155**.

Nutile, G. E., and Nutile, L. C. (1947). Effect of relative humidity on hard seeds in garden beans. *Proc. Ass. Off. Seed Anal.* **37**, 106.

Olsen, K. J. (1964). Variety testing of beets by means of laboratory and field plot methods. *Proc. Int. Seed Test. Ass.* **29,** 909.

Overaa, P. (1962). A new germination apparatus designed for alternating temperature and light exposure. *Proc. Int. Seed Test. Ass.* **27,** 742.

Porter, R. H. (1938). Uniform techniques for the analysis of small-seeded grasses. *Proc. Ass. Off. Seed Anal.* **30,** 133.

Porter, R. H. (1949). Recent developments in seed technology. *Bot. Rev.* **15,** 221.

Porter, R. H. (1959). "Manual for Seed Technologists." Dar Al-Kitab Press, Beirut, Lebanon.

Porter, R. H., and Leggatt, C. W. (1942). A new concept of pure seed as applied to seed technology. *Sci. Agr.* **23,** 80.

Przyborowski, J., and Wilénski, H. (1935). Statistical principles of routine work in testing clover seed for dodder. *Biometrika* **27,** 18.

Radersma, S. C. (1964). Morphological, physiological and other characters as the basis of variety testing by laboratory methods. *Proc. Int. Seed Test. Ass.* **29,** 785.

Rodewald, H. (1904). Untersuchungen über die Fehler der Samenprufungen. *Arb. Deut. Landwirtschaftsges.* **101,** 1.

Schubert, J. (1965). Vergleichuntersuchungen zur Prufung der Excised-embryo Methode an Hand des Keim-und Tetrazolium tests bei *Fraxinus excelsior, Prunus avium* und *Pinus monticola. Proc. Int. Seed Test. Ass.* **30,** 821.

Shenberger, L. C., Carter, A. S., and Quackenbush, F. W. (1946). A study of methods of seed sampling. *Proc. Ass. Off. Seed Anal.* **36,** 56.

Simon, P. W., and Kulik, M. M. (1971). Routine estimation of captan on individual sorghum seeds. *J. Ass. Offic. Anal. Chem.* **54,** 1110.

Smith, A. N., and Crosier, W. F. (1965). A comparison of methods and micro-organisms for assaying treated seeds. *Proc. Ass. Off. Seed Anal.* **55,** 104.

Smith, C. P. (1916). Studies in tolerance for purity variations. *Proc. Ass. Off. Seed Anal.* **9,** 18.

Stermer, R. A. (1968). An alternating cycle seed germinator with thermister-controlled temperature. *U. S. Dep. Agr., ARS,* **ARS 51-17,** 1.

Stevens, O. A. (1918). Variations in seed tests resulting from errors in sampling. *J. Amer. Soc. Agron.* **10,** 1.

Stout, M., and Tolman, B. (1941). Factors affecting the germination of sugar beet and other seeds, with special reference to the toxic effects of ammonia. *J. Agr. Res.* **63,** 687.

Thompson, V. J. (1941). Buckhorn plantain (*Plantago lanceolata*). *Proc. Ass. Off. Seed Anal.* **33,** 39.

Thomson, J. R., and Doyle, E. J. (1955). A comparison between the halving and the random cups methods of sampling seeds. *Proc. Int. Seed Test. Ass.* **20,** 62.

Toole, E. H. (1961). The effect of light and other variables on the control of seeds. *Proc. Int. Seed Test. Ass.* **26,** 659.

Toole, V. K. (1963). Light control of seed germination. *Proc. Ass. Off. Seed Anal.* **53,** 124.

Toole, V. K., and Borthwick, H. A. (1968). Light responses of *Eragrostis curvula* seed. *Proc. Int. Seed Test. Ass.* **33,** 515.

Walls, W. E. (1965). A standardized phenol method for testing wheat seed for varietal purity. *Ass. Off. Seed Anal., Handb. Contrib.* No. 28, p. 7.

Weir, H. L. (1959). Germination of Johnson grass. *Proc. Ass. Off. Seed Anal.* **49,** 82.

Wellington, P. S. (1965). Germiability and its assessment. *Proc. Int. Seed Test. Ass.* **30,** 73.

Wellington, P. S. (1970). Handbook for seedling evaluation. *Proc. Int. Seed Test. Ass.* **35,** 449.

West, D. W. (1952). A rapid technique for purity analysis of orchard grass seeds. *Proc. Ass. Off. Seed Anal.* **42,** 51.

Wheeler, W. A., and Hill, D. D. (1957). Insects injurious to forage seeds. *In* "Grassland Seeds," p. 77. Van Nostrand-Reinhold, Princeton, New Jersey.

Woods, A. F. (1910). The adulteration and misbranding of the seeds of alfalfa, red clover, orchard grass and Kentucky bluegrass. *U. S., Dep. Agr., Circ.* **31.**

Woodstock, L. W., and Niffenegger, D. (1971). Priorities for seed technology research. *U. S., Dept. Agr., Misc. Publ.* **1207,** 1.

Zeleny, L. (1954). Methods for grain moisture measurement. *Agr. Eng.* **35,** 252.

Zeleny, L. (1961). Ways to test seeds for moisture. *Yearb. Agr. (U. S. Dep. Agr.)* p. 443.

6

SEED CERTIFICATION

J. Ritchie Cowan

I. Field Seeds

As long as man has practiced agriculture, it has been necessary for him to give some consideration to the preservation and maintenance of seed stock. In spite of the very significant and essential role played by seeds, very little attention has been given to their maintenance, multiplication, and distribution until the current century. It is recorded that the early Romans made every effort to guard against the deterioration of cultivated races of seed. According to Virgil, a need was recognized for continued care in preventing the inclusion of variations of inferior value in seed. The first organized merchandising of small seeds, such as tall fescue (*Festuca arundinacea*), *Agrostis* species, and *Bromus erectis,* was handled by seed merchants in Darmstadt, Germany, prior to 1775. There developed in this period two other important seed merchandising centers, one in Paris, France, and the other in Edinburgh, Scotland. Initially, the firm of Keller and Sohn of Darmstadt was organized to handle seeds which were hand-harvested from meadows in the forests adjacent to the community of Darmstadt south of Hanover. Later in 1775 the L. C. Nungesser Seed Company was organized and is still in business today.

A. Cultivar

In the mid-1800s, Charles Darwin presented his well-supported theory of evolution which he called "The Theory of Natural Selection." This was followed by the famous principles set forth by Mendel and presented formally to the Naturalists' Society of Brunn in 1865. These principles remained buried in their archives for 25 years or so until simultaneously three scientists, DeVries, Correns, and Tschermak, defended and developed the basic principles of heredity. All these developments led to the field of endeavor which is recognized today as plant breeding. This is a means whereby scientists can isolate genetic material from within a given species to perform a specific production requirement. Such material when so identified was initially known as a variety as it moved in commerce. On an international basis, it is now recognized as a cultivar. The term *variety* (*cultivar*) denotes an assemblage of cultivated individuals which are distinguished by any characteristics (morphological, physiological, cytological, chemical, or others) significant for purposes of agriculture forestry, or horticulture, and which, when reproduced (sexually or asexually) or reconstituted, retain their distinguishing features. (See Fig. 1.)

A cultivar may be isolated or developed by two procedures. One is through natural selection, under a given set of conditions. Such a cultivar would be considered to be an ecotype. The other, and more common approach, involves a very sophisticated and coordinated research program under the direction of a plant breeder which may involve many disciplines such as genetics, cytogenetics, plant breeding, plant physiology, cereal chemistry, and nutrition. The goal of such a development will be to provide a cultivar which will perform a specific job for a specific purpose. The plant breeder will make an inventory of those character-

Fig. 1. Research trials at Oregon State University, Corvallis, Oregon. Synthesizing a new variety is a very exacting science. The plant breeder is the architect and the seed certification official guarantees its preservation.

istics that would be desirable in a new cultivar or cultivars to solve certain problems affecting the production of the crop under consideration. Then a search would be made for parent material which would provide the genetic basis to effect such improvement. This parental material would be combined in various ways in order to obtain eventually a combination of the most desirable characteristics coming from the different plant sources. It might be necessary to evaluate the progeny for disease resistance, insect resistance, quality, yield, etc. This will require establishment of large experimental plantings of the material and the taking of extensive observations as a means of identifying the most desirable progenies. (See Fig. 2.) These will of necessity need to be tested over a wide range of conditions. It may require very sophisticated laboratory analysis in order to select the right materials. Depending on the nature of the kind, as to whether it is self- or cross-pollinated, different procedures will be required to stabilize the genetic makeup. This scientific approach can be very time-consuming and expensive. Development of a new cultivar may require as much as 10 or 15 years and conceivably an expenditure of 50,000 to 300,000 dollars. Thus, it becomes extremely important that, once a new cultivar is produced through research, every effort is made to maintain and multiply it in desired quantities. As development of plant varieties became more complex and sophisticated, it naturally followed that responsibilities of those involved in seed certification became much more complex as well.

FIG. 2. Evaluation trials to determine eligibility of new forage varieties for Plant Breeder's Rights in the United Kingdom, conducted by the National Institute of Agricultural Botany, Cambridge, England.

B. History

Exactly where the concept of seed certification originated is not clear. However, all records appear to indicate that the credit should go to the Swedish workers. Although agriculture had been practiced since the Stone Age in the part of the world now known as Sweden, significant developments in agriculture did not take place until about 1840.

Cereals were one of the main early crops in southern Sweden. They were exported in a large part to neighboring countries in northern Europe. It was soon discovered that high quality of seeds being sold was not being maintained. A very discerning farmer, Birger Welinder, living near the village of Svalof in southern Sweden was responsible for initiating a movement which later was to become one of the most significant factors in multiplication and distribution of improved seeds. In April 1886, he and some of his neighbors were responsible for organizing what subsequently became known as the Swedish Seed Association. The aim of this association, as reflected in the first section of its constitution was "to work for the cultivation and development of improved sorts of cereals and other crops and for the utilization of these sorts in Sweden and other countries" (Newman, 1912). From this beginning two rather significant units developed—an outstanding and world-renowned plant-breeding research station, still in existence today in the community of Svalof, as well as a seed distribution organization. Great stress was placed on the necessity of checking carefully variations within the composition of a given variety and employing techniques, in those days referred to as "plant improvement" and today would fall in the classification of "plant breeding," to make sure that stock seeds were available in reasonable quantities of high genetic quality and free of contaminants.

In the late 1800s, Mr. E. Hellbo of Sweden spent some time in Denmark and subsequently returned to Sweden early in 1900. He set the stage for the seed testing station which was the primary basis of initiating concern for certification of varieties and making sure of varietal purity in Denmark. Originally, the Danish certification program was primarily concerned with fodder root crops. About this time, Johannasen, a Danish worker, advanced the pure line theory, particularly as it pertained to self-pollinated crops. This theory was based on the fact that once the material was genetically homogeneous, it should be possible to maintain it in a stable state for an indefinite period as long as care was exercised in preventing admixtures.

Dr. James W. Robertson, the Commissioner of Agriculture and Dairying for the Canadian Department of Agriculture from 1885 to 1905, conceived the idea that an organized system of seed selection conducted systematically on a large number of Canadian farms could not only increase productivity but could also stimulate the farmer's pride and interest in his work. As a result of Robertson's vision and guidance, the Canadian Seed Growers' Association was established. Subsequently, the Association sent Dr. L. H. Newman to Sweden, Norway, and Denmark to study systems of seed multiplication and maintenance. When Newman returned from Sweden in 1912, he promoted the establishing of a class of seed known as "elite stock seed." This was a class of stock seed which was multiplied under careful supervision and did not move in the channels of commerce. It provided the base for subsequent multiplications of this variety for certification. Today this particular class would be known as *foundation* (Newman, 1912).

In 1919, when the International Crop Improvement Association was formed, Newman chaired a committee that enunciated the following fundamental concepts of seed certification:

1. Pedigree of all certified crops must be based on lineage.
2. The integrity of certified seed growers must be recognized.
3. Field inspection must be made by qualified inspectors.
4. Verification trials to establish identification and the usefulness of varieties and strains certified must be conducted.
5. It must be recognized that there is necessity of keeping proper records to establish and maintain satisfactory pedigree of seed stocks.
6. There should be standards for purity and germination established.

7. The principle of sealing seeds to protect both grower and purchaser must be approved.

8. Species of farm weeds which would be included within the meaning of noxious weeds, as listed by the International Crop Improvement Association, must be defined.

9. There must be a standardization of nomenclature used in describing the classes of pedigreed seed.

In Denmark, it was also a farmers' organization that was responsible for initiating a comprehensive seed testing program. The Federal Seed Testing Station was responsible for checking the genetic purity and stability of all material sold as seed stock. This was the beginning of seed certification and maintenance and multiplication of cultivars for distribution in Denmark. At present Denmark has one of the most complete and modern research and service centers for seed testing. In Holland the seed certification program originally started from farmer seed fairs. Subsequently, the farmers began inspecting fields. The present program in Holland is under the Nederlandse Algemene Keuringsdienst (NAK).

After the U.S. Department of Agriculture was established, in 1862, one of its main activities was distributing seeds. During the late 1800s and early 1900s, many farmers received seeds of new varieties from members of Congress. Although some of these seeds were multiplied and further distributed, generally they benefited only those farmers who received them.

Because of obvious difficulties in increasing and distributing seeds of new varieties, agronomists at several state agricultural experiment stations began to help growers by inspecting their seed-increase fields prior to harvest. In 1913, Wisconsin started a field inspection program for members of the Wisconsin Agricultural Experiment Station. Montana followed with a similar service in 1915, Minnesota and Missouri in 1916, and Ohio in 1919. A pure seed association was formed in Arizona in Maricopa County in 1914, at the time of the introduction of Pima cotton. Oregon began a seed certification program in 1916 for potatoes, in 1918 for cereals, and in 1924 for forage seeds. Seed certification received legal status in Oregon in 1937. These early efforts laid the foundation for today's seed certification program.

The Land Grant Colleges came into being in 1862. As these institutions came of age, one of their important early contributions to agriculture was the development of new varieties of plants. There was little interest by the commercial seed trade in merchandising of varieties. Their prime concern was the merchandising of kinds. Therefore, the plant breeders who had developed new varieties sought assistance of interested farmers in multiplying these varieties. The plant breeders, in turn, inspected the multiplication of these new varieties in the field to ascertain their purity and assist in roguing offtypes or undesirable plants. From this cooperative effort, crop improvement associations were developed for multiplying and distributing new varieties which had been developed by the state agricultural experiment stations. Subsequently, it became important to have a more formal type of organization to maintain genetic purity and quality of new varieties. As a result the crop improvement associations in many states assumed leadership in organizing seed certification programs which consisted of field inspections, seed analysis, and appropriate labeling so that the seed could be identified as to variety and minimum mechanical quality.

On July 11, 1919, six persons, one from the Province of Ontario, Canada, and one from each of the states, Michigan, Minnesota, North Dakota, South Dakota, and Wisconsin of the United States, met at St. Paul, Minnesota, to discuss the feasibility of forming an organization to strengthen the efforts of individual certification agencies. This conference led to organization of the International Crop Improvement Association (ICIA) in December of 1919, in Chicago, Illinois. The first comprehensive set of minimum standards for certification of seeds was prepared by the ICIA in the early forties and published in 1945. These standards, which have been revised from time to time, became the reference for certifying agencies in the United States, Canada, and some other countries. In 1969, the ICIA changed its name to the Association of Official Seed Certifying Agencies (AOSCA).

When the ICIA was initially organized, there were no federal nor state seed laws in the United States. The organizers felt that it would not be reasonable to distribute seeds of high genetic quality which were contaminated by seeds of lesser quality having low germinating capacity or containing weed seeds or other kinds. For these reasons minimum standards were established which assured the consumer that he was buying seeds pure as to variety with a minimum level of germination and free of contaminants. This total comprehensive feature of the American and Canadian certification program has remained intact for 50 years. In the meantime, federal seed laws have been enacted which are based on truth of labeling. State laws have evolved over the same period of time placing certain requirements on the quality of seeds that can be merchandised within states. Legislation has now been developed whereby minimum standards have been set up on a national basis in the United States for certification as to genetic purity. It is still the prerogative of the individual certifying agencies to establish additional requirements within the framework of their particular state and area of jurisdiction if they should so desire. Traditionally certification in Europe has been based primarily on genetic purity.

In the United States and Australia, seed certification is handled by state agencies. India is considering organizing on this basis (Douglas, 1969). In the United States it is carried out as a legal function of state governments. A current amendment to the Federal Seed Act provides for recognition of seed certification on a national basis in the United States providing for minimum genetic standards. In Australia, it is a voluntary service conducted by the state departments of agriculture and coordinated nationally by a committee of state department officials. For the most part all certifying agencies in the United States are members of AOSCA and by virtue of such membership adhere to the same minimum standards. In a number of states, seed certification is the responsibility of a crop improvement association, in others, the Land Grant University, and in some the State Department of Agriculture. In 95% of the cases, the seed certification program is a function of the Land Grant University. This is very desirable and advantageous inasmuch as the certification operation is a step subsequent to the completion of research by the plant breeders. Involvement is essential as they can provide research talent in genetics, cytogenetics, plant pathology, etc., and can help in developing appropriate standards and regulations for seed certification. In all other countries where seed certification programs are in operation, they are a function of the national or federal government.

C. Generation System

Initially in certifying of self-pollinated crops, it was possible to re-certify seeds for an indefinite time. However, as plant improvement work extended into cross-pollinated crops, it was recognized that in order to maintain genetic purity, some limitation had to be placed on the number of generations allowed for multiplication from the original stock seeds which had been developed by the plant breeder. It was necessary, of course, to have enough generations so as to permit practical buildup of a given cultivar, but at the same time not to have so many that there would be opportunity for extensive segregation in subsequent generations. The "generation system" was developed. It was first introduced in 1946 by the ICIA for use with cross-pollinated crops, primarily forage. In 1968, ICIA put into effect a similar program for cereals. Initially, it was designated as three generations from breeder seed. There were four classes of seed; breeder seed, foundation seed, registered seed, and certified seed in the United States. In Canada there were breeder seed, select seed, foundation seed, registered seed, and certified seed. The select seed class is used only for cereals. In the OECD certification scheme, there are basic seed, certified seed–first generation, and certified seed–second generation.

Breeder seed is seed directly controlled by the originating or sponsoring plant-breeding institution, firm, or individual and is the source for the production of seed of the certified classes.

Select seed is unique to the Canadian certification system. It is the approved progeny of breeder or select seed produced in a manner to insure its specific identity and purity by those growers authorized by the certifying agency for the production of this class. Select seed is not a seed of commerce.

Foundation seed is the progeny of breeder, select, or foundation seed, handled to maintain specific genetic purity and identity. The production must be acceptable to the certifying agency.

Registered seed is the progeny of breeder, select, or foundation seed, handled under procedures acceptable to the certifying agency to maintain satisfactory genetic purity and identity.

Certified seed is the progeny of breeder, select, foundation, or registered seed so handled as to maintain satisfactory genetic purity and identity and which has been acceptable to the certifying agency.

Basic seed is equivalent to either the foundation or registered class as described above.

Certified–first generation seed is one generation removed from basic seed. (See Table I.)

TABLE I
CLASSES OF SEED USED IN TWO MAJOR CERTIFICATION SCHEMES

AOSCA[a]	OECD[b]
Breeder	Breeder
Select[c]	Breeder
Foundation	Basic
Registered	Basic
Certified	Certified–first generation (blue tag)
	Certified–second generation and subsequent generations (red tag)

[a] Association of Official Seed Certifying Agencies.
[b] Organization for Economic Cooperation and Development Certification.
[c] This class used in Canada for cereals.

It was found in the United States that when the generation system was initially introduced, frequently very satisfactory commercial buildups could be achieved without using three generations beyond the breeder seed class. Thus, when a cultivar is released, the plant breeder often stipulates that there be no registered class in the increase of this particular cultivar. The purpose behind the deletion of one of the classes was to provide seeds for commercial use which were fewer generations removed from the breeder seed. This would provide for greater genetic stability, particularly in cultivars of open-pollinated kinds. The adoption of the use of the limited generation system for self-pollinated crops was primarily to overcome contaminations due to admixtures. Frequently, because of new plant breeding techniques, many self-pollinated crops, particularly some of the cereals, were producing a higher percentage of the plant population which were capable of outcrossing. Thus, if some slight contamination occurred early in the multiplication process, it could become magnified appreciably if the production procedure was not required to go back and start with the basic stock or breeder seed at rather regular intervals.

D. Establishment

Proper planting stock must be used in the production of certified seeds. The grower of certified seeds must retain labels of stock seed for identifying the kind, variety, and class of seed sown. Checking of these labels is an important step in the official documentation performed by the seed certification official. Once the grower has obtained proper stock seeds, it is extremely important to plant them on an area that is free from contamination. Thus, the history of the field to be used for certified seed production is imperative. In fact, documentation of the field history of the site on which certified seeds are to be grown is probably one of the most important documentations which the seed certification official will make. Contamination can result from pollination of plants of the same kind growing in the immediate vicinity. Isolation is extremely important. However, seeds of other kinds, the same kind, or weeds in the field that may germinate can all serve as possible sources of contamination. Initially certification rules were devised to provide for a time interval in years during which a variety of the same kind as that previously grown on a given location could be planted. This interval of time does not necessarily provide adequate means or assurance that undesirable seeds which may be in the soil will not subsequently germinate.

Rampton has shown that seeds of some forage species, for example, will remain viable in the soil for many years longer than the time interval normally required for certification purposes as is shown in Table II. Wiesner (1971) has shown that variations between varieties within a species also exists.

This difficulty was overcome for many years by moving to new areas

TABLE II

PERCENTAGE OF WHOLE SEEDS RECOVERED OF 400 SEED SAMPLES BURIED IN 1961[a]

Year	Annual ryegrass	Perennial ryegrass	Orchard grass	Kentucky bluegrass	Chewings fescue	Colonial bentgrass
1962	30	0.1	8	30	7	41
1963	4	0.1	0.3	4	2	20
1964	4	0	0	1	0	13
1965	0	0	0	0	0	17
1966	0	0	0	0	0	13
1967	0	0	0	0	0	12

[a] Rampton (1961).

for production of certified seeds. However, this procedure became increasingly impractical. The work of Lee (1965) has provided a new and unique approach to meeting this particular cultural step in certified seed production. Herbicides can be used in establishing new plantings, but careful and well-timed land preparation is required. Best results have been obtained by summer fallowing to destroy established perennial plants prior to seeding. The field should be worked immediately following the previous crop harvest. Late fall cultivation of the field to be seeded is advised to provide good results. The key to this particular cultural practice is spraying fall-tilled fields with either a combination of iso-propylcarbanilate plus 2,4-dichlorophenoxyacetic acid or paraquat in late November or December after the maximum weed seed germination has been attained. The fields must not be tilled following the spraying. Usually seeding can be done sometime between the middle of February and the middle of March. This procedure, of course, is limited to those areas where winter farming is permissible. Drilling is preferable to broadcast seeding. This method of seedbed preparation and establishment has provided pure stands of a given variety and kind for production of high-quality seeds during the first year of harvest. Although successful, this procedure has some drawbacks. It is rather limited in flexibility and can only be used where limited winter farming operations are possible. However, this research by Lee has been responsible for evolving an entirely new concept in the production of certified seeds.

Lee (1969) subsequently developed a technique which involved preparing a site for seeding in the fall and harvesting of a seed crop during the following season. The main difference in this procedure from the one previously discussed is that it employs herbicides to eliminate all plant growth at the time of their application. A narrow band of activated charcoal is then applied, either in the seed furrow during the seeding or over the seed row on the soil surface after seeding. The charcoal deactivates the chemicals. This provides a narrow band in which the seeds may germinate and not be affected by the blanket chemical application. There is the possibility that some contaminants might germinate within the row, but this source of contamination is minimal. This method is based on the premise that there may be undesirable seeds in the soil, but as long as the soil is not disturbed, then the possibility of these seeds germinating is rather remote. Once a seed crop is established by this technique, all subsequent maintenance for control of undesirable plants must be performed by some selective herbicide application. If the surface is broken or cultivated, there would be an opportunity for contaminants to germinate. This practice enables the certified seed grower to switch from one variety of a given kind to another variety of the same kind on the same piece of land in a relatively short period of time. Prior to the development of such establishment procedure, it was necessary to wait at least 4 or 5 years if the grower wished to plant back a variety of the same kind on a piece of land which had grown another variety of that kind.

Table III shows the result obtained for one herbicide on one crop from

TABLE III

PERCENT STAND OF RYEGRASS (*Lolium multiflorum*) PROTECTED BY BAND-APPLIED ACTIVATED CHARCOAL ON DIURON-TREATED SOIL
(3.4 KG/HECTARE)

Method	Applied		Control, untreated
	Preplant	Preemergence	
No charcoal	2	0	100
Charcoal over row	83	93	100

a major research effort initiated in 1967. The preplant herbicide applications were made September 27, 1967; the ryegrass was planted on September 28; and the preemergence applications were made September 29. The "charcoal method" is now in commercial use, and there is sufficient evidence to show that this method is based on a sound crop production principle. Many herbicides have been evaluated. However, more research is needed to identify the best procedures and materials for different crops, varieties, and locations.

E. Isolation

Seedling inspection is used to determine the presence of contaminants. An inspection at this time will reveal the presence of old perennial plants not destroyed in seedbed preparation which could be a major source of contamination. If there is germination of volunteers showing up between the rows, then this also provides a means for the seed certification official to identify possible sources of contamination. For this reason an increasing number of seed certification agencies are requiring seedling inspections. These can also provide the seed certification official with a formal record that a given planting has been established for anticipated production of certified seeds. Isolation from other kinds within the immediate vicinity of the field which is producing certified seeds is very important.

In the case of self-pollinated crops, the isolation distance must be adequate to prevent mechanical mixtures. In the case of cross-pollinated crops, the minimum distance can vary, depending upon whether or not these crops are insect- or wind-pollinated. Although it is known that pollen can be carried great distances by the wind, it has been difficult to determine exactly how far pollen might be carried and still be viable. Studies by Griffiths (Parsons *et al.,* 1961) at the Plant Breeding Station, Aberystwyth, Wales, demonstrated that the amount of outcrossing in a seed field is reduced by the protective effect of pollen released by the variety itself. He observed that contamination in seeds from the first and sixth rows of several plantings of perennial ryegrass spaced at varying distances from a contaminating source was 42 and 18% at 25 ft isolation, 6 and 2% at 99 feet, and 0.8 and 0.6% at 396 ft, respectively. Similar studies by Knowles (1966) and Copeland (1968) had substantiated this principle. Typical isolation distances are reflected in Table IV. These

TABLE IV

ISOLATION DISTANCE FOR FIELDS OF 5 ACRES OR LARGER

	Isolation distance (ft)		
Class of seed produced	Kentucky bluegrass	Red fescue	Colonial bentgrass
Foundation	165	900	900
Registered	165	300	300
Certified	16	150	150

distances apply when there is no border removal. If the farmer removes a 9-ft border (after flowering) the isolation distance can be decreased to 600, 225, and 100 ft for cross-pollinated species, and to 30, 15, and 15 ft for apomictic pollinated species, respectively. Removal of a 15-ft border allows a further decrease to 450, 150, and 75 ft for cross-pollinated species, respectively. The isolation requirements for all crops must be enough to keep outcrossing at a minimum but sufficiently realistic to permit efficient production of certified seeds.

Contamination from neighboring fields or from volunteer plants along ditch banks and roadsides can be serious. Therefore, the seed grower must make certain that all such plants near the production field are removed or suppressed from producing pollen at the time that the seed field is pollinating.

The seed field may be located on the leeward side of natural barriers, such as canyons, wooded areas, major drainage ditches, or canals, and derive some isolation in this manner. However, these barriers can also harbor natural vegetation which might contaminate the crop. Therefore, the grower and seed certification official will take special note of the vegetation in the immediate vicinity of the seed field so that it will not be a source of contamination. In general, most plants growing in undisturbed areas tend to flower later than the same kind of plants growing under cultivated conditions. Thus, there is a built-in escape mechanism in this case. The good seed grower is alert to this problem and puts forth every effort to see that the contaminants are either clipped or sprayed well in advance of heading to avoid contamination by pollen which might otherwise blow into the production field. The smaller the seed field, the greater is the possibility of outcrossing. The isolation requirements for many cross-pollinated crops thus is relatively greater for smaller fields.

F. Field Inspection

Field inspection provides opportunity to check plant material when it is in an advanced stage of development. Self-pollinated annual crops generally are inspected once about maturity. Forage crops, such as red clover, alfalfa, and grasses, are usually inspected once during the full bloom stage. Potatoes may be inspected several times. Hybrid corn requires several inspections to determine that pollination is controlled as required for a hybrid cultivar. The field inspector's report as to the presence of other crops, varieties, diseases, weeds, isolation distances, and pollination control is compared with appropriate standards to determine if the crop meets requirements of certification.

G. Harvesting

Two primary precautions must be exercised when harvesting certified seed. The combine thresher equipment must be thoroughly clean before moving into a certified field. It is quite possible that every effort has been made up to this point to be certain that there are no contaminants in the production. However, if harvesting equipment contains contaminants prior to initiating the harvest of a certified field an admixture can result. It is not customary to have harvesting equipment inspected by an official of the seed certification program. However, spot checks are made from time to time to assure the certification office that the grower is making every effort to guarantee that no possible contamination might result from carelessness in not having the equipment properly cleaned before harvesting a new kind and variety of certified seed. The other essential precaution is to make certain that seeds are kept properly identified as they are moved from the field to the processing plant. During this stage, seeds are commonly handled in bulk. As a result the bulk containers, whether they be a truck or tote box, must be properly labeled and covered to prevent contamination. Some certification agencies require that a plan for appropriately identifying seeds (while in transit and until processing) be submitted before such seeds can be considered eligible for certification. In some instances, if the seeds must be transported for an appreciable distance, an interim temporary tag is attached to the seed container to aid in identification.

H. Processing

Contamination can also occur during seed processing. Certifying agencies require that processing plants meet certain standards before they can process certified seeds. Some of the requirements are that the plants have conveyors, legs, ducts, etc., that can be readily cleaned. They must show evidence of taking special precautions when shifting from one variety of seed to another to assure that all equipment is carefully cleaned. The building surrounding the cleaning equipment must be maintained in a sanitary condition so that there is no chance of material which has been lodged on beams, supports, etc., becoming a possible source of contamination through vibration. Every effort must be made to make certain that the variety of seed placed in the processing equipment is exactly the same at the completion of processing, less chaff and other undesirable components which have been removed in the cleaning process.

Processors wishing to handle certified seeds must apply to the certifying agency which has jurisdiction where their services are to be performed. In granting approval for processing certified seeds, each agency stipulates that the following requirements be met by the processors:

1. Facilities are available to perform processing without introducing admixtures.
2. Identity of the seeds must be maintained at all times.
3. Records of all operations relating to certification must be complete

and adequate to account for all incoming seeds and the final disposition of the seeds.

4. The processor must permit inspection by officials of the certifying agency of all records pertaining to certified seeds.

5. Processors approved to handle seeds for certification must designate an individual who will be responsible to the certifying agency for performing such duties as may be required.

6. Approval must be requested on an annual basis.

I. Analysis of Seeds

Once seeds have been processed, they must be carefully stored in the warehouse so as to retain their identity until approved for official labeling and sealing. Inspections are made to determine whether certified seeds are of reasonably good planting quality. Representative samples of clean seeds are submitted to the certifying agency. From these samples analysts determine percentage of mechanical purity, weeds, other crops, inert matter, and germination percentage. Each certifying agency has standards that define minimum requirements for certified seeds as determined by seed inspection. Usually it is virtually impossible to determine varietal identification by seed characteristics.

The standards for the Association of Seed Certifying Agencies which provide a guide for the state certifying agencies, require that seeds of most crops show a germination of at least 80%. Maximum tolerances for weed seeds, seeds of other kinds, and inert matter are somewhat different for each crop. The level of such impurities is held to a practical limit consistent with what is acceptable for planting seeds.

After processing, a final check is made of seed quality. Procedures are standardized for all official laboratories. Representative seed samples are evaluated for purity, germination potential, and undesirable factors such as the presence of weeds and disease. Traditionally analytical work has been based on hand separation. All components within a seed sample must be accounted for when it is analyzed.

It is rather difficult to devise equipment to do the seed analysis in a manner required by the seed analytical rules and regulations which have been developed and coordinated by AOSA (Association of Official Seed Analysts, 1965) and ISTA (International Seed Testing Association, 1966). These regulations provide uniformity of interpretation both nationally and internationally. In recent years mechanical devices, such as the vibratory separator (Hardin and Grisez, 1967), the continuous seedblower, and the microscopic inspection station, have been developed. These devices permit more accurate analysis of seeds to be accomplished more efficiently than by hand separation. Larger samples can be examined in less time than required for the standard hand analysis. This advancement in seed technology expedites the merchandising of certified seeds and enhances the opportunity of the customer obtaining high-quality seeds.

Seed analytical work has been of real concern to agriculturalists for many years. The AOSA was organized in 1908 (Munn, 1941) as a result of a study committee that had been set up by the 16th Annual Convention of the Association of American Agricultural Colleges and Experiment Stations in 1904. It was organized by a group of representatives from sixteen states, the United States Department of Agriculture, and the Canadian Department of Agriculture. This association which was formed to perfect and promulgate uniform methods for seed testing has made invaluable contributions in assuring that seeds of high quality are available to growers.

J. Labeling

The label on a seed container is of prime importance since it indicates to the buyer that the seed lot has met all the requirements of certification. Labels should be as conspicuous as possible so that they can be easily recognized in warehouses. In promoting seed certification in the United States, the purchase of "blue tag seed," the certified or commercial class of seed, has been emphasized. White tags are used for foundation seed, and purple for registered seed. The label reflects the class, variety, and kind of seed and provides an identification number for the grower, seed lot, and state. In the event that a given lot of certified seeds is of low quality, the purchaser can trace its origin. Certification labels are extremely important to the purchaser and should be filed for future reference. Too frequently these important records are discarded at the time of purchase or use of the seeds. (See Fig. 3.)

Certified seeds were originally packaged and merchandised in cotton or burlap bags. In recent years, various other packages, such as paper and large laminated cardboard containers, have been used. Although there have been problems in labeling paper containers, labels that can be affixed with glue have been developed or special paper bags marked with the official label of the certifying agency have been used. Paper containers may bear a certification label that is glued or cemented to a container across the opening in such a way that it must be torn when the package is opened. The certifying agency determines the effectiveness of sealing devices for special containers. Sometimes a certifying agency authorizes imprinting of certification labels on the containers. In Great

Britain some commercial concerns market certified seeds in bags that have been dyed the same color as the identify tags for a given class of seed.

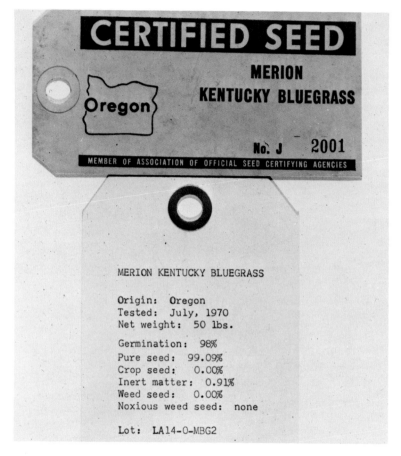

FIG. 3. Seed certification label reflects class, place of origin, variety, kind, and lot number. Seed analysis label reflects mechanical quality and germination potential.

Two or more agencies may perform the services required for certification of a seed lot. Special procedures have been developed to cover such interagency certification. For example, certified seeds might be produced, processed, tagged, and sealed by a certification agency in State A. The seeds are moved through trade channels to State B where the seed law might require recleaning of the seed lot. The certification agency in State B supervises the breaking of the original seals, removes all labels, and supervises reprocessing. When the seeds meet certification requirements of that state, they are retagged and resealed without prior permission of State A.

Seed certification labels are retained by the certification agency until they are to be attached to the container in which certified seeds are packaged. They are serially numbered to facilitate record keeping. For OECD seed certification, tags for basic seed are white; tags for certified–first generation seed are blue, and the tags for certified–second generation seed or successive generations are red. On all red labels the appropriate generation must be stated.

K. Sealing Containers

Once the seed containers have been labeled, the final step in the process of certification is sealing. The file is examined carefully for the particular variety under consideration. Field records reflecting the various field inspections, the documents indicating the stock seed which was used for establishing the crop, and the results of the seed analysis are all taken into account when making a decision on whether or not a given lot of a given variety is eligible for certification. If all requirements have been met, the certification official permits affixing of the appropriate seed certification labels to containers of the seed under consideration and sealing same containers. Field seeds usually are packaged in burlap or cloth bags. The twine with which they are sewn or tied is sealed with a metal seal that cannot be removed or reused. The seal is the property of the certification agency, and its use by any other agency is illegal. The seal protects the bag or container from being opened and prevents removal of the official certification tag. The records become extremely important documents for any given lot of seeds of a given variety. They are retained by the certification agency for an indefinite period. If questions are raised about a given lot of certified seeds, it is possible to trace the source of the difficulty by sending in the label to the certifying agency.

L. Growing out Tests

Some certification agencies require postcontrol testing of certified seeds. This is not a requirement of the system used by member agencies of AOSCA. Where it is used, it is made up of an actual growing out test to determine whether or not impurities exist. It is used in various ways. In some countries the growing out test is used to examine critically certified lots of stock seeds. If a growing out test reflects some serious admixtures or departure from the description of the variety, the source of seeds being checked cannot be used for further multiplication. In some certification programs, attempts are made to have growing out tests of every lot that has been certified. In the case of cereals, some programs require that a sample of stock seeds planted for production must be entered into a growing out test at the same time. As a result, it is possible for the certification official to determine if major problems exist in stock seeds before the commercial production would be considered for certified status. In the case of forage crops, it is impossible to have such information prior to actual seed labeling. However, if it is found that there is a producer for a certain lot of seeds who consistently gives results not in keeping with the variety description, then future certification is denied. In countries in which Breeders' Rights are a legal entity, the information obtained from the description of varieties submitted for Breeders' Rights is used by the certification officials to determine the stability of such varieties. This provides an opportunity to detect seed lots which might not properly represent a given variety as a source of stock seeds.

II. Tree Seeds

Improvement of forest production by specific seed production is relatively new. Forest geneticists concern themselves with identifying superior trees or by developing superior types through crossing two or more desirable parents. Collecting of forest tree seeds is largely from wild stands, but increasing quantities are being gathered from specially established plantations or seed orchards. Beginning in the 1950s, seeds have been collected in seed production areas, that is high-quality stands specially treated to foster heavy production. In the South, there have been seed orchards made up of vegetatively propagated material representing superior selected trees. More and more seeds of forest trees are expected to come from such special stands and orchards. In wild stands, changes in elevation can be very significant in the seeds produced. Elevations will indicate different ecotypes within a given species.

Certification of tree seeds is relatively new. However, it has rendered an invaluable service as this infant seed industry has become established. It has provided a means whereby official identification can be given to the seed as to genus, species, and location of harvest. There has been an attempt to apply similar technology in handling of certified seeds to that used for field seeds. This has not always been a satisfactory approach and as a result certain adjustments have had to be made.

There are four classes of seed recognized. *Certified seed* means seed from within a seed zone or a portion thereof, and within a 500-ft elevation increment, and from trees of proven genetic superiority which were either in a seed orchard or were plus trees with controlled pollination which is produced in a manner assuring maintenance of genetic identity. Such seeds are labeled with a blue label stating "certified seed." The certification agency is required to examine the orchard or trees and records pertaining to this production prior to collection of cones and conduct appropriate field inspection, plant and warehouse inspection, and audit.

Selected seed which is the second class, is seed from within a seed zone or portion thereof, and from within a 500-ft elevation increment and further from rogued stands or from individual trees that have promise but no proof of genetic superiority. This seed is labeled with a green label stating "selected seed." The certifying agency examines stands or trees and records prior to the collection of cones and conduct field inspection, plant and/or warehouse inspection, and audit.

Source identified seed means seed from within a seed zone or portion thereof from within a 500-ft elevation increment. The trees from which the seeds are collected are assumed to be indigenous. There are two subclasses: (A) personally supervised collections; and (B) procedurally supervised collections.

Audit certificate seed means that the applicant's record of procurement, processing, storage, and distribution states that the seed was collected within stated seed zones or described portions thereof from 500-ft elevation increments. For areas not delineated, containers of seed identified as "audit certificate" carry serially numbered brown and white labels. All records of the applicant for this class of seed are subject to audit.

The certification agency provides a documentation of evidence showing that the seeds as labeled have come from a specific source. In the instance of tree seeds, it is important to have some provision of checking those who collect cones, for example, and the areas in which they work, etc.

Obtaining an accurate measure of viability of tree seeds is frequently very difficult. Dormancy is a significant feature of many tree seeds. Thus techniques, such as tetrazolium and X-ray, have contributed substantially in getting some means of predicting the potential of viability. This has been of real concern to tree seed certification officials because the value of these seeds is only as great as their viability when planted.

Real progress has been made in developing a useful national basis on which to certify tree seeds. There are situations that are peculiar to various parts of the country, but in general, the same principle is being used throughout. The Organization for Economic Cooperation and Development (OECD) has now evolved a tree seed certification scheme to aid in the international movement of tree seeds.

III.   The Organization for Economic Cooperation
and Development Certification Schemes

In 1952, at the VIth International Grassland Congress, considerable discussion revolved around the problem of having adequate supplies of genetically pure seeds of new forage varieties to move in channels of international trade. In many countries there were no specific certification programs. In others, there were certification programs which had many different classes of seed. There was little uniformity in requirements from country to country. As a result of this discussion, committees were created to study the possibility of an international certification scheme for herbage. In 1958, the Organization for European Economic Cooperation (OEEC) assumed responsibility for developing a herbage certification scheme. Originally, the herbage certification scheme was confined to member countries in Western Europe. The United States and Canada were observers at some of the earlier meetings. Eventually, the United States, Canada, Japan, and New Zealand became a part of this total program. The United States and Canada officially became involved in 1962. The most recent member nation is Australia. By this time the OEEC had been changed to the Organization for Economic Cooperation and Development (OECD). Thus, this international certification program has become commonly known as the OECD Scheme for Certification. It was designed, initially, to certify herbage or forage crops, but since its establishment schemes have been developed for cereals, tree seeds, and sugar beet seeds. The objective of these certification schemes was to provide a means of expediting and facilitating international movement of certified seeds. The United States has been a participating member of the Herbage Scheme since 1962. The Cereal Scheme was established in 1966, the Tree Seed Scheme in 1969, and the Sugar Beet Scheme in 1970. The government of each participating country identifies a designated authority which has the responsibility for implementation and operation of the schemes within that country. The designated authority in the United States is the Crops Division of the Agricultural Research Service (ARS) of the U.S. Department of Agriculture (USDA). The USDA has, in turn, executed a memorandum with the certifying agencies of those states which have an interest in participating in the OECD schemes. In the memorandum, it is agreed that the certifying agency shall adhere to the rules and regulations as set forth by OECD for various certification schemes.

The Crops Division of ARS, USDA, is responsible for (a) arranging for the approval from the country of origin for the multiplication of their varieties in the United States according to the OECD rules, (b) receiving

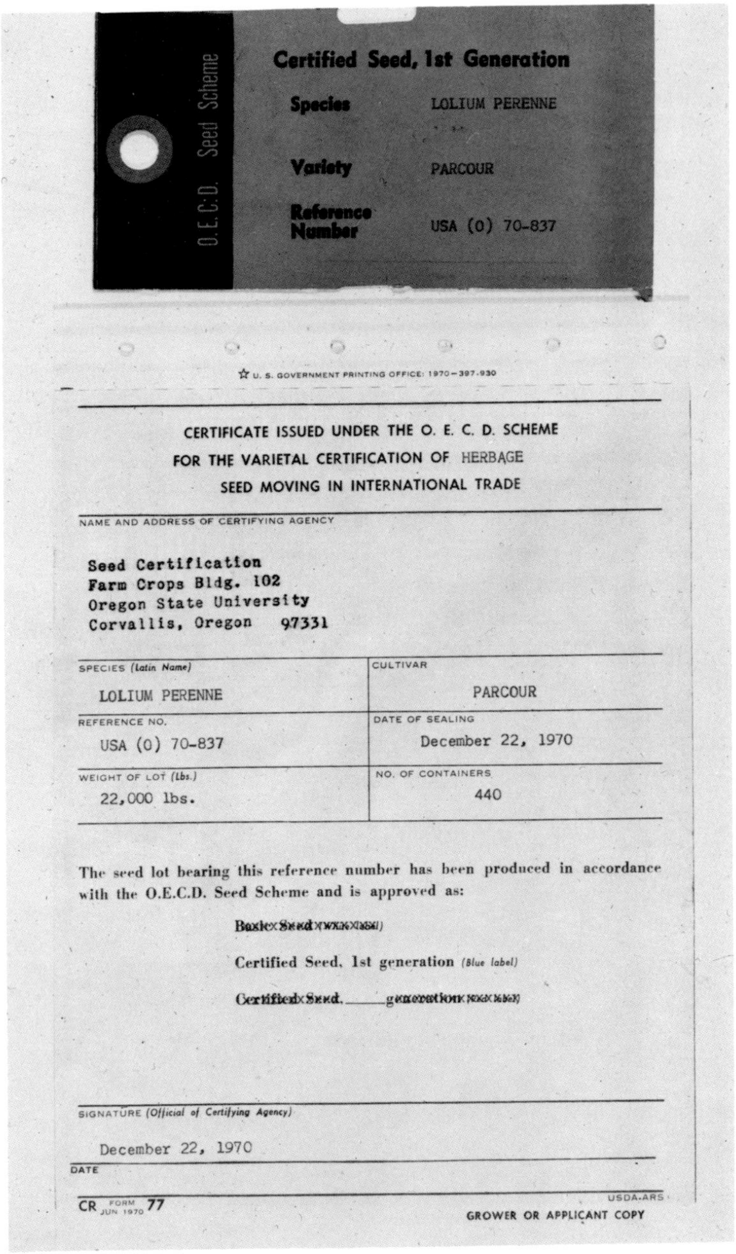

Certified Seed, 1st Generation

O.E.C.D. Seed Scheme

Species LOLIUM PERENNE

Variety PARCOUR

Reference Number USA (O) 70-837

☆ U. S. GOVERNMENT PRINTING OFFICE: 1970—397-930

CERTIFICATE ISSUED UNDER THE O. E. C. D. SCHEME
FOR THE VARIETAL CERTIFICATION OF HERBAGE
SEED MOVING IN INTERNATIONAL TRADE

NAME AND ADDRESS OF CERTIFYING AGENCY

Seed Certification
Farm Crops Bldg. 102
Oregon State University
Corvallis, Oregon 97331

SPECIES *(Latin Name)*	CULTIVAR
LOLIUM PERENNE	PARCOUR
REFERENCE NO. USA (O) 70-837	DATE OF SEALING December 22, 1970
WEIGHT OF LOT *(Lbs.)* 22,000 lbs.	NO. OF CONTAINERS 440

The seed lot bearing this reference number has been produced in accordance with the O.E.C.D. Seed Scheme and is approved as:

~~Basic Seed~~

Certified Seed, 1st generation *(Blue label)*

~~Certified Seed, ____ generation~~

SIGNATURE *(Official of Certifying Agency)*

December 22, 1970

DATE

CR FORM JUN 1970 77

USDA-ARS

GROWER OR APPLICANT COPY

FIG. 4. Organization for Economic Cooperation and Development label and certificate for a specific lot of certified seeds.

the country of origin approval of a generation system for each variety and the number of seed crops being harvested from the stand, and (c) maintaining certification records, conducting field inspection and issuing appropriate OECD certificates, labeling and sealing seeds and conducting pre- and postcontrol tests. Those seed-certifying agencies that have executed the appropriate memorandum of understanding to conduct certification under the OECD scheme have responsibility for maintaining certification records, carrying out appropriate crop inspection, reviewing crop inspection reports for adherence to field standards, maintaining certification records, and issuing appropriate documents. Labels and certificates that can be used under the schemes are prescribed by the OECD and are readily identifiable as belonging to the OECD schemes (1966, 1967). (See Fig. 4.)

The schemes are not intended to replace the domestic certification systems of participating countries but, rather, are designed so that they can operate either independently or in coexistence with the domestic systems. The United States interest in the schemes to date is mainly in the multiplication of foreign forage varieties for export to the country of origin or to some other market. Seed exporters of the United States play a very important role in multiplication of these foreign varieties since they have the facilities to make the initial contact with the owner of the variety and also for contracting, processing, and exporting seeds.

Two categories or classes of seed are recognized in the schemes; namely, basic seed and certified seed. *Basic seed* is equivalent to the foundation or registered class of the AOSCA, and *certified seed* is equivalent to the certified seed of AOSCA.

The nomenclature used for the different classes of certified seed by OECD is becoming widely accepted in many countries which are developing certification schemes. Japan is moving in the direction of using only the OECD Scheme for Certification. Many states in Australia are giving it consideration. The OECD publishes a list of eligible varieties for certification each year. Policy relative to rules and regulations for the OECD certification schemes is evolved by an advisory committee made up of representatives from member countries.

IV. Future Trends in Seed Certification

Seed certification is used primarily to protect genetic qualities of a cultivar. In the early days of crop improvement most cultivars had easily identifiable phenotypic characteristics. At present most cultivars are syntheses of many complex genotypic characteristics which are not necessarily expressed in a phenotypic manner. Thus, seed certification officials have a much more demanding assignment to work out appropriate procedures to provide protection of genetic purity but not to impose restrictions that will make seed certification impractical.

A second requirement of seed certification is maintaining mechanical quality of seed — germination percent and purity (freedom from weed seeds, other crops, inert matter, and disease). This aspect of certification perhaps has been given more relative weight in the past because, when seed certification was first employed, there were no seed laws. As seed laws were enacted, both on a state and national basis, mechanical quality of certified seed became less important. This does not in any way detract from its overall importance in the merchandising of high-quality seed. As a result of action taken by AOSCA in 1970, the minimum standards for certification in the United States will be based on genetic purity only. With the change in minimum standards which have always included genetic standards (varietal purity plus seed standards, such as germination, purity, weed content, etc.), the official standards of AOSCA will be based on genetic considerations and apply to variety purity only. Seed standards as to germination, purity, etc., will not be a part of the official minimum standards but will be published separately as a guide to the production of high-quality seeds. Although this may appear to be a relaxation of standards, it is not. Each individual state certifying agency will have its own minimum seed quality standards for merchandising seeds within the state and the federal seed law requiring truth of labeling. Hence, seeds that move in interstate commerce must have a certificate indicating the quality of seeds in a container. Under this program, the user of certified seeds must study carefully both the seed certification label and the mechanical quality label.

Seed certification documents provide a guarantee of the product developed by the plant breeder's researches. The seed certification official cannot devise a series of rules and regulations which will serve satisfactorily for an indefinite period. He must be prepared to be flexible in keeping with the techniques of plant breeders. In the seed certification process, the plant breeder holds the key position and has a major responsibility. Development of a variety is not accomplished by the plant breeder alone but rather through a cooperative effort involving, usually, other plant breeders, agronomists, plant physiologists, cereal chemists, seed technologists, seed certification specialists, plant pathologists, etc. The plant breeder must be recognized as the team leader or coordinator of this effort. If he is a competent scientist he knows more about the potential of his cultivar developments than anyone else. Thus, his advice and guidance in how cultivars should be multiplied are essential. He also has the important responsibility to describe carefully and in detail the products of his researches (cultivars). Unfortunately, plant breeders have not always exercised this responsibility. The only way in which seed certification officials can carry out their responsibility is by having a well-documented description of the cultivar being certified. A certification official must have an understanding and appreciation of plant breeding. However, he must not attempt to be a plant breeder, or by so doing he would alter the genetic composition of the material that was released by the plant breeder.

The process of seed certification must have flexibility which will permit quick multiplication of new cultivars. Much valuable time can be lost if seed certification is not permitted to proceed promptly once the plant breeder has identified promising experimental materials. In many instances seed certification is not brought into the picture until final testing of the experimental material has been accomplished. If there is a serious emergency, it may be wise to increase a promising line on an experimental basis during the last year or two of its evaluation. If this seems to be an advisable procedure, the certification official should be in a position to accept such an experimental line on a basis of "intent to certify" made by the plant breeder. If the line should then prove to be satisfactory, in the final year or two of testing, there will be a substantial quantity of stock seeds available which will meet all certification requirements for a rapid buildup upon release.

As more regions of the world develop expertise and specialization for making seed production a primary enterprise, there will be an increased movement of seeds from a region of production to regions of use. Comprehensive research programs will be needed on a continuing basis to study improved production techniques and monitor possible genetic shifts if varieties are produced in environments somewhat different than their origin. Research on isolation and seed production relative to maintaining a high level of genetic purity will be essential. For such reasons the duties of certification officials will become increasingly complex and sophisticated. The seed certification official of the future will have to be highly specialized. He must be knowledgeable in fundamentals of plant breeding and in goals of plant breeders. Expansion of specialized advanced training, specifically in the area of seed certification, will be needed to provide trained personnel. In the past, too many rules and regulations in seed certification have been established on an arbitrary and empirical basis. In the future this will no longer be acceptable. Regulations must be established on the basis of sound research and with a view toward revision as new information based on continuing research becomes available.

In 1961, a convention was held in Paris relative to the protection of new varieties of plants. This is known as Plant Breeder's Rights. The agreement reached at this convention was signed by eight countries and as of this date has been ratified by four. Certified seed moving in international trade will be influenced by the activity of those countries that have ratified the Paris convention. This will be particularly true in the case of seeds produced in a country that has not ratified the Paris convention and expects to export to one that has. Also, countries in the European Economic Community may establish requirements for certified seeds which will be in addition to those of the OECD schemes. Legislation on variety protection in the United States which has been proposed to Congress can have a significant bearing on the future.

REFERENCES

Association of Official Seed Analysts (1965). Rules for testing seeds. *Proc. Ass. Off. Seed Anal.* **54,** 2.
Copeland, L. O. (1968). Measurements of outcrossing in *Lolium* spp. as determined by fluorescence tests. Ph.D. Thesis, Oregon State University, Corvallis.
Douglas, J. E. (1969). Seed improvement in developing countries building a comprehensive program. *Proc. Int. Seed Test. Ass.* **34,** 427.
Hardin, E. E., and Grisez, J. P. (1967). Vibratory separator—a new research tool. *Agron. J.* **59,** 384.
International Seed Testing Association. (1966). International rules for seed testing. *Proc. Int. Seed Test. Ass.* **31,** 1.
Knowles, R. P. (1966). *Ann. Rep. Int. Crop Improvement Ass.* No. 48, p. 93.
Lee, W. O. (1965). Herbicides in seedbed preparation for establishment of grass seed fields. *Weeds* **13,** 293–297.
Lee, W. O. (1969). Oregon State University (personal communication).
Munn, M. T. (1941). "History of the Association of Official Seed Analysts." Ass. Off. Seed Anal., Geneva, New York.
Newman, L. H. (1912). "Plant Breeding in Scandinavia." Canadian Seed Growers' Ass., Ottawa, Canada.
OECD Scheme. (1966). "OECD Scheme for Varietal Certification of Herbage Seed Moving in International Trade," new, rev. ed., OECD Publ. No. 2. Rue Andre Pascal, Paris.
OECD Scheme. (1967). "OECD Scheme for the Varietal Certification of Cereal Seed Moving in International Trade," OECD Publ. No. 2. Rue Andre Pascal, Paris.
Parsons, F. G., Garrison, C. S., and Beeson, K. (1961). Seed certification in the United States. Seeds. *Yearb. Agr. (U.S. Dep. Agr.)* No. 394.
Rampton, H. H. (1961). Oregon State University (personal communication).
Wiesner, L. E. (1971). Temperature preconditioning of ryegrass (*Lolium* sp.) seed dormancy. Ph.D. Thesis, Oregon State University, Corvallis.

AUTHOR INDEX

Numbers in *italics* refer to the pages on which the complete references are listed.

Wojciechowska, B., 123, *143*
Wojterska, H., 102, *140*
Wollenweber, H. W., 199, *245*
Woodroof, J. G., 161, *245*
Woodroffe, G. E., 296, *300*
Woods, A. F., 304, *370*
Woodstock, L. W., 353, *370*
Worley, S., 13, *52*
Wright, R. H., 276, *300*

Y

Youngman, B. J., 187, 189, *245*

Z

Zachariae, G., 262, *300*
Zeleny, L., 303, 309, 311, 320, 324, 325, 330, 338, 339, 350, 354, 355, 356, 360, *367*
Zink, E., 163, 164, 240, *245*

SUBJECT INDEX

A

Abiotic pollination, 4, *see also* Pollination
Abnormal seedlings, 150, 315, 346–350
Abscission, 28, 151, 152
 buds, 29
 cones, 28
 cotton bolls, 31, 33
 flowers, 29
Absorption of water, 150, 159, 341, 343,
 see also Hydration, Imbibition
Achene, 96, 104, 110, 127, 129, 309, 327
Acrylic polymer emulsion, 87
Activated alumina, 228
Activated charcoal, 381
Adsorption, 210, 271, 272
Adulteration of seeds, 303, 304, *see also*
 Contamination of seeds, Impurities
Adulteration Seeds Act, 304
Aeration, 206, 338, 340
Aerosol OT, *see* Dioctyl sodium sul-
 fosuccinate
Aging of seeds, 152, 155–158, 167, 206, *see*
 also Senescence
Aggregate fruit, 110
Air–oven method, 236–238, 355, *see also*
 Moisture content
Air–tight storage, 270, 271, *see also* Storage
 of seeds
Aleurone layer, 106, 131
Alfalfa seed chalcid, 37, 39, 43, 46
Alfalfa leafcutter bee, 1, 17–21
Alkali bee, 1, 15–17
Allergy, 259
Alternate host, 46
Aluminum foil, 231
Aluminum phosphide, 221
Amide groups, 210
Amino acids, 90
Amino groups, 210
Amphitropous ovule, 88, 89, 113
Ammonia, 226
Anacampylotropous ovule, 88, 89, 104

Anaphylaxis, 69
Anatropous ovule, 88, 89, 94, 96, 98, 100,
 106, 108, 110, 111, 113, 115, 117,
 119, 121, 123, 125, 127, 129, 131,
 132, 134, 136
Anaerobiosis, 259
Anemophily, *see* Wind pollination
Anthophily, 3
Anthropology, 58
Antioxidants, 213, *see also* Oxidation
Ants, 3
AOSA (Association of Official Seed Ana-
 lysts), 85, 305–308, 328–331, 341,
 346, 352, 354, 355, 361–364, 386,
 387
AOSCA (Association of Official Seed Cer-
 tifying Agencies), 377–379, 389, 394,
 395
Aphids, 26, 28, 41, *see also* specific types
Aphrodisiac, 276
Apical beak, *see* Stylopodium
Apiculture, 14
Archeology, 66, 85, 86, 167
Arginine, 91
Aril, 87, 96
Artichoke plume moth, 47
Assassin bug, 43
Association of Official Seed Analysts, *see*
 AOSA
Association of Official Seed Certifying
 Agencies, *see* AOSCA
Attractants, 5, 267, 276
Audit certificate seed, 391, *see also* Certi-
 fication of tree seeds
Autogamous varieties, 7
Autoxidation, 213, *see also* Oxidation
Awn, 127

B

Bacteria, 211, 259, 275, 342, 347, 354
Bacteriacide, 352
Bald head, 153